FUNDAMENTALS OF
FIGHTER DESIGN

FUNDAMENTALS OF
FIGHTER DESIGN

Ray Whitford

CROWOOD

First published in 2000 by Airlife Publishing

This edition published in 2004 by
The Crowood Press Ltd, Ramsbury
Marlborough, Wiltshire SN8 2HR

www.crowood.com

British Library Cataloguing-in-Publication Data
A catalogue record for this book
is available from the British Library

ISBN 1 86126 664 2

Typeset by Servis Filmsetting Ltd, Manchester

Printed in England by The Cromwell Press, Trowbridge, Wiltshire.

Preface

FOR THE FIRST fifty years of fighter aircraft, each new generation was marked by a significant increase in speed over its predecessor. This was the Holy Grail of aeronautical engineering. At first it was a matter of getting enough forward velocity to make the aerofoil principle work. Even in its infancy, military aviation depended on speed as the prime element of air superiority, and until the 1970s this remained so. Thus when the jet engine made possible an almost overnight speed increment of 100kts, where previous progress had been in small fractions of that value, there was no holding back. On the contrary, there were intensive efforts to research and develop even more powerful engines, to reduce drag and to solve the numerous stability and control problems inherent in the new aircraft shapes.

In airframe design, which had been refined to give the lowest drag for the 350kts regime, new trends were started involving radical changes of shape, flight control systems, equipment, structures and weapons. Swept, low-aspect-ratio and thin wings, powered control surfaces, buried antennae, and heavy-gauge tapered sandwich structures were among the solutions to problems. The advances in speed were achieved only by changes in propulsion and airframe shape: the World War One rotary-engine biplanes could reach 80kts; monoplanes with retractable undercarriage powered by in-line liquid-cooled engines in the mid-1930s achieved 250kts; supercharged piston engines with constant-speed propellers increased speeds to over 300kts. The introduction of jet propulsion in the mid-1940s enabled 450kts to be exceeded with straight wings, swept wings made transonic flight possible, and by the mid-1950s Mach 2 fighters with thin swept wings and afterburning turbojets were being developed. Only two specialised types of combat aircraft forged ahead to beyond Mach 3 and fighters settled at a maximum Mach number of 2.5.

This does not mean that development has stood still; on the contrary, there is more technology available today than can be utilised in an affordable aircraft. What changed was the emphasis: from the pure 'get up and go' interceptor, whose cruising efficiency, manoeuvrability, airfield performance and handling qualities were sacrificed on the altar of maximum speed, to versatile aircraft efficient in many phases of flight. More recently speed has become even less of an issue so that fighters can go about their business in the black world of stealth in order to remain hidden at speeds of no more than Mach 2. It may be that this century is the one that has seen the beginning, and from which can be seen the end, of manned fighter aircraft development.

Throughout the history of the fighter, technological battles have been fought to pursue the goal of air superiority, and some fighter design skirmishes are ongoing:

Stable aircraft versus unstable aircraft
One-seat versus two-seat
Thin aerofoils versus thick aerofoils
Biplane versus monoplane
Complexity and low drag versus simplicity and high drag
Air-cooled engines versus liquid-cooled engines
Geared propellers versus variable-pitch propellers
Turbosupercharger versus supercharger
Machine-gun versus cannon
Propeller versus jet
Centrifugal versus axial compressors
One engine versus two engines
Straight wing versus swept wing
Swept wing versus delta wing
Safe-life structures versus fail-safe structures
Fabrication versus integral construction
Gun versus guided missile
Turbojet versus turbofan
Ground control versus autonomous operations
Dogfighting versus beyond-visual-range combat
Fixed wing versus swing wing
VSTOL versus CTOL
Few complex aircraft versus many simple aircraft
Strength/weight versus stiffness/weight
Metals versus composites
Automation versus 'seat of pants'
Mechanical linkages versus electrical signalling
Inherent stability versus autostabilisation
Spin avoidance versus spin protection
Central computers versus task-dedicated computers
Conventional configuration versus canard layout
Turbofan engine versus variable-cycle engine
Stealth design versus conventional design
Hydraulics versus electrics
Manned fighters versus uninhabited vehicles

The chapters that follow are hopefully able to convey some of the spirit of the times and the background against which design decisions had to be made, often under the duress of wartime conditions and amid fierce competition among the leading protagonists.

Ray Whitford
Shrivenham, UK

Acknowledgements

I WOULD FIRSTLY LIKE to thank Malcolm English, the editor of *AIR International*, for his suggestion for me to write a series for his magazine and his constant encouragement. It was originally planned to be in eight-parts but gradually grew as the research progressed. I wish also to thank other colleagues, namely, Professor John E. Allen, Dr Tony Hewitt of Rolls-Royce Bristol, Mike Hirst of British Aerospace Warton, Lt-Col Steve Brandt of the USAF Academy, Professor Oleg Samoilovitch of Moscow Aviation Institute, Alistair Doig and Peter Moss of RMCS, Professor Ed Obert of Delft University and Stephen Webb, all of whom were very helpful with information, suggestions and corrections. I also want to thank the following who have greatly helped in the provision of photographs: Professor Yury Shustrov of Moscow Aviation Institute; Alexei Mikheyev; Professor Zdobyslaw Goraj of Warsaw Institute of Aeronautics and Applied Mechanics; Ryszard Jaxa-Malachowski; Jerzy Grzegorzewski; Andrew Bunce, Julia Sevin and Karl Moore of British Aerospace Warton; Geoffrey Lee; Cliff Knox; Dr James Young and Freida Johnson of AFFTC/HO at Edwards AFB; Bob Ficcarra and Denny Lombard of Lockheed Martin Skunk Works; Dr Patrick Berry, Anna Emanuelsson and Eva Jacobson of Saab; Jan Rogers and Mike Taylor of Matra BAe Dynamics; Linda Purses of The Boeing Company at St Louis; Andy Head and Eileen Patrick of Raytheon Systems; Heinz Vogel; Phil Boyle of British Aerospace Dunsfold; Walt Bylciw and Lindy Shambaugh of Pratt & Whitney Large Military Engines; Capt. M.A. McClelland of US Marine Corps HQ; Paul Hermann of Eurojet; Dr K Dang-Tan of ONERA; Hua Liping of Sheyang Aircraft Industry; Professor Phillipe Lacomme and Bernard Hovine of Thomson-CSF; Monique Navarra of Dassault Electronique; Jean-Pierre Robillard of Dassault Aviation; Mike Tull of Boeing Aircraft and Missile Systems, of The Boeing Company Seattle; Jane High of Marconi Electronic Systems; Public Information Department of the Japan Defense Agency; Yukiharu Terakawa of Mitsubishi Heavy Industries; Cdr Tony Croke and CPO Paul Cope of the Ministry of Defence, Gary Atkins and Andrew Siddons of Rolls-Royce Derby; K. Venkatesh of Hindustan Aeronautics Ltd; George Lambe of British Aerospace Airbus; David Gibbings of GKN Westland; John Measell of Sander; Ray Hancock of GEC Marconi; Richard Bayliss of Smiths Industries; Patrick Darphin of Sextant Avionique; MSgt Scott Haberer of 185th Fighter Wing (ACC/ANG) Soux City; Public Affairs Office of 113rd Fighter Wing (ANG) Andrews AFB; Garland Cougar of NASA Langley; MSgt Russell Maheras of 51st Fighter Wing (PACAF); Jack Martin of Northrop Grumman; Dr T. N. Prakash of Indian Aeronautical Development Agency; Patti Heady of IAT; Brian Birlison; Mark Richardson of Cranfield University; Michael Buxton; Public Relations staff of Flight Refuelling; SSgt James Rush of 338th Fighter Wing Hill AFB; Chris Grant; Matt Jevons.

Many readers of the magazine series from around the world contacted me to congratulate and criticise. I have hopefully paid heed to their welcome comments, though the final responsibility is entirely mine.

Contents

Contents

Contents

Fighter design has come a long way in 85 years, with these two aircraft showing the changes of the last half century. Probably the best fighter of its day (1943), the Supermarine Spitfire IX (shown in the markings of 32 Sqdn) was powered by a 1,600hp Rolls-Royce Merlin 61. The Eurofighter Typhoon is much larger with an empty weight almost four times that of the Spitfire and powered by two Eurojet EJ200s in which Rolls-Royce has the major involvement. *(British Aerospace via author).*

Chapter 1

Requirements

The aeroplane was not initially regarded as a serious aggressive instrument but more of a sportsman's toy, and any improvement between 1908 and 1912 was not the result of military demands. Since their conception during World War One, fighters have been required to develop in a number of ways as technology has advanced and roles have changed. Turning points occurred, for example, with the advent of the turbojet, airborne radar and the change in threat from high-altitude bombers to terrain-hugging attack aircraft.

Changing Roles

It appears that Italy was the first country to employ aircraft in actual hostilities when, during a campaign in Libya in 1911, they were used for reconnaissance and bombing.

When in 1912 the Royal Flying Corps (RFC) was formed, aircraft started to be tested with machine-guns and light bombs. In the early stages of World War One, aircraft were predominantly used for aerial reconnaissance, hence their being called scouts. Reconnaissance, far from being a luxury item in a general's bag of tricks, became, instead, vital. An elaborate logistics and communication network grew up to determine photographic targets, launch the missions, develop, print and interpret the films and send them to unit commanders for use. Dogfights were extremely rare, and when they did occur the outcome was largely dependent on the initial engagement tactics.

However, as the war progressed and aircraft performance improved, pilots began exploiting the fighting capabilities of their aircraft. Machine-guns were fitted to two-seat reconnaissance aircraft for defence; guns were strapped onto the sides of single-seat scouts and tactics to engage or evade hostile aircraft developed almost naturally. The first dedicated fighter was arguably the Fokker EI, a monoplane fitted with a synchronised machine-gun firing through the propeller arc. By the end of World War One the originally unarmed scout had become a highly manoeuvrable fighter with two synchronised machine-guns. The best aircraft had a service ceiling over 20,000ft and could fly at 120kts at low level and 110kts at 10,000ft, which they could reach in fifteen minutes.

By the outbreak of World War Two, the need for fighters to counter hostile bomber threats and protect friendly bomber forces was clearly understood. During the Battle of Britain in 1940, information from the United Kingdom's radar chain, which extended from Dundee in Scotland to the Isle of Wight off the southern coast of England and from the Observer Corps was plotted on the operations and filter tables in the RAF's Fighter Command HQ. Approaching air formations were identified and the information passed simultaneously to the groups and sectors. Fighters were then guided by the sector controller

Arguably the first dedicated fighter, the Fokker EI was fitted with a synchronised machine-gun firing through the propeller arc. The combination of manoeuvrability, speed and forward-firing gun was so effective that the latter quarter of 1915 became known as that of the 'Fokker Scourge'. This is a replica of the more powerful variant, designated the EIII. *(Philip Jarrett).*

by radio telephony (R/T) to the point of interception, where they were dependent on making visual contact with the enemy. Hawker Hurricanes and Supermarine Spitfires would encounter large groups of bombers escorted by fighters, often in a 3:1 ratio. Ideally, while Hurricanes dealt with the bombers, Spitfires handled the fighters, some of which might often be stepped up in grids to 30,000ft (9,000m). As the war turned in the Allies' favour and was taken to Germany, the American strategy of daylight bombing required longer-range North American P-51 Mustangs and Lockheed P-38 Lightnings as escort fighters. Ironically, the P-51 had been designed to meet British needs and was evaluated by the United States Army Air Corps (USAAC) as a close-support dive bomber, though it eventually excelled as an escort. The P-38 was designed as a high-altitude interceptor, but performed best against the Bf 109 below 18,000ft (5,500m).

With the advent of the turbojet, altitude became a vital factor in cruise efficiency and therefore in range capability. Cruise altitude was suddenly boosted by about 10,000 to 15,000ft (3,000 to 4,500m) and speed by about 100kts (185km/h). Further increases came with turbojet improvement. This presented new survival problems for designers and aircrew, including pressurisation and low-temperature, high-speed escape provision. The Focke-Wulf Fw190 was used to test German compressed-air ejection seats, and several Axis aircraft were so equipped during the later stages of World War Two. The British Air Ministry guidelines considered that no form of ejection seat was required for the Hurricane and Spitfire, only for the Meteor and other jet aircraft then coming into service. The first live test in the UK was made successfully in 1946 from a Gloster Meteor. By 1961 the first zero-altitude, zero-speed seat had been successfully tested, giving an acceptable ride with acceleration limited to 15g.

There was also a need for elaborate search and fire control equipment to accomplish the mission under conditions of reduced manoeuvrability and a vastly expanded battlespace. The lag of military altitude requirement behind capability is illustrated by the fact that the world record at the start of World War Two was 56,000ft (17,000m); six years later no combat aircraft had been designed even to approach that height.

Superiority in speed was essential in the gun-armed dogfighting era: the initiative was for the swiftest to pursue the enemy, engage, or evade by flying out of range. During the Korean War a piston-engined P-51 could wrestle with turbojet-powered MiGs and not be touched, but it could not touch them either. Following experience with the North American F-86 Sabre in Korea, the USAF requested a lightweight fighter. Fortunately, the idea never really caught on (although the direct result was the Lockheed F-104 Starfighter), since subsequent conflicts have often demonstrated the importance of a large fuel capacity to confer sufficiently long range and endurance. In more recent conflicts inflight refuelling has added a vital tactical element.

By the 1960s computer-controlled interceptions were possible: the airborne intercept (AI) radar picked up its target at a range of many miles, locked on and followed it automatically. In the case of the Convair F-102 Delta Dagger its fire-control system was linked via the autopilot to the flight control system to steer the aircraft onto a suitable engagement path. Flight need not be towards the target in the previous curve of pursuit. An onboard computer capable of predicting future positions worked out collision-course or other types of flight path.

'Speed', it is said, 'is life'; by increasing an interceptor's speed the engagement point can be moved further away from the attacker's target. In 'vis-ident' interceptions of intruding high-flying reconnaissance aircraft, speed can be exchanged for height at constant energy to enable transient flight well above the sustained flight ceiling by zoom-climbing. In future air-to-air, beyond-visual-range (BVR) combat, fighters will attempt to engage their targets with medium-range air-to-air missiles that demand long-range AI radar and foolproof identification systems for their effective use. They must then manoeuvre hard to minimise an enemy missile's chance of a kill by flying outside the

Following the turn in the air war during World War Two, the primary role of Allied fighters changed from point defence to escort. The long-range capability of the North American P-51 Mustang made it ideal for supporting 8th Air Force bombing raids from England. *(AIR International Collection).*

Celebrating 75 years of 29 Squadron, the three Mach 2 interceptors that have seen service with the RAF show some distinctive differences in design philosophy, though each has two engines. The Lightning (top, F.6 version) was always short on fuel volume, having been designed in the early 1950s as a 'get up and go' interceptor and had to be given increasingly larger external tanks. The McDonnell Douglas F-4 Phantom (middle, FG1 version) was designed in 1954 as a carrier-based fleet-defence fighter to have long endurance and featured a short vertical fin in order to get below decks. The Tornado ADV (Air Defence Variant) was a British-developed fighter derivative of the shorter Panavia Tornado swing-wing interdictor/strike aircraft, and easily identified by its tall fin. *(BAe via author).*

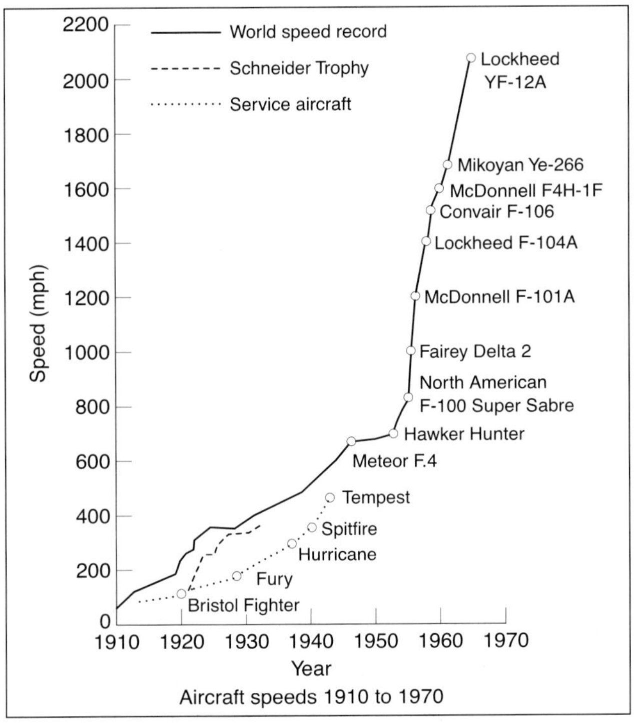

Illustration of increases in aircraft speeds from 1910 to 1965, beyond which there has been no known increase.
(Pete West/AIR International).

missile's engagement envelope, by forcing it to manoeuvre near the end of its flight, just when it has minimum manoeuvring capability.

High supersonic speed has received renewed interest for the BVR role, but a law of diminishing returns applies. Only the speed which is attained during acceleration from radar contact to missile launch is useful (to impart maximum energy to the missile). However, costs increase sharply if operation at speeds significantly beyond Mach 2 is required, and since the time spent at such speed is likely to be in the region of only 0.1% of the total service life this is difficult to justify. In most fighter roles other than BVR, speed is less important than manoeuvrability and the ability to change energy by accelerating or climbing rapidly.

With modern agile missiles there is no escape from close combat by virtue of speed. Once engaged, only superior manoeuvrability or deploying countermeasures (flares, chaff, jamming) will win the day. Indeed for most fighter-to-fighter air superiority engagements, except for slashing attacks on an unwary foe, speed is an embarrassment because rate of turn is restricted by g limits (structural or physiological). Even if a head-on pass is made at supersonic speed at high or medium altitude, speed must be reduced to exploit maximum turn rate while manoeuvring for a firing opportunity.

Weapon System Concept

With fighters getting more complex, the traditional method of designing them from the outside inwards had involved installing usually 'off-the-shelf' equipment wherever it would fit. More often than not this meant it was difficult to access and maintain, and the packaging density was low. When studies started for the F-102 supersonic interceptor in 1950, the USAF realised that a fundamental rethink was necessary. Every factor affecting combat performance was to be integrated with every other item at the earliest possible stage of design. It developed the project management concept of the completely integrated weapon

The Convair F-102A Delta Dagger was the first aircraft to be designed as part of an integrated weapon system. Powered by a single Pratt & Whitney J57-P-23 rated at 16,000lb in afterburner, the F-102A could reach Mach 1.175 and had a combat ceiling of 52,000ft, both figures less than originally specified. Almost as long as a Lancaster bomber, the F-102A owed its inception to the then over-ambitious objective of combining the traditional night-fighter's adverse-weather combat capability with the performance of the single-seat day-fighter. It was the first USAF interceptor to dispense with a gun and this one (53-1809) operated from Truax Field, Wisconsin. *(via author).*

system in which, not just airframe and engine, but avionics, armament, test and support equipment, training devices, tools, trials, etc. were designed as part of the whole package. Allied to this was the Cook–Craigie plan, an idealised production cycle for the new weapon system concept. From the drawing-board, production tooling would be installed with a capacity for production at the maximum rate envisaged, and it would then be used for manufacture of a progressively accelerating flow of aircraft. Such a scheme was intended to permit intensive flight testing with an increasing number of aircraft to uncover all the major bugs before embarking on full production. By 1952 these ideas were guiding many USAF weapon systems, but the F-102 was the first to be tied to them from the beginning. Thanks to the Cook–Craigie plan only twelve aircraft had been built when in 1954 the YF-102 ran into trouble with excessive drag (see Chapter 2).

It was unfortunate, perhaps due to a misconception of what it meant or mere lack of experience with advanced types of aircraft, that the weapon system concept was looked upon with some disfavour in certain UK quarters. However, English Electric, with its Lightning project, appreciated from the outset that the mere installation of two guns and a gunsight was incompatible with the achievement of an efficient combat system. Accordingly, the whole basis of its programme was broadened. It marked a welcome change of thinking, which it was hoped would prevent repetition of the unhappy histories of some earlier RAF aircraft. It is pertinent to note that contrary to a previous decision made by the USAF, the RAF considered that the full requirements for all-weather interception could be met only by a two-seat aircraft. English Electric felt that since the ultimate aim (as stated by the UK government in its Defence White Paper of 1957) was the elimination of the human crew altogether, it was logical to minimise the crew as far as possible and make automatic equipment perform all the essential functions of the weapon system.

Development Times

Little emerged during World War One which can be regarded as an advance in production technology. Post-war aircraft were still being designed within six months by draughtsmen and built by carpenters, fitters and riggers from spruce, metal tubes and fabric. This remained the case up to the end of the 1920s. By the 1930s the Air Ministry's F.7/30 Specification called for a single-seat monoplane fighter; that led to the Spitfire, for which Specification F.37/34 was issued. A two-year period covered the design to first flight of the aircraft.

North American's experience with its P-51, F-86 and F-100 Super Sabre showed that the development cycle was taking around four years. The trend of manhours and calendar time for aircraft development from the start of design to acceptance of the 200th unit is shown in Table 1.

Table 1 The trend of development time

Aircraft	*Engineering manhours*	*Years*	*Time*
P-51	250,000	2.0	(1940–42)
F-86A	1,700,000	4.3	(1945–49)
F-86D	2,800,000	3.7	(1949–53)
F-100A	4,800,000	4.6	(1951–55)

Notice that while the engineering manhours increased from 1.7 million for the F-86A to 4.8 million for the F-100A, the difference in elapsed time between the two basic types was only a matter of months. In comparison, the Hawker Hunter took four years from design start to first flight (1951) and a further three years to enter service. The English Electric P.1, originally planned as a technology demonstrator for high-speed flight, took five years to fly (1954) and another five years to get into service as the Lightning. Such was the size of the programme that it was decided to order what was, by UK standards, an unprecedented number of prototypes. An intermediate development design, called P.1A, accounted for two complete aircraft and one airframe for static testing, and a later development, the P.1B, required three prototypes and twenty pre-production aircraft. Such a quantity was considered adequate for the programme of research and development (R&D) facing English Electric. For the Convair F-102A, a total of fifty-two aircraft was engaged in the intensive flight test programme that enabled it to enter service eighteen months after first flight.

More recently it took less than three years from VFX concept definition to the first flight of the Grumman F-14A in 1970. A total of twenty development aircraft was used to get the F-14A into service by the end of 1972. The number of development aircraft allocated to the Eurofighter Typhoon programme (4,700 flight hours over six years) is seven, the same number as allocated to the Boeing F/A-18E/F Super Hornet derivative aircraft, whose flight test programme was scheduled to last three years. Comparison of late 1970s US and European aircraft showed that, typically, European aircraft had between 60 and 100% more parts per main component (wing and fuselage) than US aircraft, and for similar production runs overall US manufacturing manhours were 35% less than European.

Operational Requirements

The preliminary design of a new fighter must be preceded by the establishment of the aircraft's requirement. This is extremely important as it drives the design and is the yardstick by which the success of the aircraft is measured. However, the design is not normally based on a completely defined specification but on what is termed a 'cardinal point specification'. This is a list of what is considered by the customer to be the most salient requirements and will include a measure of the weighting applied to each. For example, low radar signature may be more important than the top speed.

The processes of fixing the major design parameters in order to arrive at the optimum configuration and developing a workable specification for a weapon system are mutually influential. Furthermore, this period of definition now takes so long that there are sometimes profound changes of opinion about how the aircraft should best be used. Other, and just as radical, changes of view take place following a type's entry into service.The operational lifetime of a modern combat aircraft type can be thirty years or more. For example, the McDonnell Douglas F-4 Phantom first entered service in 1960 and will remain in service into the twenty-first century. Aircraft like the Dassault Rafale, Eurofighter Typhoon and Lockheed Martin F-22, planned before the end of the Cold War and due to enter service in the next decade, should be operating well beyond 2040. Combined with a lengthy project gestation, such extended operational lives can mean that a requirement has to be defined some twenty years before it is put to the test of combat.

The odds against accurately foreseeing what will be needed are thus very great. They are lengthened still further by the current rapid development of the weapon and sensor

Even in the late 1940s the signs suggested that single-seat fighters would get larger. This photo, taken at Lockheed's Palmdale plant, shows two of 'Kelly' Johnson's designs. The P-80B Shooting Star is dwarfed by the XF-90 that was designed as a long-range penetration fighter. Though supersonic in a dive it was underpowered even with two Westinghouse J34 turbojets and lost in competition to the McDonnell XF-88 that was developed into the F-101 Voodoo. *(Lockheed Martin via author).*

True to tradition, the Sukhoi Su-27 'Flanker' has demonstrated the Russian philosophy of taking a good simple design (in this case exceptionally good) and improving it. The photo shows a Su-27K (now known as Su-33 by Sukhoi) operating from the *Admiral Kuznetsov* aircraft carrier. 'Navalising' required structural reinforcement, addition of arrester hook, wing fold, two-segment double-slotted flaps and drooped ailerons, foreplanes and a shortened tailcone. *(Avexceu Muxeeb via author).*

technology which might be deployed by and also against any new aircraft. These developments, in today's unstable world situation, make it doubly difficult to predict the nature, strength and location of likely threats.

Design

The design of an aircraft is a very large undertaking. It requires the team efforts of many engineers, scientists and mathematicians with expertise in a wide spectrum of fields including aerodynamics, propulsion, structures, materials, flight control, electronics, performance and operational analysis, human factors, weights, manufacture and costing. The designer's task is not necessarily to invent some completely new concept, but rather to determine in what respects existing combat aircraft are inadequate and to judge which of the emerging technologies are worth pursuing. With shrinking defence budgets and the uncertainty of the threat it is increasingly important that the design allows for wide versatility and

In the 1960s the defence of the US Navy's aircraft carrier battle groups against stand-off missiles called for a long-range interceptor. The radar and air-to-air missiles were already in hand, namely the Hughes AWG-9 fire control system and AIM-54 Phoenix, intended for the General Dynamics/Grumman F-111B. When it became evident the latter would fall well short of specification, Grumman designed a new fighter around them. First flown in 1970, the two-seat F-14 can carry six AIM-54s and a General Electric M61 six-barrel Gatling gun mounted just below the forward cockpit on the left side. Its complexity led to MMH/FH of over 50 and earned it the sobriquet 'Hangar Queen'. Shown is an F-14A of VF-84 'Jolly Rogers'. *(via author).*

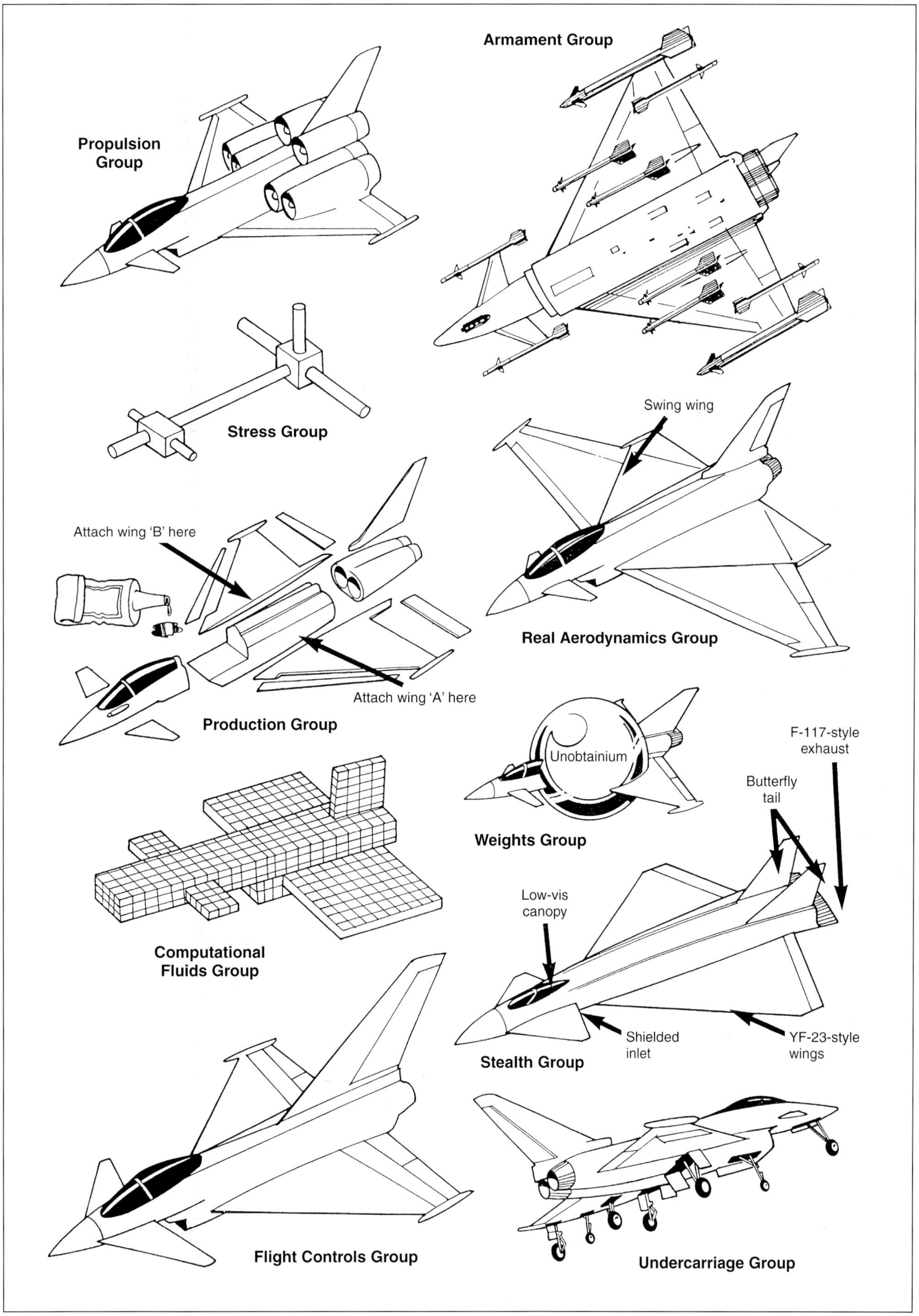

Lighthearted look at how specialists would have designed Eurofighter Typhoon, if given a free hand. *(Pete West/AIR International).*

adaptability of the final weapons system. The absolute minimum the designer must attempt to avoid is undue sensitivity to a change in the emphasis applied to operational phases.

It is important to realise that preliminary design of an aircraft, which might occupy its first one or two years, is an activity that absorbs much less than 1% of the total life-cycle cost of the aircraft. However, this is the period in which decisions are taken that have the largest impact on life-cycle cost.

Table 2 Summary of the main design features required for operational versatility

Operational phase	*Primary design requirement*
Short take-off	High afterburning thrust/weight ratio (T/W), high flap lift, low wing loading (W/S), thrust vectoring
Economical transit/loiter	Low throttled specific fuel consumption (sfc), high cruise lift/drag (L/D)
High penetration speed	High dry T/W, low store drag, good ride quality
High combat agility	High dry T/W, high control power, good L/D at high g, high usable lift, thrust vectoring
High combat persistence	Low combat sfc, good L/D at high g, versatility of weapon carriage
High combat mobility	High afterburning T/W, low drag at all Mach numbers
High survivability	Stealth (to give first shoot/first kill), threat awareness and countermeasures, robustness/redundancy
Short landing	High flap lift, low W/S, flareless, good retardation

Design Requirements

Lethality

This is a function of the destructive power of the aircraft's weapons, which must be easy to use, reliable, non-counterable and effective. The lightest, cheapest, easiest to use, hardest to counter and most effective air-to-air weapon is the gun. The 20mm cannon is adequate for dogfighting with other aircraft, though ideally it should be complemented by the short-range infra-red (IR) guided missile, which prevents disengagement. Interceptors operating against less manoeuvrable aircraft are armed with medium- to long-range missiles which home in on radar returns, either from the parent aircraft – semi-active homing – or the missile's own seeker – active homing.

Manoeuvrability

Also often known as agility, this is the ability of an aircraft to change position and velocity rapidly in order to gain an advantage in air-to-air combat. The parameters used to assess an aircraft's turning performance are load factor (normal acceleration in g), turn rate and turn radius, with turn rate regarded as the most important. Combat superiority is seen as depending on the product of specific excess power (SEP is a measure of the ability to regain energy by climbing or accelerating), sustained turn rate (maximum turn rate without loss of speed) and instantaneous or attained turn rate (maximum achievable turn rate with transient loss of speed).

Handling qualities

The aircraft should have handling qualities that permit the pilot to use the full extent of the manoeuvring performance envelope with ease and safety. Indeed, superior handling qualities are essential if the full performance potential of the aircraft is to be realised, and carefree manoeuvring is now a prime objective.

Radius of action

The ability of an aircraft to reach and attack its target and return depends on its radius of action. This in turn is influenced by the nature of the combat in which the aircraft is engaged. For the air superiority fighter the main arena may be close at hand, with combat consisting primarily of linear and turning accelerations. The best aircraft is the one that can perform most combat-relevant manoeuvres at a given radius of action, or achieve the longest radius of action for a given combat requirement.

Persistence

This is the ability of an aircraft to stay in combat while continuing to achieve superior manoeuvring performance. It is expressed in units of time, assuming a specified radius of action and fuel load for given flight-to-combat profiles. The ability to do without afterburning greatly improves an aircraft's capacity to stay and fight. An adequate dry thrust/weight (T/W) ratio is made all the more important by the fact that any attempt to break off close combat is extremely dangerous if the opponent has any unused armament. The ability to carry a mix of armament and not to be rigidly constrained by today's weapons is also a factor in persistence.

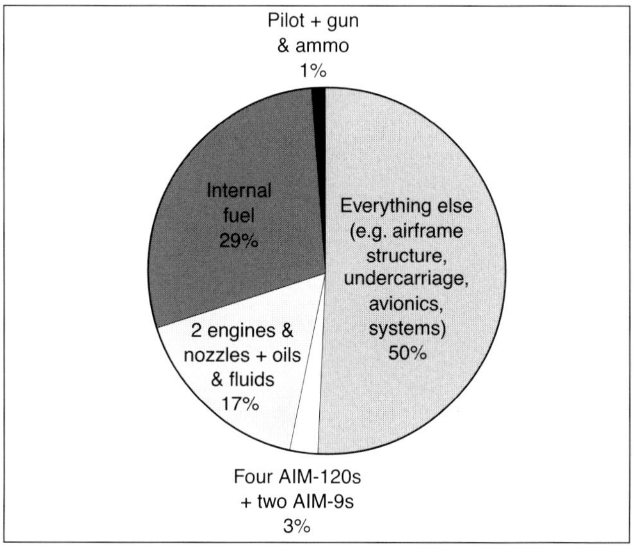

Estimated weight breakdown of the Lockheed Martin F-22A Raptor. One of the design criteria that has been uppermost in the minds of USAF planners is the aircraft's empty weight. This has turned out to be higher than originally hoped and will have an effect on the F-22's ability to supercruise. The fuel fraction, hoped to be well over 30%, has slipped to yield a figure (29%) less than that of the F-15C.

Visibility

This is the ability to detect the target first and keep it in view by whatever means. This is true of interceptors which operate against targets well beyond the visual range. For air superiority fighters, however, unobscured forward and aft vision is vital for target detection and warning of rear-hemisphere attack.

Stealth

This is the ability of an aircraft to attack its target with the maximum amount of surprise by denying long-range detection. It can be achieved by reducing the aircraft's visual, radar and infra-red signal strengths. All three are minimised by keeping the aircraft small, and may be further reduced:

VISUAL Eliminate the smoke trail with and without afterburner. Camouflage the aircraft by painting it in the colours predominating in its mission environment, and reduce the size, number and visibility of identification markings.

RADAR Reduce radar cross-section (RCS) by: avoiding surfaces at right angles to each other (to limit the number of corner reflections), designing for a minimum of radar spikes (by minimising the number of airframe angles), shielding the engine compressor face, using radar-absorbent materials, carrying stores internally, treating the cockpit canopy and radar cavity, minimising energy emissions from the aircraft's own sensors, providing comprehensive electronic countermeasures and decoys.

INFRA-RED Less use of afterburner, with its enormous IR signal, since the very short afterburners now employed have increased the angular detection range. Shield and/or cool engine exhaust. Use decoys (i.e. flares).

Table 3 Some of the compromises for stealth made in the F-22

Aerodynamics/ Propulsion	*Stealth*	*Compromise*
Requires big horizontal tail for supersonic manoeuvrability	Requires no horizontal tail	Close-coupled layout and thrust vectoring
Requires variable geometry air inlet to give high intake efficiency	Requires shielded inlets and exhaust nozzles	20 to 35% loss of thrust due to fixed and highly stealthy inlet and 2-D nozzles
Requires a smooth cross-sectional area distribution and high sweep for low-wave drag	Requires small radar energy spike angles	Similar but not equal LE and TE sweep angles, very high taper and highly loaded wing tips

Resilience

This is the ability of a force of aircraft to return repeatedly to the battle arena after the first engagement. The availability of large numbers of aircraft can provide this, though maintainability, survivability and reparability are also important. With the increasing complexity of fighters in the 1960s there was a rapid rise in the number of maintenance manhours per flight hour (MMH/FH). This brought about a strong response from customers who demanded reduced downtime, and as a result reliability and maintainability became two of the most important parameters in subsequent fighter programmes. The ability of an aircraft to survive battle damage depends on how well its four most vulnerable

Very careful attention to detail of the production models of the Lockheed Martin F-22's airframe has been necessary to ensure a low RCS. In comparison with the YF-22, a dedicated airbrake has been dispensed with (by using differential rudders), constant-chord wing control surfaces are used and a reduced number of inspection hatches. Because of concerns hanging over from problems with early F-117s, the YF-22 had relatively large 'vertical' tails, those on the F-22A have been shortened. *(Pratt & Whitney via author).*

components – crew, powerplant, fuel and flight control systems – are protected. Threat awareness on the part of the crew is the first line of defence, and this can be assisted by all-aspects detectors and the maximum use of head-up and/or helmet-mounted displays in the battle zone.

Costs

Accuracy in delivery of the warload became a critical problem only in the early 1950s. Twenty years previously the design and installation of equipment necessary for accuracy was a negligible part of the whole aircraft. It involved a few relatively simple navigation instruments and an elementary sighting device for gunnery and bombing. Even during World War Two not that much relative progress was made through the addition of radio navigation and a gyro-stabilised gunsight. However, with the rapid performance increase of fighters that began after World War Two, much more elaborate avionics equipment was necessary to maintain tolerable target detection and weapon lethality.

By the early 1950s, aircraft were becoming ever more complex with a tremendous increase in gadgetry, ranging from very complex, for the time, thermionic valve-driven radars (much of it in its early development stages, difficult to install, hard to maintain and in many cases unreliable) to small servomotors required to work at unprecedented speeds. Additionally it created serious space, weight and cooling problems, resulting in compromises of other design aspects. However, since the precision attained in the delivery of the weapon is one of the most vital factors in determining the size of the force required to perform a given mission, the penalty had to be accepted. The design, construction and installation of such equipment was in some instances more than a third of the total aircraft cost. With the passage of time, the proportion of the total aircraft cost devoted to avionics has grown to over 50%. Current efforts are underway to explore the possibility of using commercially off-the-shelf (COTS) electronics to reduce costs.

It is in the field of electronics where the most spectacular developments have taken place, bringing a completely new concept to aircraft systems. The introduction of the microprocessor, miniaturisation of many other components and a better understanding of the capabilities of digital data processing have allowed a completely integrated approach to all aircraft systems. One of the major disadvantages of 1970s aircraft was the escalating cost of avionic systems, because though the capability was available and the need was there, the equipment had to be specially developed for a particular aircraft. With a limited market the cost was inevitably high.

In the immediate post-World War Two period, the prototyping of a great number of single-purpose aircraft was the order of the day to ensure the highest-quality aircraft in service inventories, but many of those aircraft never went into production. (The number of first flights of new fighter/attack aircraft declined by a factor of almost five between the 1945–49 and 1970–74 time periods.) Aircraft became larger and more expensive, because each aircraft needed more effort to design, build and test and also because of the incumbent need for them to stay in service for much longer. Between the end of World War Two and 1980, technical manhours/lb of aircraft increased eightfold, even for simpler types like the British Aerospace Hawk and Dassault Alpha Jet the increase was 4.5 times. Over the same period the real cost/lb increased by a factor of 11, while direct manhours/lb increased by a factor of 2.5. The large increase in capital value relative to direct manhours (4.4:1) is surprisingly large, but reflects increased indirect charges associated with modern production techniques and the inclusion in the aircraft of expensive materials and very-high-value equipment. Allowing for an increase in weight by a factor of four results in technical manhours increasing by a factor of 30 and cost increasing by a factor of more than 40.

The growth in fighter flyaway cost is shown in the graph below which does not pretend to be wholly accurate but merely

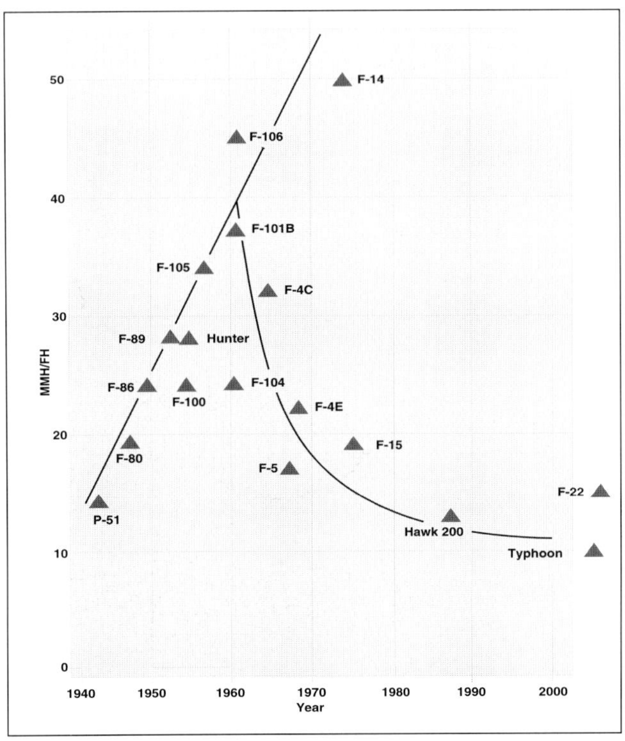

Plot of MMH/FH against time showing the dramatic rise up to around 1970, and the effect of the steps taken to combat this.

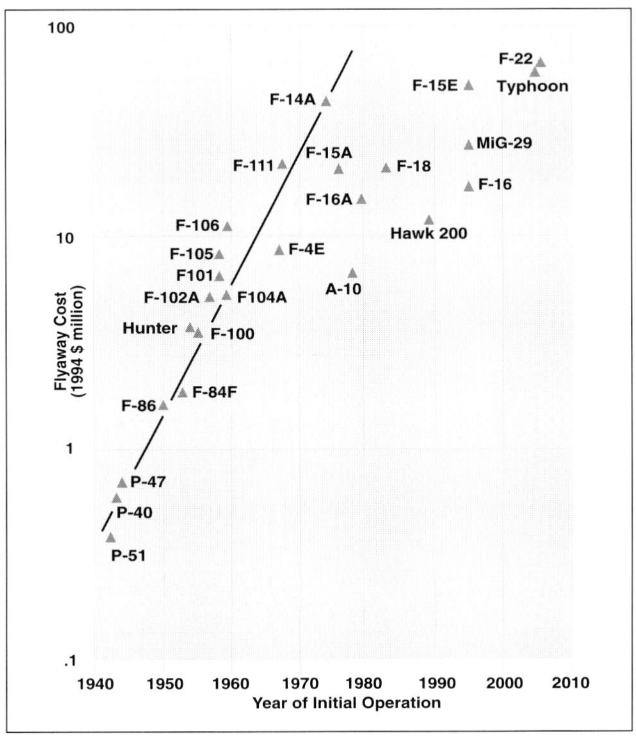

Plot of flyaway cost against year of initial operation. Note the logarithmic cost scale.

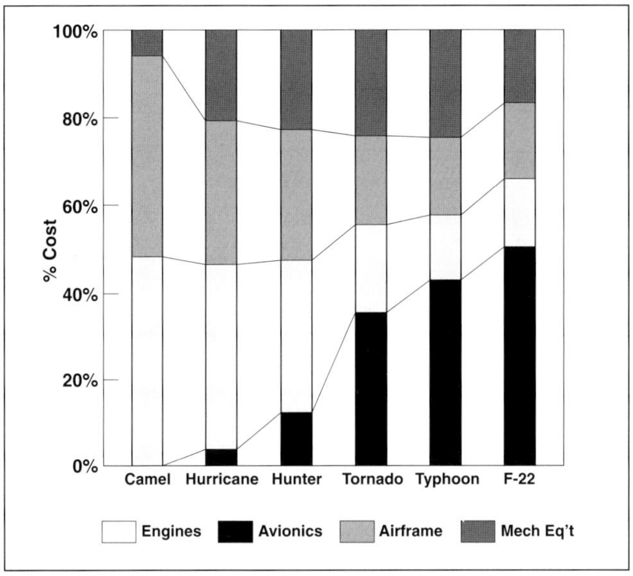

Cost breakdown of a range of fighters, showing the increasing contribution of avionics.

illustrates the trend. The cost of engines alone, excluding inflation, between 1950 and 1973 was increasing such that cost/lb of thrust was doubling every twenty-five years. On the basis of the early 1970s trends in flyaway cost and US Gross National Product (GNP), it was predicted that by 2150 an entire year's GNP would be required to buy just one fighter. With this dire prediction, procurement agencies across the world began to insist that cost be given equal weighting to aircraft performance. The divergence of recent aircraft, from the unit cost line shown connecting the Curtiss P-40 Warhawk to the F-14, may be a demonstration that cost awareness has paid dividends. On the basis of the original trend even the Northrop Grumman B-2A is cheap!

With increasing unit cost and diminishing total spend on new equipment, it should therefore come as no surprise that the numbers of aircraft in the world's air forces have declined (frontline strengths in 1980 were only one seventh that of 1950) and service life has been extended. At present there is more technology available than can be included in any single aircraft if cost is to remain within bounds. This has meant that the 1950s philosophy of 'doing what can be done' was gradually replaced by 'doing what should be done', whereas today's philosophy is one of 'doing what can be afforded'. It has been suggested that the only way out of this quandary is to state the cost first and make everything else subordinate to it. The other approach is to have fewer types and adequately long production runs, this being one of the aims of collaborative projects. It still appears to be true that aircraft are bought by weight, and flyaway cost per empty weight for the upcoming Typhoon, Rafale and F-22 ranges between $2,500 and $3,000/lb in 1996 US dollars. The only certain thing about the coming generation of fighters is that they will have to do everything that the current generation is capable of, but better, and at a comparable overall programme cost.

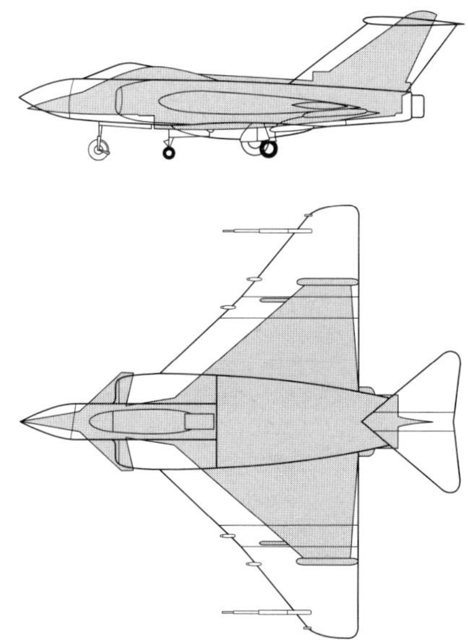

	Typhoon: Javelin FAW.8
Maximum Mach number	2:1
Empty weight	0.8:1
Maximum take-off weight	1.07:1
Wing loading (Maximum weight/Wing area)	1.85:1
Thrust loading (Maximum thrust/ Maximum weight)	1.52:1
Missile stations	2:1
Internal fuel volume	0.87:1

Adverse weather fighter evolution over 40 years: comparison of the Gloster Javelin FAW 8 with the Eurofighter Typhoon. Despite its lower empty weight the Typhoon's large external load-carrying capability gives it a larger maximum take-off weight. Moreover, its integral fuel tanks hold almost 90% that of the big winged Javelin. Hidden, of course, is the Typhoon's vastly more capable avionic suite.

With costs escalating, long gone are the days of the dedicated-mission fighter. The two-seat Dassault Rafale B shown here demonstrates the multi-role capability of modern fighters. It is carrying two Matra Apache stand-off weapon dispensers, two tip-mounted Matra Mica IR air-to-air missiles and three 2,000lt subsonic drop tanks. *(F. Robineau via author).*

Chapter 2

Aerodynamics

The biplane layout was the most common type for fighters into the 1920s with Germany leading the field of aerodynamics. Despite the vision of several experts, the value of aerodynamic cleanliness, to take advantage of increases in available engine power, was largely ignored in the UK until the early 1930s. America pursued streamlining to benefit its burgeoning airline industry and helped, with Germany, to inspire the monoplane fighters with which the warring nations entered World War Two. The efficient exploitation of the turbojet via swept and delta wings was begun in Germany and quickly spread among the Allies following the end of hostilities.

Glossary

The source of aerodynamic loads: The aerodynamic loads imposed on the skin of an aircraft in flight are due to the distribution of pressure. These arise from the wide range of local flow velocities of the air as it passes over the aircraft. The link between pressure and velocity is that where flow velocity is high, the local pressure is low, and vice versa. In essence, total pressure = static pressure + dynamic pressure, where total pressure is the pressure measured at a point in the flow where the velocity has been reduced to zero without any losses; static pressure is the pressure measured at a point moving with the flow; dynamic pressure is the difference between the total pressure and static pressure and contains the velocity term.

This photo of the Gloster Gauntlet of 1932 demonstrates the reluctance of designers to fully embrace streamlining. The aircraft does have a Townend Ring to reduce engine cylinder drag and this particular one has spats over the wheels. However, by the time that the aircraft was beginning its brief career, the first of the modern monoplane fighters (Bf 109) had flown. What the Gauntlet did though was to herald a radical improvement in firepower as it was designed to carry six guns (two in the fuselage and four in the wings). By the time it entered service in 1935 however armament had declined to the standard two fuselage-mounted machine-guns. *(via author).*

The Messerschmitt Bf 109 first flew in 1935 as an air superiority fighter. Only a relatively small fuel load was envisaged and it was decided that performance could be optimised by the use of small wings whose high wing loading could be offset by leading-edge slots and trailing-edge flaps. The fuselage was slim and the cockpit cramped, this emphasis being carried over to the undercarriage, which was of narrow track. These design features were later to become the major drawbacks of the aircraft. *(Cliff Knox).*

Angle of attack (AoA): Angle between wing chord line and freestream direction.
Boundary layer: The thin layer of air next to the surface in which the velocity changes rapidly from zero at the surface to the freestream velocity a small distance from the surface.
Drag: The total drag of a wing is made up of lift-induced drag, profile drag and compressibility drag. Induced drag is primarily a function of the planform shape of the wing for a given speed and wing loading and is not influenced by aerofoil shape. Conversely, profile drag, for a given speed and AoA, is a function of the profile shape and surface roughness, and can be subdivided into skin friction drag and pressure drag. Wave drag is due to the loss of momentum as the air flows through shockwaves set up around an aircraft in transonic and supersonic flight.

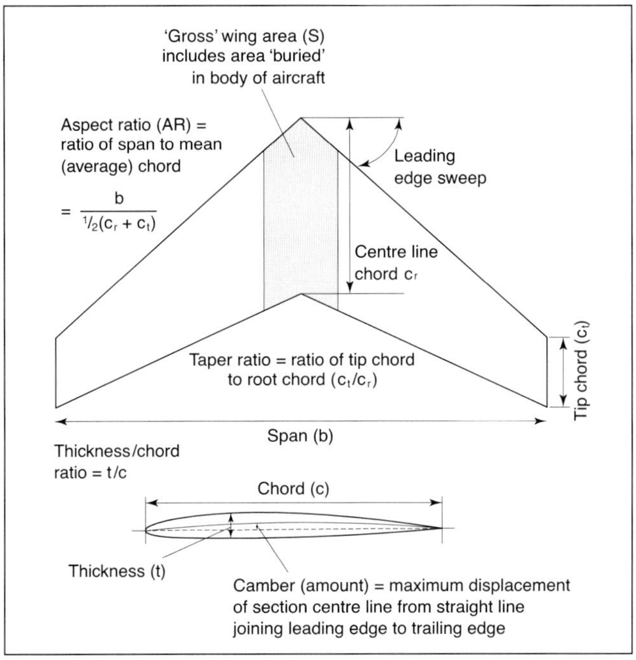

Wing geometry.

Drag equation: Drag $= C_D \times$ dynamic pressure $\times$ wing area

$$= [C_{D_0} + (C_L^2/\pi Ae)] \times \frac{1}{2}\rho V^2 \times S$$

where: C_{D_0} is the zero lift drag coefficient (comprising skin friction drag and pressure drag) and $(C_L^2/\pi Ae)$ is the lift-induced drag coefficient

C_L is the lift coefficient $\left(\text{lift force} \div \frac{1}{2}\rho V^2 S\right)$

ρ is the air density
V is the velocity
S is the wing area
b is the wing span
A is the aspect ratio (b^2/S)
e is the wing planform efficiency factor

Drag can be alternatively expressed using the above:

$$\text{Drag} = \frac{1}{2}\rho V^2 S C_{D_0} + \frac{2\left(\frac{W}{b}\right)^2}{\rho V^2 \pi e}$$ where W is aircraft weight.

The second term on the RHS shows clearly the dependence of lift-induced drag on span loading squared, $(W/b)^2$.
Mach number (M=V/a): The ratio of the speed (V) of the aircraft to the local speed of sound (a).
RAF: Royal Aircraft Factory (Farnborough).
Reynolds number $\left(\frac{\rho V l}{\mu}\right)$**:** The ratio of inertia forces to viscous forces. It is a relationship between the flow conditions over a scale model used in wind-tunnel testing that needs flow similarity to be made as close to the full-scale aircraft as possible, and where l is a typical length (e.g. wing chord), and μ is the viscosity (stickiness) of the air.
Speed of sound: For air, this varies only with temperature, so that in standard conditions at sea level (15°C) the speed of sound is 660kts (1,220km/h), whereas at 36,000ft (11,000m), where the temperature is -57°C, the speed of sound is 570kts (1,060km/h).
W/S (wing loading): Aircraft weight/wing area.

Pre-World War One

The Wright brothers used the biplane because it doubled the wing area for a given aircraft size. Precise doubling of the lift with a two-fold increase in wing area was not achieved, however, due to aerodynamic interference between the upper and lower wings. Monoplanes were not initially uncommon, and in 1910 Nieuport built a single-seat one with fully enclosed and streamlined fuselage. It was derived from the Blériot XI but had fewer bracing wires for its wings than earlier ones. It could achieve 65kts with a 28hp engine, and even with a 20hp engine it was as fast as biplanes with twice the power. It was a Deperdussin monoplane of 1913 that held the world airspeed record of 110kts. Nevertheless, 1912 saw the ascendancy of the biplane due to its lift, strength/weight and compactness, compared with the large, braced monoplane, at a time when thin wing sections were considered obligatory. Because of the misunderstood causes of early monoplane crashes, there was official resistance in Britain to monoplanes. Consequently, until the beginning of the 1930s, the type was comparatively rare in the UK.

World War One

Aerofoil sections

Prior to a series of wind-tunnel tests conducted by Ludwig Prandtl in 1915 at the University of Göttingen, the only available aerofoil data was for Reynolds numbers considerably less than those actually encountered in flight by aircraft. For example, the Wright brothers' wind-tunnel tests of 1901 were performed on very small model aerofoils of just one inch (2.5cm) chord so that with the maximum speed of their tunnel being 27mph (43km/h) this corresponded to a Reynolds number of only 20,000, whereas their first flights were all made at greater than one million. Their tests, and later ones carried out by others at a Reynolds number of around 70,000 (still too far from full-scale), showed that maximum lift and lift/drag (L/D) ratio both increased as the aerofoil was made thinner and leading-edge radius was reduced. This merely reinforced the almost intuitive but mistaken notion that thin aerofoils would have less drag than thick ones. Consequently, almost

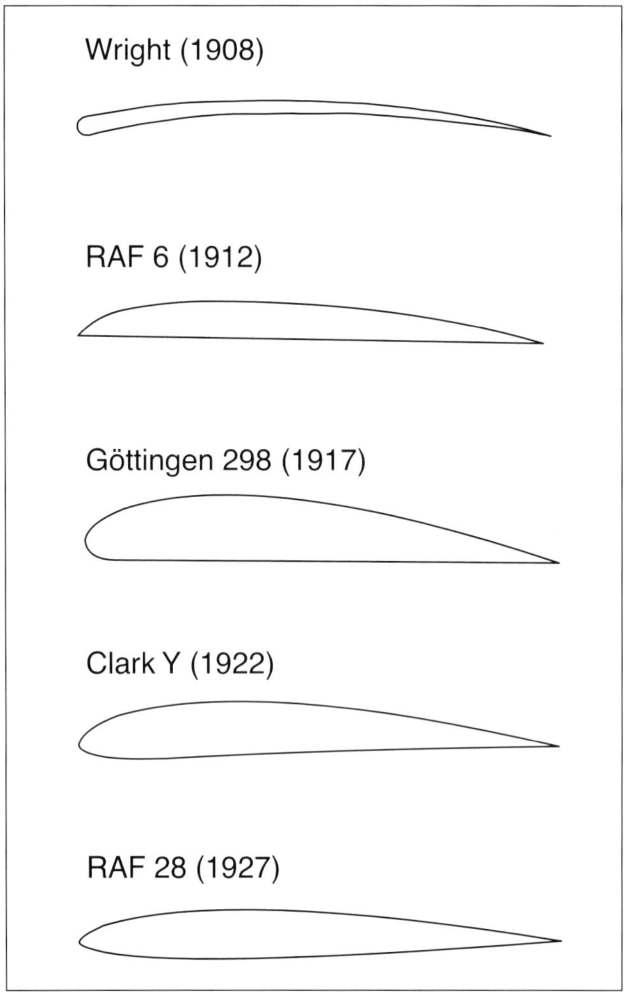

Historical development of aerofoil sections. By 1917, German research was showing that contrary to previous tests (and the almost intuitive but mistaken notion that thin aerofoils had less drag) thick ones were better from both drag and lift points of view. The Göttingen 298 section was used with great effect on the Fokker DVII. *(Pete West/AIR International).*

all World War One biplane fighters had very thin wings (thickness/chord ratios no more than 6%) with relatively sharp leading edges.

However, Prandtl's research was revolutionary since it showed that as the Reynolds number approached one million a thicker aerofoil with a blunt leading edge developed much greater maximum lift and a better L/D. This led to a German family of sections that were of distinctly new form having much greater thickness than hitherto. The Fokker Dr I Triplane (as flown by the 'Red Baron', von Richthofen) was the first to use the Göttingen 298 (a 13% thick aerofoil section with a generously rounded leading edge). There were two major benefits: sufficient room within the wing to enclose the structure without the need for conventional wire bracing, and hence lower drag; and its higher maximum lift coefficient gave it increased manoeuvrability. The same aerofoil was used on the Fokker DVII designed in 1917 which had excellent stall characteristics. These not only made the aircraft easy and safe to fly by inexperienced pilots but permitted it, in skilled hands, to totally out-manoeuvre the opposition and made it the most feared of all German fighters. For example, it was able to 'hang on its prop', and its rate of climb was far better than the Sopwith Camel and the Spad XIII, its principal opponents, which used thin aerofoils. When the British captured their first Fokker DVII in June 1918 they made a 1/20 scale model and tested it in a wind tunnel, but because of the rampant scale effects at the low Reynolds number (70,000) of the tests the benefits of its thicker and blunter wing section were completely hidden. The conclusion was therefore that the thicker section had been used for structural strength, and definitely not for aerodynamic reasons.

Table 4 Use of early aerofoils

Aircraft	C_{D_0}	*Aerofoil*
Nieuport 17	0.0491	Thin with a relatively sharp leading edge
Spad XIII	0.0367	Thicker with a more rounded leading edge
Fokker Dr I	0.032	Thick with an even blunter leading edge

Biplane configuration

At the start of World War One both biplanes and the rarer monoplane were appearing in cleaned-up form with enclosed fuselages and fineness ratios (fuselage length/diameter) of about 4. However, with the mass of bracing wires and fixed undercarriage, speeds were insufficient to demonstrate much value of any streamlining. Thus the biplane, supplemented by the triplane, was the accepted configuration.

It was realised that the greater the vertical gap (G) between the wings, the smaller the lift interference. Whereas, at best, a monoplane might have an L/D of 17, a biplane's L/D was 14, and that of a triplane was 12, all at around 3 to 4° angle of attack (AoA). The gap was usually about equal to the mean chord (c); with G/c greater than 1.0, the structure was weaker, and with G/c less than unity, excessive interference occurred. Consequently, any departure from G/c=1, was usually for reasons of visibility.

A further aerodynamic disadvantage of biplanes was that in addition to lift interference, the trailing vortices from each wing interacted with one another. Thus additional drag was produced on each wing by the presence of the other and had the same effect as lowering the aspect-ratio. However, it should be remembered that in those days aircraft could be made relatively cheaply and modified quickly. Hence, despite the lead given on lift-induced drag by Lanchester in 1915, the drag of 3-D wings was treated as an entirely empirical subject throughout World War One. Designers generally arrived at the 'best' aspect-ratio (between 3 and 5) for the prevailing conditions and did not greatly feel the need for a precise means of calculating the induced drag.

The use of overhang was regarded as a bad design feature and spans needed to be equal to keep the induced drag to a minimum. The increased induced drag, by reducing the aspect-ratio of one wing, tended to offset the reduction in interference drag between the two wings. The 'equal winged' solution proved to be the best. With a gap of 15% of the span, the equal-winged biplane had about 20% less drag than the monoplane of like span and total area.

Stagger (meaning the position forward of the upper wing

The Bristol M.1C Bullet which first flew in 1916 was an aerodynamically clean monoplane and achieved 114kts with only 80kW from its le Rhône rotary engine. However, it succumbed to technical opposition by virtue of the monoplane's supposed lack of structural integrity. Only 125 were built, all relegated to operations in secondary theatres of war. *(Philip Jarrett).*

In spite of its uncowled Jaguar radial engine, fixed undercarriage, open cockpit and externally braced wings, the Armstrong Whitworth Siskin was regarded in its day (introduced in 1924) as a high performance fighter. This particular aircraft is a Siskin IIIA operated from Tangmere in 1930–31 by 43 Sqdn 'Fighting Cocks'. It displayed an unusual wing arrangement in that the top wing had three times the area of the lower wing. *(via author).*

One of the most influential aircraft evolved in Germany during the early 1930s was the Heinkel He 70. Among its many aerodynamic refinements was a smooth, oval section monocoque fuselage, retractable undercarriage and an elliptical planform wing. The latter was chosen to minimise induced drag. It was so aerodynamically clean that Rolls-Royce used one to test the drag of engine radiators. *(Philip Jarrett).*

in relation to the lower) was usually dictated by structural considerations or field of view for the pilot. Its effects on lift and drag were of secondary importance and insignificant compared to those that appeared beyond the stall, where its influence on stability and spinning characteristics was much more serious. At high AoA the upper wing was shielded (negative stagger increased shielding, and vice versa).

Engine power during the war increased greatly, but the fastest fighters of 1918 did only 125 to 130kts (233 to 240km/h) even though they had more than twice the power of pre-war aircraft that flew at 115kts (210km/h). Aerodynamically, the efficiency of the aircraft had improved little, especially in the UK. It was remarked in 1923 by Professor Sir Leonard Bairstow, vice-chairman of the Aeronautical Research Committee, that 'Our war experience showed that while we went forward with regard to horsepower, we went backwards with regard to aerodynamic efficiency.'

Between the Wars

Aerofoil sections

Following the end of World War One the two lines of development being pursued by the Allied nations and Germany began to draw together. However, since the British tendency was to favour the biplane, the UK continued to use the relatively thin RAF15. It was only the potential for increased speed that forced the development, in the UK, of thicker and lower-drag aerofoils such as the RAF28 (1927) and RAF32 (1939) for monoplanes.

The UK's reluctance to dispense with the biplane was based on some good arguments, though with false assumptions. For instance, it was clear that the monoplane wing would be heavier than the biplane wing. The biplane structure is deeper, structurally more efficient and could be made lighter. In addition to reservations regarding its structural integrity, early development work on the monoplane was held back because of the misleading wind-tunnel data. High drag coefficients were found from tests because all the available wind-tunnels had much higher turbulence than the atmosphere in which an aircraft flew, although this was not generally known at the time. So if the drag of the biplane's struts and wires was compared with the drag of the monoplane's thicker wing, it was assumed that there would be little to choose between the two. It was concluded that for a monoplane to be more aerodynamically efficient than a biplane, it would have to have a smaller wing area (i.e. higher wing loading). If the wing area was made such that the top speed occurred at the AoA for maximum L/D, then the minimum permissible flight speed would result in longer take-off and landing runs and thereby limit its usability, and be so high as to make landings very hazardous. Thus, designers were forced to use a larger wing area for the sake of safety.

At the same time, the full-scale wing surfaces were often rough or inaccurate in shape, with heavily ribbed or stretched fabric, corrugated light alloy, wrinkled plywood or a mass of snaphead rivets along lapped joints. It was later realised that the drag of the bracing struts, wires and interference of the biplane had been underestimated and that the drag of the wing itself had been overestimated, thus giving a quite misleading picture of full-scale drag. In many cases it could, and it was proved that monoplanes, especially low-wing ones, had no advantage over biplanes. However, it was discovered in the 1930s that this was caused by interference at the wing/fuselage junction. This caused not only additional drag but sometimes extremely unpleasant stalling characteristics.

High-lift devices

As a means of increasing lift for landing and take-off, trailing-edge flaps were first used on the SE4 (1914). The value of such flaps on biplanes was limited due to the thin aerofoil sections they featured and the leading-edge stall from which they suffered. This was sudden and usually vicious. Using a trailing edge flap increases the leading-edge suction peak and thereby increases the tendency to leading-edge flow separation. It was as a result of these undesirable characteristics that biplane pilots of the day were often reluctant to use their flaps.

A crucial development was that of the slotted wing. The slot effect was discovered independently in quite different ways between 1917 and 1919. Handley Page carried out tests in 1917 on a wing with a spanwise slit near the leading edge through which air could flow from the lower surface to the upper surface. With a good deal of effort to produce an efficient nozzle shape, from lower to upper surface, the results were very encouraging. Not only was the stalling angle increased but the maximum lift was raised by 50 to 60%. Almost simultaneously Lachmann in Germany made the same discovery, and both men subsequently collaborated on developing the concept. Use of the slot on the thin aerofoils used by biplanes was especially beneficial. For normal flight the slot would remain closed, and when needed for high AoA flight the leading-edge slat would move forward to open the slot. At first a control mechanism was needed to move the slat forward, though it was eventually realised that the resultant aerodynamic force on the slat changed with AoA in such a way that automatic opening and closing could be obtained. Though the slatted slot received appreciative notice from many quarters during the 1920s it was not flattered by adoption.

While the slot had been originally conceived to increase high AoA performance, lowering landing speeds were not a pressing matter because wing loadings were very low. In any case, slots were not very useful as a means of reducing landing speeds. Because they benefit high AoA flight only, an aircraft would have required a very long landing undercarriage to take full advantage of them. At the other end of the speed spectrum, the profile drag of contemporary aircraft, strutted and wire-braced, with fixed landing gear and bristling with other paraphernalia, was such a big component of total drag that any benefit of slots to reduce wing area was insignificant. Furthermore, the cumbersome operating mechanisms for the slats were not popular with pilots, and precluded their use as a safety device, despite there being a pressing need.

The better an aircraft is made from the point of view of low drag and high speed, the more difficult it is to land. The reason for this is twofold. Firstly, the gliding angle is reduced by the low drag so the aircraft comes in on a less inclined flight path and travels further over the ground before it touches down. Secondly, the low drag means less deceleration during the ground run. Aircraft of the 1920s were easy to land by virtue of their low stalling speeds and high drag. The need for 1930s aircraft was to make them temporarily as inefficient as those of the previous decade.

Trailing-edge flaps were very effective, for not only did they produce high maximum lift so that landing speeds could be reduced, they also increased the drag.

Another difficulty was that of take-off. High drag is not required and is a hindrance to acceleration, whereas lift increase is advantageous since it reduces the take-off speed. The split flap was used at angles of about 15° to give appreciable extra lift without much increase in drag. However, the slotted flap developed by Handley Page was better because it gave much more lift with very little increase in drag and was still as effective as the split flap for landing when deflected to 60°.

Aerodynamic cleanliness

In terms of aerodynamic efficiency, as measured by L/D ratio, the best that had been achieved by a biplane in 1912 was about 7:1. Even by the late 1920s little improvement in service aircraft had been made and an L/D of about 10:1 was typical for an aircraft with a water-cooled engine (those with air-cooled engines were slightly worse). The increase in top speed from 65kts in 1910 to 310kts in 1929 was due more to an increase in available engine power than to cleaner aerodynamic design.

By comparing the drag of contemporary aircraft with an ideal aircraft having only skin friction and induced drag, Professor Melville Jones in the UK was able to show in 1929 that nearly all contemporary aircraft had two to three times more drag than was inevitable. He showed conclusively that some aircraft would need only a third of the power with which they were actually provided, *if* they conformed to his ideal aeroplane. Only Lindbergh's Ryan NYP monoplane *Spirit of St Louis* came anywhere close to the ideal. Designers, well aware of the drag associated with external wing bracing, exposed undercarriage and uncowled engines, were inclined to the view that the penalties of weight and the complication of eliminating each of these drag-producing items would more than offset the advantages. What they failed to appreciate was that if all the improvements were made simultaneously, the total gain would not be some insignificant percentage of the whole but might reduce power requirements to a fraction of what was actually being used. However, many British designers were reluctant to accept the inevitable design complexity which was needed, with the consequence that both America and Germany had streamlined aircraft in series production long before the UK.

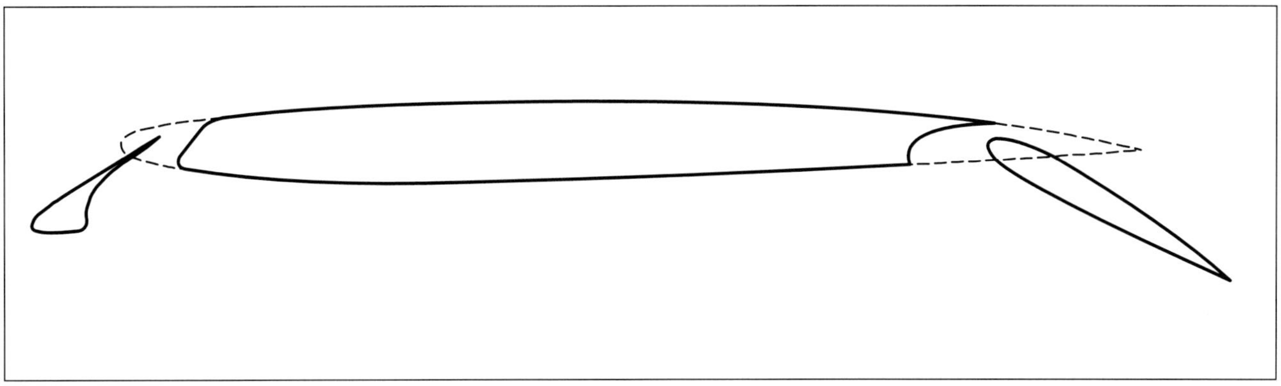

Wing with slat and Fowler flap, showing how camber and wing area are increased, together with boundary layer control via the slots.

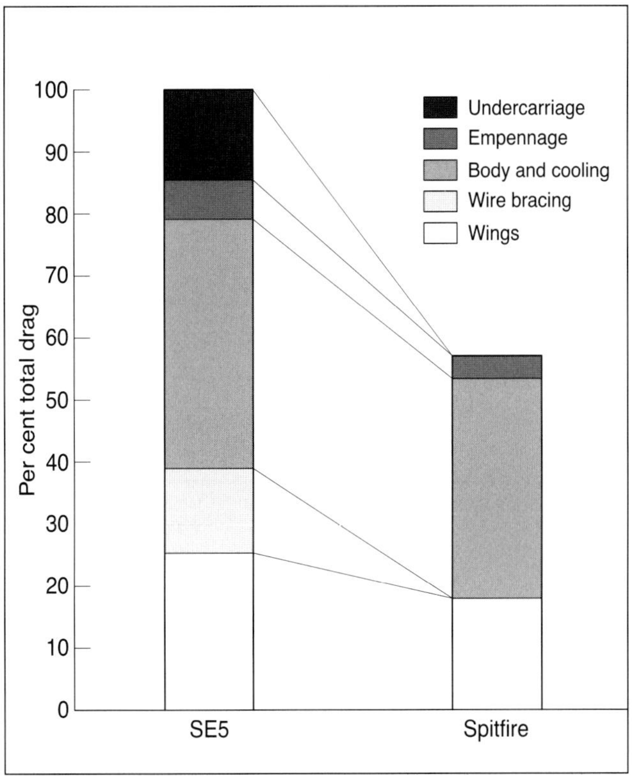

Drag comparison between the SE5 and Spitfire at the same speed (60kts) in a wind-tunnel. The figure shows the significant drag reduction (over 40%) obtained by moving from the braced biplane with fixed undercarriage to the cantilever monoplane with retractable undercarriage.

Wing/fuselage aerodynamic interference

Considerable investigation had been carried out by the early 1930s on the aerodynamic interference generated at wing/fuselage junctions. In particular, the low-wing configuration was found to be by far the worst arrangement from the point of view of drag. Putting a circular fuselage on top of the wing has the effect of producing a pair of rapidly diverging surfaces which steepens an already adverse pressure gradient and almost guarantees flow separation and its attendant drag. Not only that, but the lift of the wing was reduced and, more important still, the separated flow would likely strike the tailplane and cause stability and control problems. Why, then, did the low-wing design become so popular? The discovery of the beneficial aerodynamic effects of a fillet or wing/body fairing, at CalTech in the USA, largely overcame the problem. What little inferiority remained for the low wing was more than compensated for by its benefits, which included provision

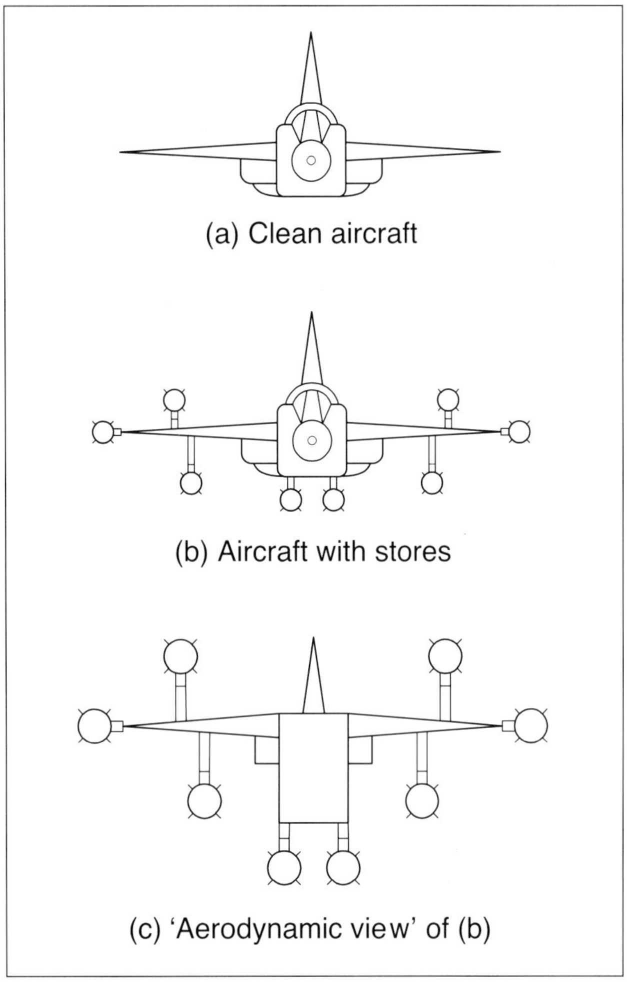

Designers need to minimise adverse aerodynamic interference and nowhere is this more important than when considering external stores. The figure contrasts the true geometric view of an aircraft carrying a large load of stores with the 'aerodynamic view' in which the frontal area of each component has been scaled up in proportion.

for a short and easily retractable undercarriage and the cushioning ground effect on landing with a wing close to the ground.

Laminar flow

Prandtl proposed the concept of the boundary layer in 1904, though in UK literature the term does not appear until 1925. The existence of a laminar boundary layer (which gives minimum skin friction drag) was unsuspected in the UK until 1924 when experimental results demonstrated its reality. With the eventual overthrow of the biplane in favour of the monoplane with retractable undercarriage, drag reduction of the basic configuration concentrated on detail design, and during the late 1930s much attention was paid to this. Tests at the National Physical Laboratory (NPL) on a wing of 20cm (8in) chord at a Reynolds number of seven million had shown that excrescences of about 0.025mm (0.001in) resulted in a drag increase over the smooth aerofoil of about 70%. On a full-scale wing of 2.13m (7ft) chord this corresponded to the roughness of fabric-covered wings of the time. These results ultimately led to more care over aircraft construction. In addition to keeping a smooth leading edge where flush riveting had to be employed, advantage could be gained by retaining a smooth surface back to the trailing edge, since the rear half chord of the wing was found to produce about a third of the drag.

Even though wind-tunnel results showed little chance of much laminar flow on aircraft wings, Melville Jones carried out flight tests with a Hawker Hart in 1937 at realistic Reynolds numbers and found it. To increase the extent of laminar flow it was found that the position of maximum wing thickness should be around 40 to 45% chord and that the section should be thin. Reducing the thickness of the aerofoil has the added benefit of minimising pressure drag, caused by the distortion of the flow around the section due to the effects of air viscosity. These results undoubtedly inspired US researchers to develop the NACA 6-series laminar flow aerofoils with pressure minima far back along the chord. Although these sections have been widely used (first on the P-51 Mustang), the full benefits of laminar flow were never realised. This was because laminar flow requires

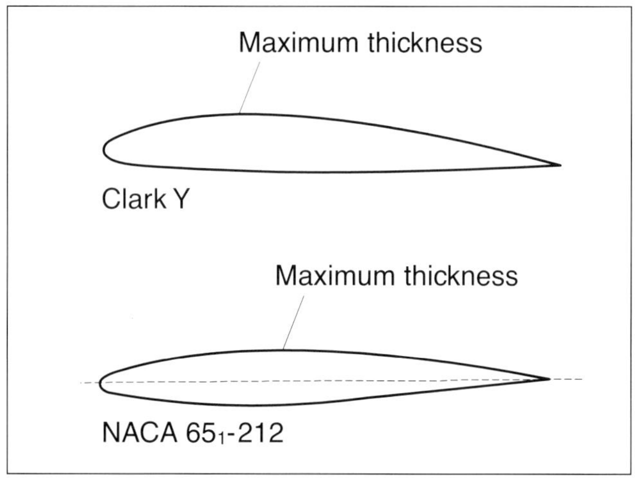

Comparison of a NACA 6-series laminar flow aerofoil with maximum thickness at around 40% chord with a typical older type, the Clark Y with maximum thickness at around 28% chord.

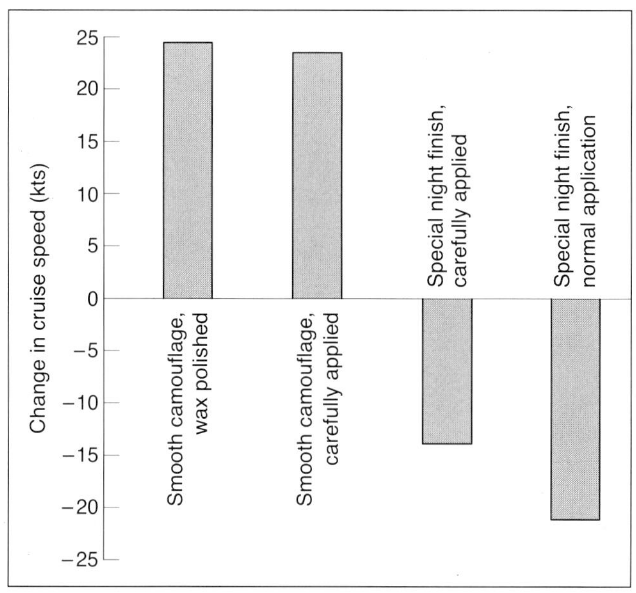

Effect of surface finish on cruise speed. Analysis carried out at RAE Farnborough in 1943 on a North American Mustang showed the following effects on cruising speed (306.5kts as received from the RAF): A general cleaning up from a normal service condition gave a 24kt increase in the cruising speed and the night finish caused a 45.2kt reduction in speed compared with the clean condition, equivalent to a drag increase of 34 per cent on the smooth camouflage condition.

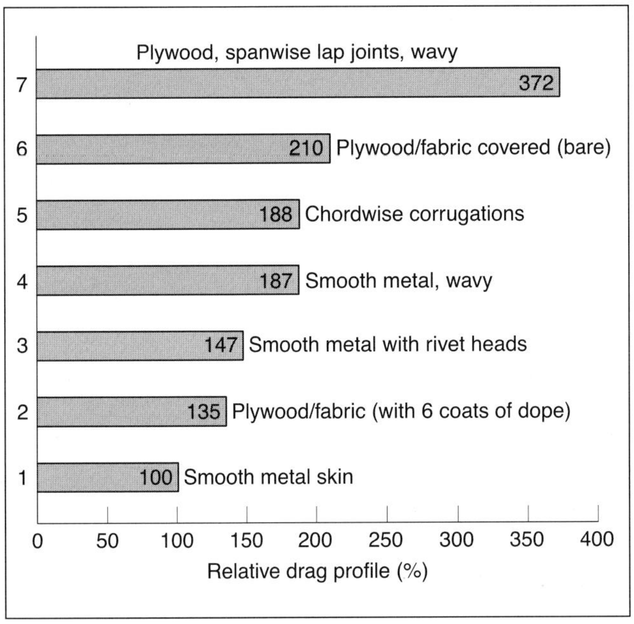

The effect of surface condition on profile drag: result of flight tests in 1928 on a number of typical wing surfaces on a Junkers aircraft having a thickness/chord ratio of 17%.

a wing surface to be manufactured and maintained in a very smooth condition. This could easily be contaminated by the build-up of insects or dirt on the leading edge, and wings without roughness and waviness, though achievable on modern sailplanes, have never been available for production fighters. However, they did have the virtue of a high critical Mach number (delaying the onset of the drag rise), wherein lay their greatest practical benefit.

World War Two

Compressibility

Up until midway through World War Two, compressibility effects on wings of service aircraft had not been serious and aerofoil sections with a low minimum drag had been satisfactory. As speeds increased, consideration had to be given to compressibility and thinner aerofoils of lower camber with the maximum thickness at about 40% chord were needed. This created installation volume and structural strength problems since the best solution of these was plenty of thickness well forward.

The first Lockheed YP-38 Lightning flew in 1940, and a year later the USAAC's first 350kts aircraft ran into severe problems in a high-speed dive. As the aircraft accelerated in a dive, the control column shook violently, control forces became very heavy and it was impossible to pull the control column back to counteract a nose-heavy tendency. This was the first reported flight experience of severe compressibility. The P-38 designer ('Kelly' Johnson) knew that the aircraft with its high speed could readily get into compressibility in a dive and was not surprised by the flight report. Its wing had been designed so that in level flight there would be a reasonable speed margin before compressibility struck. This had been done with what experimental data were available, mostly 2-D aerofoil tests combined with the limited amount of theoretical understanding of the flow phenomena. However, this work had concentrated on estimating how aerofoil drag would vary with Mach number, but, what kinds of forces and consequences would be encountered as the aircraft exceeded the estimated critical Mach number of 0.68 were unknown. The impact of shockwave/boundary layer interaction and trailing-edge separation and its effect on control hinge moments had been waiting in the wings, yet to be studied, tested and understood.

The birth of sweepback

The use of sweepback to reduce drag at supersonic speeds was first proposed by Adolf Busemann (a German aerodynamicist) at the Volta Congress for High Speed Flight in Rome in 1935. The concept was promptly classified secret by the *Luftwaffe* even though its application to aircraft was not then regarded as a practical proposition (Britain had yet to put into service the Gloster Gladiator biplane with a top speed of 220kts, only one third the speed of sound). By 1939, Albert Betz of Switzerland, another aerodynamicist, had shown that sweepback could also be used to delay the onset of compressibility problems arising from local regions of supersonic flow terminating in shockwaves in the transonic flight regime. By joining together a jet engine, which has acceptable propulsive efficiency when it is flying fast enough, with a swept wing that could fly fast without encountering the transonic drag rise, one could produce a high-speed aircraft with a good payload/range capability.

German aircraft with sweepback appeared in the early 1940s, notably the Messerschmitt Me 163 Komet tailless rocket fighter. In this design, however, sweepback was needed principally to give an acceptable margin between the centre of gravity and aerodynamic centre, and to increase the moment arm of the control surfaces on the wing's trailing edge. Sweepback's high-speed aerodynamic benefits were seen as largely incidental. The twin-turbojet Me 262 was originally designed with straight wings, but because it was one of the first high-speed aircraft not to have the forward weight concentration of piston engines, it turned out to be tail-heavy. In order to restore the balance between the lift and centre of gravity the wing outboard of the engine nacelles was swept back. Eventually the aircraft was given a leading-edge sweepback of 18½°, a modest level of sweep that had little effect on drag.

The Jet Age

Following in the wake of the Allied occupation forces, research teams found vast quantities of data which showed the enormous advances made by German research into high-speed aerodynamics. Sweepback was incorporated into many of its future jet fighter designs. Though the significance of sweep was immediately clear, by 1945 several completely new families of transonic wings were being discussed without anyone having sufficient practical knowledge to recommend which type should be chosen. Nevertheless, sweepback was quickly put to use by the United States (North American F-86 Sabre and Boeing B-47, both of 1947), the Soviet Union (MiG-15 of 1947) and Sweden (Saab 29 of 1948). The reluctance of UK authorities to adopt the swept wing in 1945–46 lost the country a great opportunity that subsequent British aircraft failed to recover until the late-1950s.

The major benefits of sweepback were recognised as being postponement and reduction of subsonic drag rise,

Europe's first swept wing jet fighter to enter service was the Saab-29, aptly nicknamed Tunnan or Flying Barrel, the first of which flew in 1948. A swept wing, having a significant spanwise component of velocity, encourages a thickening of the boundary layer with the likelihood of premature flow separation, possibly leading to pitch-up. A wing fence produces a vortex outboard of its location to resist the loss of lift towards the wing tip. Shown are Saab 29Fs of Flottilj 3 from Malmslätt in 1963 armed with Rb 24 (AIM-9B) Sidewinders. *(I.Thuresson via author).*

reduction of transonic trim changes, and more gradual change in aerofoil lift across the transonic regime. There was also the promise of significant drag reduction at supersonic speeds. For all its useful attributes, though, experience was to show that sweepback also brought with it some serious problems. These included reduced lift at a given AoA and reduced effectiveness of trailing-edge flaps, both of which resulted in increased take-off and landing speeds. Because, simultaneously, wing loading was rapidly increasing, the problem was compounded, though some relief was obtained by the reintroduction of the leading-edge devices. Other repercussions of sweep are: increased lift-induced drag, increased rolling due to sideslip, tip stalling and pitch-up, and structural (aeroelastic) problems.

Aerofoils

With its 13% thick root and 7% thick tip the wing of the Spitfire was atypical and enabled it to reach a very high speed in dives, the highest recorded being Mach 0.92. Early jet fighters had thickness/chord ratios of 12 to 14%, typical of their more common piston-engined predecessors. German research during the 1940s showed a clear appreciation of the value of reduced thickness for high-speed flight. The thinner sections not only had lower drag coefficients but also delayed the onset of drag rise due to shockwave formation. The former was the main argument in favour of thin sections, which, together with sweepback, has been successfully employed on all subsequent high-speed aircraft.

A typical example of early high-speed aircraft design philosophy was that adopted by North American for its F-86 Sabre. This had 35° leading-edge sweep with the thickness ratio held at 12%. It was felt that the delay in the drag rise did not warrant the structural weight penalty of a thinner wing. In addition, a wing of conventional thickness, permitted by the use of sweepback, maintained reasonably high maximum lift and provided room for both leading- and trailing-edge devices, together with useful fuel volume and stowage for the undercarriage.

At higher speeds, thick sections are impractical because the excessive airflow acceleration over the surface produces velocities which exceed the local sonic velocity and lead to premature shockwave formation. This can be delayed by more sweepback, since 30° sweep plus 10% thickness is roughly equivalent at subsonic speeds to 40° sweep plus 12% thickness. Since the wave drag due to thickness is proportional to thickness ratio squared, the demand for aircraft capable of penetrating the fully supersonic flight regime caused thickness/chord ratios to tumble. The successor to North American's F-86 was its Mach 1.2 F-100 Super Sabre with a wing of 45° leading-edge sweep and 6% thickness. The alternative was to ignore the swept wing and to reduce thickness even more. The Mach 2 Lockheed XF-104 Starfighter had stubby unswept wings and a thickness ratio of 3.3% (giving a maximum thickness at the root of barely 10cm), with such a small leading-edge radius that it could only be regarded as sharp.

The reduction in thickness demanded by the need for a limit on wing and trim drag (from the tailplane) at supersonic flight produced wing sections whose leading-edge radii and camber were also reduced. Consequently, trailing-edge boundary-layer separation, typical of the older, thicker and more cambered sections, gave way to leading-edge separation during subsonic manoeuvres. This caused wing buffet at relatively low AoA, so that at higher lift, buffeting became so severe as to render the aircraft ineffective as a gun or missile platform. The large regions of separated flow also led to poor handling, with consequent restrictions on allowable AoA, and prompted the development and use of manoeuvre flaps in the late 1960s.

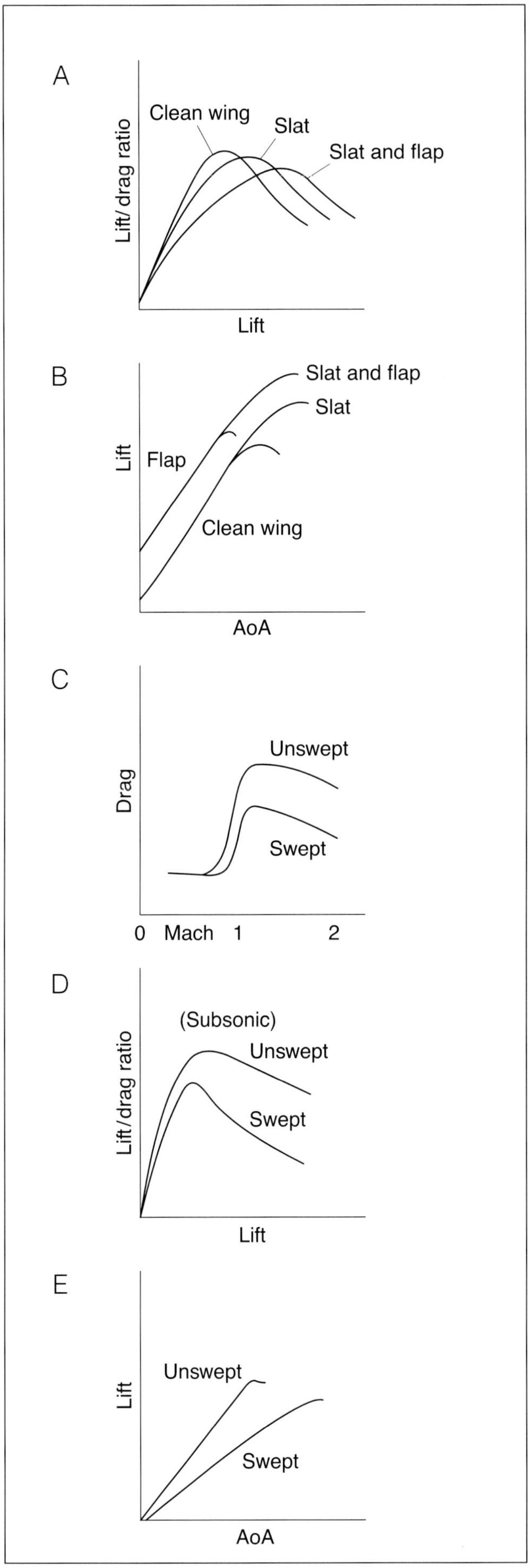

(A) At high AoA the lift/drag ratio of a clean wing decreases with lift. Some of the loss can be recovered by the use of high lift devices.

(B) Use of a trailing edge flap causes the lift curve of a clean wing to move upwards to the left. This increases maximum lift but decreases stalling angle. By employing a leading edge device (e.g a slat), further increase in lift can be obtained and the stalling angle recovered.

(C) Use of sweepback both delays the drag rise near Mach 1 and reduces the peak drag.

(D) At subsonic speeds the use of sweep has the penalty of reduced lift/drag ratio. This is more than compensated for by the lower drag at supersonic speed though.

(E) Another penalty of sweep is the lower slope of the lift-versus-AoA curve. This implies flight at higher AoA to achieve the required lift and increased lift-induced drag.

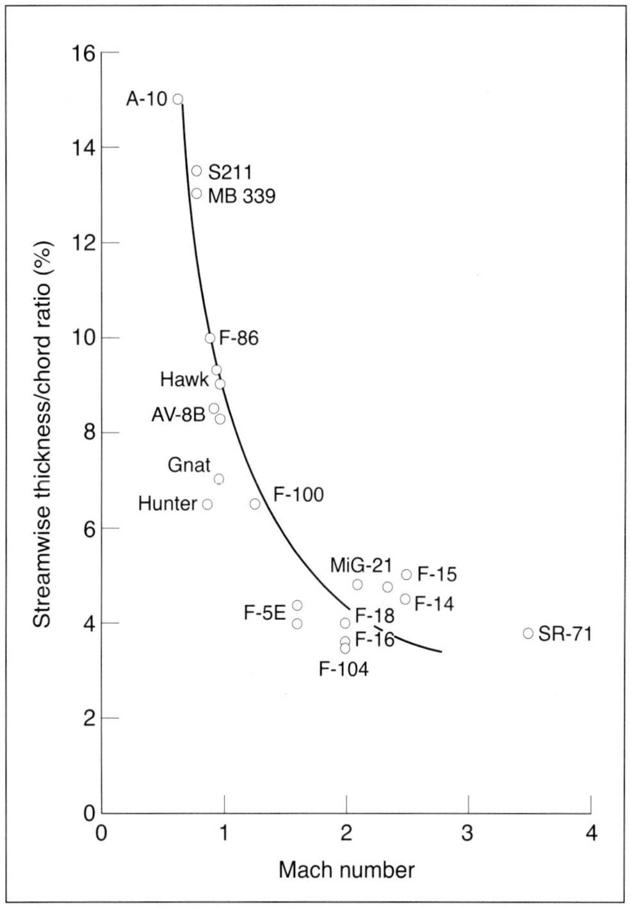

Reduction of thickness/chord ratio with Mach number.

Delta wings

German theoretical and wind-tunnel data on transonic wing design included studies of delta-wing planforms. The delta planform was seen to offer structural and fuel volume benefits, a more gentle transonic drag rise and, with its usually lower W/S, improved manoeuvrability, especially at high altitude. Many countries experimented with the delta planform for both subsonic and supersonic designs, and it became a common configuration during the 1950s in both tailed and tailless forms.

High supersonic performance led Dassault to opt for the delta planform in 1952. At that time the French company was working on a light supersonic fighter and expected that

Swept wings do not lift as well as straight ones. This is an acute problem for shipboard aircraft that need the slowest approach speed for safety. One solution was that adopted by Chance Vought for its supersonic F8U Crusader of 1955. The aircraft had full-span leading-edge flaps, trailing-edge flaperons, but more remarkably, a variable-incidence symmetric section wing (+8° for take-off/landing). So well did the combination work that, on at least one occasion an F8U took off with its wings folded, though not from a carrier, and was recovered safely. This F-8C belonged to VF-84 'Jolly Rogers' operating from USS *Independence. (US Navy via author).*

a development of its classic swept-wing Mystère would suffice. More sweep and a lower thickness/chord ratio were required for Mach 2, so much so that the conventional swept wing came out with a maximum thickness of only 10cm (4in). It would have been extremely heavy, being almost solid metal, and would have had virtually no room for fuel. Consequently Dassault opted for the delta-wing Mirage III instead, which had a 60° leading-edge sweep and 5% thickness ratio. Its increased chord length permitted a maximum wing thickness of around 30cm (12in).

The Area Rule

In 1952 the National Advisory Committee for Aeronautics (NACA), now NASA, published the Area Rule developed at NASA Langley by a team under Richard Whitcomb. This is a simple empirical law that, if properly applied, could result in minimum transonic drag. It relates the build-up of cross-sectional area along the length of the aircraft to its drag at transonic Mach numbers. This had a profound effect on the outcome of the USAF's ambitious programme to produce the first supersonic interceptor, the delta-wing

First flying in 1951, the Douglas F4D Skyray or 'Ford', as it was known, was designed by Ed Heinemann to meet a 1947 requirement calling for an interceptor to reach 40,000ft in 5 minutes or less. It had a very fast roll rate that enabled the Skyray to change direction quickly. One major shortcoming though was lack of fuel and it became known as the 10-minute killer, and large external (300US gal) tanks were routinely carried as shown. Unfortunately it also suffered transonic 'Mach tuck' characteristics with a nose-down trim change on acceleration that could be handled but produced a vicious nose-up tendency on deceleration. To combat these effects it was fitted with a transonic trim compensator. *(via author).*

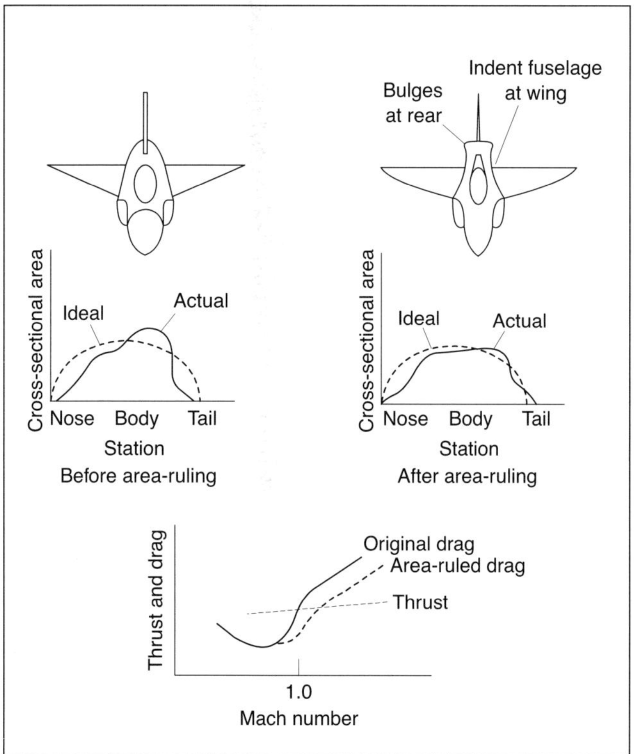

With higher drag than predicted, the original Convair F-102 was unable to go supersonic. Applying the newly formulated Area Rule to reduce transonic drag involved major revision of the fuselage profile to give reduced cross-sectional area over the wing, and filling in the area diagram by the addition of large bulges flanking the jet pipe and exhaust nozzle.

Convair F-102A. Its development from the YF-102, which did not conform to the Area Rule and the improvement that resulted, became a classic in the annals of fighter design.

The flight test programme of the YF-102 in the spring of 1954 had progressed normally until the flight envelope was extended towards the transonic region. It became clear that wind-tunnel results had been optimistic; transonic drag of the aircraft was greater than the available thrust and Convair's supersonic interceptor was subsonic. This put the whole programme in jeopardy. The US Air Force had stipulated a supersonic Mach number in the original specification, and complete failure of so fundamental a nature was sufficient grounds for cancellation. The aircraft was equipped with the most powerful flight-cleared engine in the Western world and somehow drag had to be reduced, and by a wide margin.

Changes to reduce transonic drag involved major revision of the fuselage profile to give reduced cross-sectional area over the wing, and filling in the area diagram by the addition of large bulges flanking the jet pipe and exhaust nozzle. The wings were given conical camber (progressive increase in leading-edge camber from the root to the tip, giving a marked negative AoA over the outboard portion) to reduce drag at high altitude and to improve behaviour at high AoA. The fuselage was lengthened and the nose was sharpened and lowered to reduce drag and improve visibility. The canopy was redesigned with a modified windscreen and a narrower hood with flat side panels. The air intakes were cut back to improve pressure recovery, and behaviour at the highest Mach number and a higher thrust model of the Pratt & Whitney J57 engine was installed. The first YF-102A flew supersonic in a climb on its second flight in December 1954.

Variable-sweep wings

Swing wings had been investigated in the USA in the early 1950s with two aircraft, the Bell X-5 (of Messerschmitt P.1101 origin) and Grumman XF10F-1. Similar British studies were based on wind-tunnel and free-flight model testing, mainly by Barnes Wallis of Vickers-Armstrong. It was recognised that, although variable sweep had promise, a practical means of employing the technique had yet to be found. The major obstacle was that adequate longitudinal stability for satisfactory handling at low sweep meant excessive stability at high sweep. The real breakthrough came in 1952 when NACA proposed the outboard pivot. It was shown that the lifting area of the wing could be divided between fixed and movable portions in such a manner that the movement of the centre of lift, caused by sweep variation, need not present a serious excess-stability problem.

A combination of three performance requirements stimulated enormous interest in swing wings in the 1960s. This arose primarily because, by 1957, the USAF wanted, as a replacement for the Republic F-105 Thunderchief, an aircraft that would combine the conflicting requirements of Mach 1.2 low-level penetration with an unrefuelled transatlantic ferry range and the ability to operate from 1,000m (3,300ft) unpaved strips. A secondary air-to-air role demanded Mach 2.5 at high altitude. Simultaneously, the US Navy was being urged by the Department of Defense, ostensibly for reasons of economy, to consider whether a broadly common aircraft would meet its need for fleet air defence. From these requirements arose the Tactical Fighter Experimental (TFX) programme that spawned the General Dynamics/Grumman F-111A for the USAF and F-111B for the USN (eventually cancelled).

Good subsonic performance requires a minimum-sweep design, that is, a high-aspect-ratio wing of relatively thick section generating high lift and of low span loading giving low induced drag. The great value of variable sweep lies in the fact that it allows the wing planform to be changed in order to obtain, within limits, the optimum sweep for each Mach number. Provided that the cruise range required was long enough, the improved aerodynamic efficiency was seen to offset the penalties of increased weight and cost, and reduced fuel volume.

For high-supersonic flight and the transonic low-altitude strike role, the wings need to be fully swept, so that the shockwave drag due to volume (that part of the wave drag due to the bulk of the aircraft, which is independent of lift) is minimised. Furthermore, if the crew are to have a smooth ride and airframe fatigue life be acceptably long, the aircraft must be insensitive to the vertical gusts likely to be met in the turbulent atmosphere at low altitude. The most powerful means of achieving this is via a combination of reduced aspect-ratio and increased sweep. For operation from limited-length runways or aircraft carriers, maximum lift must be generated. This is best achieved by thick-section, unswept wings equipped with high-lift devices.

The use of variable sweep wings on the F-111 allowed the apparently irreconcilable demands of the three missions to be met in a single design. It was a solution adopted for several other projects of the 1960s and 1970s, including the Panavia Tornado, Grumman F-14 Tomcat and several Russian fighters.

Exceptionally clean and fast at low level (Mach 1.2) the Republic F-105 Thunderchief could easily outrun MiGs over Vietnam, and even their missiles in later stages of their flight. The F-105B was one of the first applications (1955) of area-ruling as shown by its 'Coke bottle' fuselage. It also featured unusual forward-swept air intakes to its Pratt & Whitney J75 afterburning turbojet. *(via author).*

The quest for higher manoeuvrability

Superior transonic manoeuvrability is a prime requisite for a modern fighter aircraft. The high turn rates necessary for combat come from flight at high AoA, which normally result in flow separations on the wing. These separations lead to increased drag, buffet, and stability and control problems that degrade combat effectiveness. Such problems afflicted most supersonic aircraft designed in the 1950s and later. The development of increasingly higher thrust-to-weight (T/W) ratio aircraft capable of high sustained turn rates expanded the flight envelope and gave great impetus to the quest for good high-AoA transonic flight characteristics. Ideally, as an aircraft is brought to its limiting AoA condition, it should still have a tendency (inherent or artificial) towards a reduction of the AoA. In all events, the aircraft should remain controllable. Although some deterioration in handling qualities is inevitable, it should be progressive rather than abrupt, and distinctive enough not to mask the fact that a limiting condition is being approached. In practice many aircraft fell short of this ideal.

In the late 1960s the US Navy launched a major effort to improve the combat effectiveness at high AoA of the McDonnell Douglas F-4 Phantom, Vought A-7 Corsair and Grumman A-6 Intruder. Since then, significant improvements to air combat manoeuvring have been made via leading- and trailing-edge devices. It is now standard practice to design the high-lift devices for use in combat, and not just to improve airfield performance. Use of high-lift devices reduces the AoA for a given lift and flattens the spanwise load distribution to postpone flow separation on the outer wing, resulting in a significant delay in buffet onset.

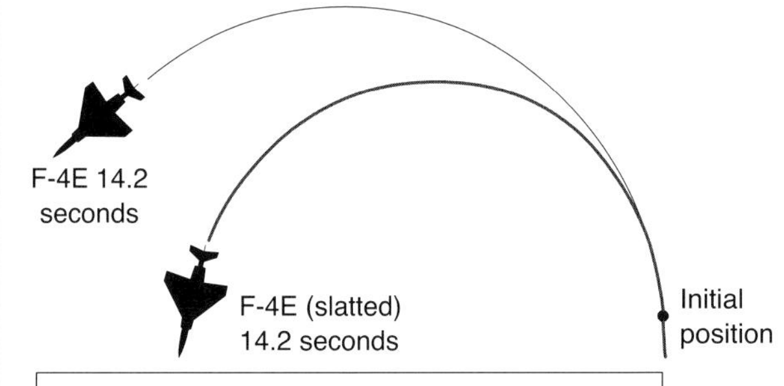

	F-4E (slatted)	F-4E
Time (180°)	14.2 sec	19.0 sec
Radius	889 m (0.48 nm)	1,850 m (0.64 nm)
Gs	4.5	3.5

Addition of leading-edge slats to the later model F-4Es greatly improved the turn rate and radii over the early models. The slats also improve stability and handling qualities.

There are exceptions to every rule. Lockheed opted for a very small straight wing for its F-104 Starfighter, rather than the delta or swept types of contemporary supersonic fighters. Among the penalties incurred were the need for flap blowing (air tapped off the engine and blown over the flap) and a very thin wing (4% thick), with a leading-edge so sharp that maintenance personnel could be injured. Shown is an YF-104A of 1955 with neither ventral stake nor tip tanks that appeared on the F-104A. *(Lockheed Martin via author).*

Vortices streaming back from the wing leading-edge strakes of this airshow demonstration F-16 at high AoA help to greatly enhance lift on the main wing panels. It was the exploitation of vortex lift that enabled this early 1970s-designed aircraft to achieve hitherto unreached flight attitudes. *(Lockheed Martin via author).*

Vortex lift

It will have been noted that, throughout the development of aerodynamics, one of the major wing design considerations has been the avoidance of flow separation. In an effort to counter adverse compressibility effects, wing sweep angles were increased and thickness ratios reduced. These changes, however, conspired to make the maintenance of attached flow more difficult, resulting in many performance, stability and control problems. It stimulated the development of a range of devices, mentioned above, designed to cope with widely disparate flow conditions. Studies in the late 1950s on slender delta wings designed to minimise wave drag for supersonic transports showed that acceptable low-speed performance could be achieved by making the leading edge so sharp as to actually provoke separation along its entire length. This, it was discovered, could produce a highly stable vortex system that would stream back over the wing and enhance the lift at low speeds and high AoA.

In the mid-1960s, Northrop began work on improving the manoeuvrability of its F-5. The company had been the leading exponent of the art of combining moderate sweep and low taper to achieve good flow characteristics and spin resistance, as typified by its T-38 Talon supersonic trainer. When the F-5A appeared in 1963 it differed from the T-38 in having a small leading-edge extension (LEX), intended basically to alter the cross-sectional distribution and so reduce wave drag for maximum transonic acceleration. The Northrop designers, aware of the vortices that the LEX would produce and the lift-induced drag penalty that would result, found that not only was the wave drag reduced, but a 10% increase in usable lift appeared at high AoA, which markedly improved instantaneous turn performance.

The greatest advance in the application of vortex lift to combat aircraft came in 1966, when Northrop began work on the F-5 derivative that became the YF-17 and, ultimately, the McDonnell Douglas F-18 Hornet. The basic YF-17 wing planform already displayed the virtues of the classic moderately swept wing. Its combination with a much larger LEX turned it into a truly hybrid wing, with the LEX generating a very strong vortex field not only on itself but also on the parent wing. The maximum lift capability of the hybrid produced a 50% increase in lift at subsonic speeds for just 10% more wing area. In its flight test programme the YF-17 demonstrated a far wider range of flying attitudes than on any hitherto comparable fighter. For the F-18, other advantages of a LEX included reduced AoA at the under-LEX air intakes and reduced transonic lift centre shift giving lower supersonic trim drag at high g. On the debit side, the separated nature of vortex flow resulted in increased drag at low AoA (though this can be rectified by the use of manoeuvre flaps) and structural fatigue of vertical stabilisers buffeted by the flowfield.

At the time Northrop was developing its YF-17, General Dynamics came to a similar conclusion regarding vortex lift for its YF-16, and the concept was subsequently applied to McDonnell Douglas AV-8B Harrier II, MiG-29 ‘Fulcrum’, and Su-27 ‘Flanker’ with excellent results.

Canard configurations

The foreplane of the canard arrangement is a destabilising surface since it is ahead of the centre of gravity (CG). However, this is not in itself significant, provided that the CG is in the correct position to take care of overall stability. The great benefit of a close-coupled canard layout of the Saab Viggen type, with highly swept surfaces, is that it can produce favourable vortex interactions between foreplane and wing at high lift. In these circumstances the foreplane acts rather like a LEX, inducing a vortex field that improves the pressure distribution on the parent wing. This lift enhancement can be as much as 20 to 30% by comparison with the maximum lift obtained by the surfaces in isolation. The benefit is felt mainly at high AoA, so that manoeuvrability is improved. Another advantage of the canard layout is that both wing and foreplane generate positive lift. This is contrary to the case of the conventional tailed layout, in which the tailplane carries a download for much of the time, particularly during high-lift flight phases. Another major application of vortex aerodynamics has been to the current plethora of fighters of canard configuration.

The newer unstable canard designs (Saab JAS 39 Gripen, Dassault Rafale and Eurofighter Typhoon) use their foreplanes predominantly for pitch control and probably do not achieve much lift enhancement because any lift they produce tends to reduce the lift on the wing. Whereas at supersonic speeds when the centre of lift has moved aft of the CG and the tailplane of the conventional layout has to carry a download to trim the aircraft, the foreplane of the canard configuration has an upload to trim. Another aerodynamic benefit of the foreplane/delta configuration arises from the lower supersonic drag due to the better aftbody lines that result from the absence of tailplane and spigot-supporting structure. Moreover, if thrust-vectoring is used, a canard layout helps to eliminate longitudinal stability and trim changes that occur by having its pitch control surface (the foreplane) as far forward as possible, as on the DASA/Rockwell X-31.

The Saab JA 37 Viggen led the way in 1977 to the later plethora of canard configurations. It is a conventionally stable aircraft of close-coupled canard delta layout to exploit favourable foreplane/wing vortex interference via the fixed foreplane with a plain trailing-edge flap. Of interest in this photo (showing aircraft from F4 Air Force Wing at Östersund) are the fairings over the elevon actuators that are collinear with the ribs that also carry the weapon pylons. *(Peter Modigh via author).*

Chapter 3

Stability and Control

The experiments the Wright brothers conducted led to the unstable biplane of canard design as the favoured layout in the earliest days of flight. European ideas differed and inherent stability became mandatory. As configurations changed to meet the demand for ever higher speeds the problems of stalling, spinning and compressibility were gradually ameliorated. Since the introduction of the jet engine these problems have needed revisiting along with new ones encountered by aircraft designed for supersonic flight. Recalling the Wright Flyer, the wheel has now turned full circle, in a sense, when one examines the crop of unstable canard-configured fighters that emerged in the 1990s.

Glossary

Stability

Static stability: An aircraft whose initial tendency is to return to its equilibrium position after being disturbed (say, by a gust) is one that is said to have positive static stability. One whose initial tendency is to diverge from its equilibrium position has negative static stability. One that is neutrally stable merely assumes its new attitude after being disturbed, with no tendency to converge nor diverge. About the pitch axis a commonly used measure of stability is the static margin. In simple terms, this is a centre of gravity (CG) margin, being the distance along the chord of the wing between the actual CG position and the position where the CG would be for aircraft to be neutrally stable (the neutral point). If the CG is ahead of the neutral point, the aircraft has a positive static margin and is stable. All of the most recent crop of fighters, especially the canard designs, have aft CGs and are unstable at subsonic speeds.

Dynamic stability: This refers to the time history of an aircraft's response to a disturbance. An aircraft has positive dynamic stability if, over a period of time, it tends to return to its undisturbed state. An aircraft that, over a period of time, tends to diverge from its undisturbed state is dynamically unstable. An aircraft has neutral dynamic stability if it has no ultimate tendency either to move towards or away from its undisturbed state, though it may oscillate through it.

Stalling: This is, in simple terms, the condition where maximum lift has been achieved. It is the result of flow separation on an aircraft's wings. For simple aircraft (e.g. those of the pre-swept wing era) the stalling angle of attack (AoA) is in the region of 15°.

Spinning: Von Karman, the eminent German aerodynamicist, was asked by Amy Johnson, 'Can you tell me in a few words what causes a spin?' He replied, 'Young lady, a spin is like a love affair: you don't notice how you get into it and it is very hard to get out of.' A spin is the autorotation of a stalled wing. In a roll, at AoA below

The Fokker Triplane's best rate of climb at 65mph was 1,000fpm. Both it and the Sopwith Camel had maximum speeds of 120mph with 130hp engines. The Camel had a slight edge in manoeuvrability, though the Triplane could fly more slowly.

stalling, the down-going wing meets the air at an increased AoA and develops more lift, while the up-going wing's lift is reduced. Both effects tend to resist and damp the rolling motion. At AoA beyond the stall, however, an increased AoA on the down-going wing produces less lift, whereas on the up-going wing, whose AoA may now be less than the stalling angle, the lift is increased. The combined effect of the two wings is now not to dampen the rolling motion; on the contrary, a propelling moment is applied. Rotation of the fuselage and fin about the flight path at high AoA gives rise to opposing (damping) moments and a steady rotation rate is reached when the two effects are equal and opposite. Rotation causes a nose-up moment due to the centrifugal forces on the nose and tail of the aircraft. This tends to fling them away from the spin axis, with the nose-up moment depending on the mass distribution of the aircraft, the rotation rate and the AoA.

Pre-World War One

The Wright brothers flew longitudinally unstable aeroplanes. Their aircraft had about neutral directional stability, negative dihedral and a roll control that introduced yawing moments. It can only be concluded that their success was due to low speed and low wing loadings, robust structures, and a lot of practice. From their own descriptions of their flights, they were over-controlling most of the time. Many flights ended when, at the bottom of an oscillation, their skids touched the ground. In these early days of flying, the flight regime was so rudimentary that control power was the key to success. The aerodynamic instabilities could be overcome by the pilot. As aircraft flew faster and became more heavily loaded, the divergence of the unstable aeroplane became too rapid for the human pilot to cope with and inherent aerodynamic stability became mandatory. This was the fundamental difference between the European approach and that pioneered by the Wrights. The Europeans insisted on inherent stability, allied to adequate manoeuvrability, the objective being safe, hands-off, continuous flying.

By 1909 the successful aircraft had: cambered biplane or monoplane wings for lifting; horizontal elevator for control in pitch, placed forward of the wings on biplanes and aft of the wings on monoplanes; a vertical rudder (aft) for control in yaw and satisfactory directional stability; and helical twisting (wing warping) of the wing tips or hinged ailerons, to provide control in roll in combination with rudder, either for preserving lateral balance or to affect banked turns.

The period 1908–12 saw the aeroplane evolve from an assemblage of disparate parts that the Wrights had made flyable and controllable, but had shown little desire to improve.

World War One

Stalling and spinning

Basic stability research was carried out by the Royal Aircraft Factory at Farnborough on the tractor BE2 (Blériot Experimental) designed by de Havilland in 1914. Small changes in CG position and in the areas and arrangement of the tail surfaces transformed the unstable BE2A into the stable BE2C. It was the first inherently stable aircraft produced by Farnborough and was used in large numbers by the RFC (1,216 being built) for a variety of roles with various armament installations.

It is a curious fact that stalling and spinning are hardly mentioned in aeronautical literature until the end of World War One, yet they were taking a distressingly heavy toll of human life. Anyone living near a busy airfield during that war would not be greatly surprised to see an aircraft suddenly falter, drop a wing and spin into the ground. Indeed this was the usual consequence of engine failure just after take-off and was accepted as one of the hazards of learning to fly. There is no doubt that the cause of these

Aircraft design is, without exception, a compromise. By changing the CG position and making slight modifications to the tail surfaces, the unstable BE2A was transformed into the BE2C (seen here). However, improvements in its flying characteristics reduced its manoeuvrability, rendering it inadequate as a fighting machine. *(Philip Jarrett).*

Far more RFC aircraft were lost in accidents during the War than were shot down. In spite of the simple mechanical flying controls fitted to contemporary fighters, Sopwith F.1 Camel 5654 came to grief when the controls jammed during take-off. *(Philip Jarrett).*

frequent stalling accidents lay in a combination of the inherent qualities of the aircraft then in use and psychological factors that are still with us today. With stalling speeds only 10 to 20kts (16 to 32km/h) below the climbing speed, prompt action was required to maintain flying speed if the engine failed in the climb. That said, the main danger arose from the tremendous temptation to turn back to the airfield rather than attempt a landing between obstructions such as trees and houses, ahead. However, provided the aircraft had reached 300ft (90m) or more, a turn was often a perfectly feasible manoeuvre and a downwind landing presented no great difficulty. Despite this, hundreds of pilots stalled on the turn and crashed, and no doubt many of them had enough height for the turn. Unfortunately, a downwind turn close to the ground looks wrong to the pilot if it is done without sideslip. It can be made to *look* right by skidding outwards. This was easily done and it resulted in a rapid loss of speed. This loss of airspeed was concealed from the pilot by his turning close to the ground, when, at the moment when he was most likely to stall, he obtained a false sense of security from the impression of increasing speed over the ground. It is not surprising that when engines failed daily, many died from this unhappy combination of circumstances. That such accidents are rare today is mainly due to vastly more reliable engines and to a much greater sense of discipline, not greatly to aerodynamic progress.

Before 1916, stalling at any height was likely to result in a spin from which no method of recovery had been discovered. Eventually, the notion was grasped that a spin was essentially a stalled manoeuvre, from which recovery could be affected by pushing the stick forward. Very soon this technique was being taught in flying schools and spinning became a recognised part of flying technique. It was often used by fighter pilots as a means of retiring from a dogfight if guns jammed or ammunition ran out. Nevertheless, despite the preaching of the spin recovery technique, accidents due to spinning continued to occur in great numbers. Of 106 flying accidents occurring in May 1918, forty-one were attributed to spinning, and of these, twenty-seven occurred with the Sopwith Camel that not only dropped into a spin very readily but called for a high degree of skill for recovery. The controls were very sensitive and any impatience on the part of the pilot in pulling out of the dive resulted in a second spin, often in the opposite direction. Although these accidents could be

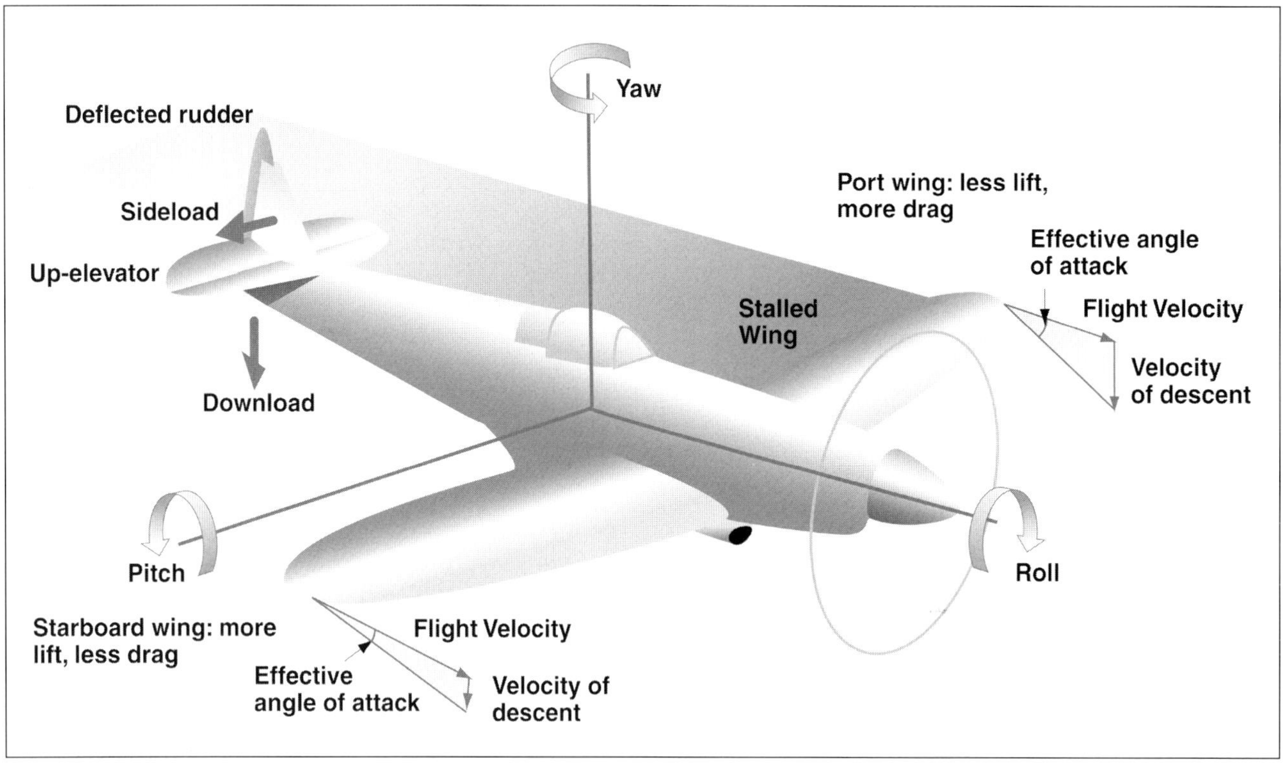

This figure illustrates the control actions (rudder and elevator) and resulting effects to induce a spin to port. Some aircraft may also need a downward deflection of the aileron on the side to which the spin is required (port down for a spin to port) to force the wing to stall. Once in a spin, higher lift on the starboard wing rolls the aircraft to port. Higher drag on the port wing yaws the nose to port causing increasing sideslip to starboard, which increases roll to port. The motion is self-sustaining. *(Pete West/AIR International).*

Not only did early aviators lack experience and the benefit of scientific knowledge, their aircraft – such as this Wright Flyer A – were inherently unstable. However, this instability and a canard layout are characteristics it shares with the latest European fighters. *(Philip Jarrett).*

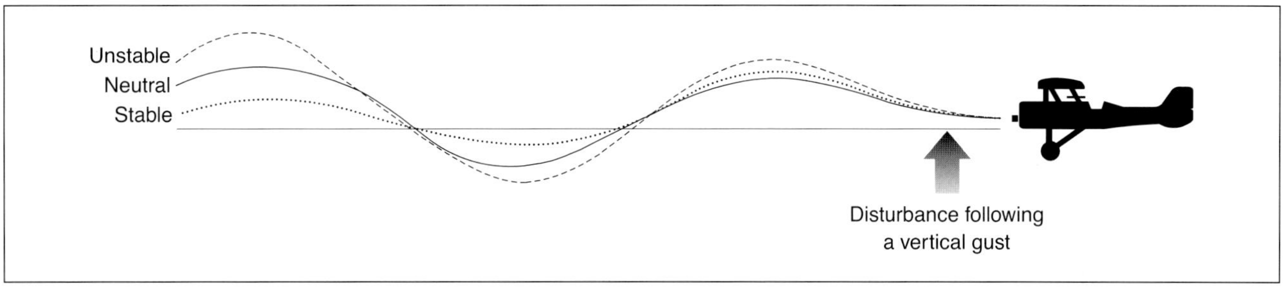

Dynamic longitudinal stability showing the resultant (in this case oscillatory) motion of an unstable, neutrally stable and stable aircraft following a vertical disturbance. *(Pete West/AIR International).*

A notable feature of many biplanes was stagger (the position forward of the upper wing in relation to the lower). This was usually dictated by structural considerations or field of view for the pilot. Aerodynamic effects were of secondary importance though not negligible. The influence on stability and especially spinning characteristics was much more serious. At high AoA the upper wing was shielded (negative stagger increased shielding and vice versa).
The Nieuport 17 was an outstanding fighter and featured pronounced stagger with a larger area upper wing than the lower. *(via Philip Jarrett).*

attributed to the inexperience of hastily trained wartime pilots, it soon became apparent that current trends in aircraft design were leading to aircraft more and more difficult to get out of a spin, even in the most skilful hands.

Control surfaces

Typical of its day, the Camel had a low wing loading of 29kg/m^2 (6lb/ft^2). Together with large, lower wing dihedral, this gave it kite-like qualities on the ground and made it very sensitive to crosswinds. With most of the weight (the 110 to 130hp (82 to 97kW) Clerget, le Rhône or Bentley rotary engine, pilot, fuel and guns) within 1.8m (6ft) of the nose, pilots had to contend with large gyroscopic engine/propeller effects. These tended to give the aircraft a mind of its own. Because of the spinning engine, aircraft tended to remain in whatever attitude or manoeuvre it happened to be in. The Camel was unstable in pitch and considerably unstable in yaw, with very sensitive elevators and rudder. When it was designed in 1916, ailerons were in their infancy, and the prevailing philosophy was if a faster roll is needed, make the ailerons bigger. The Camel had four big ailerons, requiring a large force to move them. When moved, the wing section was so degraded that the roll response was much less than, say, a Tiger Moth's. Adverse yaw arising from the aileron drag was the primary effect, aileron roll was secondary. Modern pilots who have flown Camel replicas have commented that if it was the best World War One fighter, 'God help the worst'. However, its stalls were gentle and predictable, spins quite controllable in skilled hands with low stick force/g. The Fokker Triplane, on the other hand, had very poor visibility from the cockpit and a small rudder that was very effective when its rotary engine's power was applied. The tail skid was not connected to the rudder which inevitably meant overcontrol with tail wagging on the ground. Like the Camel, the controls lacked harmony with a light and twitchy rudder and very heavy ailerons.

Between the Wars

Spinning

At least one third of the many new types of military aircraft were troublesome recovering from spins at some stage in their careers. Some types were sensitive to small variations in rigging and loading, with the result that spinning trouble would suddenly arise on aircraft that had been in service for several years.

At first a myth, which many had heard of but no one had seen, the flat spin gradually became a reality around 1919. Previously, spins involved high-pitch attitudes, with AoAs between 25° and 35°. Now, certain aircraft were found to spin very rapidly with the fuselage approaching the horizontal, with AoAs up to 45° and a relatively low rate of descent. The flat spin was more dangerous; recovery was often very slow, if not impossible, and the phenomenon underwent vigorous research by RAE and NPL. However, though these experiments led to some understanding, the crying need was for design features that would make it impossible.

To make a convincing demonstration, it was decided to use a Gloster Gamecock, a single-seat fighter, which, although popular with the RAF, had a reputation for spinning fast and flat. Apart from its reputation it had all the features known to lead to flat spins: a short, fat, round body in conjunction with wings of small stagger and a badly shielded fin and rudder. Before modifying the Gamecock allocated to this project it was necessary to establish that it did indeed have a dangerous spin. The aircraft was taken to 15,000ft, the pilot having been instructed to do a few turns in a spin and then recover. After a couple of turns the aircraft flipped into a very fast and flat spin (analysis showed a rotational speed of 50rpm) during which the aircraft made thirty-four turns, twenty-five of which were against the pilot's will. It was clear that there was a thoroughly dangerous spin to be tamed. The spin recovery technique was to push the stick forward, centralise rudder or centralise ailerons, full outspin rudder, then stick fully forward. Pilots in a spin had a natural tendency to pull the nose up to avoid the ground, but the only chance of safety was to put the nose down, so as to reduce the AoA.

The most promising design suggestion was to raise the tailplane to the top of the fin and thus unshield the fin and rudder at very high AoA. Subsequently, this was done, and exhaustive spinning tests proved conclusively that the dangerous characteristics had been eliminated. The spin was slower and recovery was rapid. However, designers were reluctant to adopt the high tailplane and bad spinning characteristics were found on many new types for years to come.

Stalling

Control at the stall was another matter engaging much attention. An especially dangerous type of stall was that in which a wing tip stalled first, resulting in the aircraft spinning. This remained a frequent source of accidents during the 1920s and 1930s. Consequently, there was a strong incentive to try to ensure that the central part of the wing stalled before the tip. A stall, if encountered, would then be straight, without abrupt wing dropping and/or loss of aileron control, and a spin would be avoided. Attention was therefore directed to the possible use of a slot at the wing tip to delay tip stall. Eventually the Handley Page automatic slat was fitted to a Bristol Fighter in 1927 and has been used in various forms ever since. However, there was a tendency to avoid its use (not just because of the many battles over patent rights), since it added mechanical complexity and it made it more difficult to secure a smooth wing surface when the slats were retracted. The heyday of the slot was in the early 1930s on thin wings; there was less call for it when thicker sections became usual and design skill improved and the tip stall was controlled by basic wing design.

The maximum lift that an aerofoil section can develop before stalling depends on its thickness and camber. In the limited ranges that designers were forced to work, thickness was the most important. For most sections the optimum thickness was found to be about 12%. Thinner sections were prone to have a leading-edge stall which was sudden and usually vicious, thicker sections tended to stall progressively and more gently from the trailing edge.

The parallel wing stalls first at the wing root, the stall spreading outwards as the AoA is increased further, so that the tips remain unstalled for a while and provide damping to a rolling motion. The situation became complicated with the increased use of wings tapered in planform as monoplanes' were. The behaviour at the stall of such wings is quite different from that of wings of constant chord used on previous aircraft. A highly tapered wing stalls first at the tips, the stall spreading progressively inboard. At a taper ratio (tip chord/root chord) of around 0.5, the tendency is

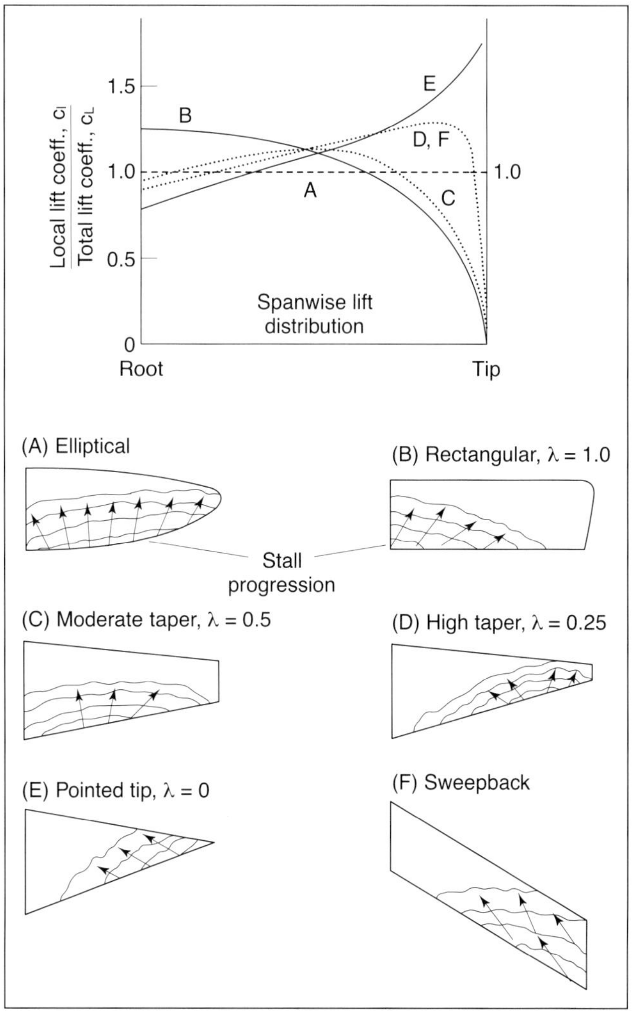

Spanwise lift distribution (top) and stall progression (bottom) for plane wings of different planforms. Notice that taper and sweep both cause higher loading to arise outboard (D, E, F) which leads to tip stalling with loss of aileron control and possible wing dropping.

for the whole wing to stall at once. This is not desirable, as the stall will occur suddenly and without warning. It is better to make the stall spread from the root outwards, as such a stall will be gentle and not accompanied by abrupt wing dropping and/or loss of aileron control. Unfortunately the critical 0.5 taper ratio was regarded as very convenient from a construction point of view, so that a compromise was necessary. Since the point at which a wing stalls is dependent upon its AoA, the position along the wing span where stalling first appears can be controlled by twisting the wing so that AoA is progressively reduced towards the tip. By judicious design, this gives a wing root stall. A drawback to this solution, particularly on high-speed aircraft, is that it creates greater drag at lower AoA. An untwisted elliptic wing (used for low lift-induced drag) formed of similar sections with the same thickness ratio will stall throughout its span at the same time. The stall of other wing planforms considered at the time (e.g. straight tapered wings) followed no hard and fast rule.

The addition of trailing-edge flaps on tapered wings aggravates the problem. They concentrate the lift at the centre of the span and thereby produce additional upwash near the wing tips. This tends to promote earlier stalling of the tips. However, with trailing-edge flaps extending only over the inner half of the span, the effect is not so serious and the usual arrangement became one with half-span flaps and conventional ailerons.

World War Two

Being able to roll out quickly from a dive is a useful tactical manoeuvre. Such a breakaway required a high rate of bank at high speed under the maximum force (about 23kg (50lb)) that the average pilot could reasonably be expected to apply. It also needed good stability and control in the turn at or near the pilot's blackout point. The time to bank through a given angle, for a given pilot's force, varies for geometrically similar aircraft, with the fourth power of the wing span. This led to the 'clipped wing' (span reduced by 11.5%) Spitfire of 1942 that could roll 150° per second at 170kts compared with 105°/s for one having a normal elliptical wing. There is a damping effect that depends on the aspect-ratio and rate of roll. For a given aspect-ratio the difficulty of providing light ailerons also varies as the fourth power of the span. The problem of aileron balance during World War Two was so delicate that comparatively small variations would make all the difference between good and bad control at high speed. Aileron hinge moments are very sensitive to small protuberances on the control surface, as well as to the curvature of the profile. Since the

In comparison with the Bf 109's heavy control forces at high speed, the Focke Wulf Fw 190A's combination of high speed and harmonised controls made it one of the most successful fighters of World War Two. Its most impressive feature was its excellent aileron control at high speed; peak roll rate occurred at around 220kts, being roughly 160°/sec. *(Philip Jarrett).*

Flight tested in France, where it made a forced landing in November 1939, and in England, Messerschmitt Bf 109E-3 W Nr 1304/AE479 was found to have a high roll rate at medium speeds. However, at speeds above 260kts aileron control was termed 'solid' by RAF test pilots. *(Philip Jarrett).*

standard of workmanship in wartime was bound to deteriorate owing to the dilution of skilled labour with semi- and unskilled workers, there was a large variation in aileron heaviness between individual aircraft of a given type.

At speeds of 130 to 170kts (240 to 320km/h) the aileron control of the Messerschmitt Bf 109E was very good. However, at speeds above this, the ailerons and elevators became rapidly heavier so that above 260kts (480km/h) the aileron control was termed 'solid' by RAF test pilots. The time to bank to 45° at 350kts (640km/h) was four seconds (11°/s), which was quite excessive for a fighter, classing it as very unmanoeuvrable at high speed. The later Spitfires' rate of roll at 350kts was 60°/s.

Pilot opinions of the Focke Wulf Fw 190 were that its handling qualities were very good, in particular its excellent aileron control at high speed (peak rate of roll was 160°/s at 225kts). Its aileron reversal speed was estimated to be about 660kts (1,220km/h). With a top speed of 350kts (640km/h) at 17,500ft (5,300m), the Fw 190A was regarded as every inch a fighter. As a result of light ailerons and elevator, manoeuvrability of the Fw 190 at high airspeeds was excellent.

Compressibility

A new and serious high-speed stability and control problem struck two new US fighters, the Lockheed P-38 Lightning and Republic P-47 Thunderbolt, during their high-speed dive tests in 1941. As the aircraft accelerated in the dives they began to shake, mildly at first, then more violently, and control forces became heavy. Even worse, the aircraft wanted to nose down further and increase the severity of the dive. In the case of the P-47, the nose went beyond the vertical and no recovery seemed possible for full-aft stick and nose-up tab, used to counteract the nose-heavy tendency. The pilots reported that upon reaching nearly terminal velocity, the aircraft that would normally develop about 7 to 8g in response to a stick pull of 18 to 23kg (40 to 50lb) failed to generate any noticeable g. The dives took the aircraft from about Mach 0.75 to 0.86 at which value it remained roughly constant for the last 5,000ft (1,500m) of the dive. This situation persisted until the aircraft descended to around 15,000ft (4,750m), when very high normal accelerations built up and pilots blacked out. If they were lucky the pilots recovered with their aircraft flying at 20,000ft (6,100m) with bent wings. The phenomenon was soon referred to as the 'frozen stick dive'. Recovery was possible because as the aircraft descended into warmer air

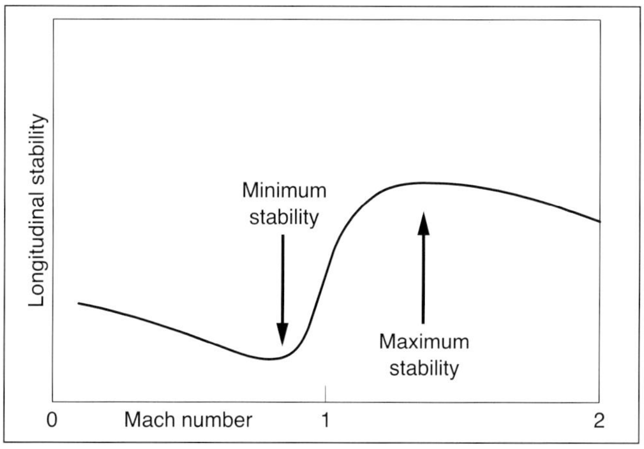

Variation of longitudinal (pitch) stability with Mach number, showing its reduction at subsonic speeds followed by a marked increase beyond Mach1.

When the Lockheed P-38 Lightning started its high speed dive tests in 1941, its control forces became heavy as it accelerated, even worse the aircraft wanted to nose down further and increase the severity of the dive. A dive flap was mounted on the lower surface of the inboard wing which, when deployed, generated more wing lift thereby increasing the downwash over the tail to produce a nose-up moment to assist in dive recovery. Being electrically driven, actuation was quick. In addition, dive trajectories were established and could be used to formulate operating restrictions for the flight manual. This is a P-38J-10. *(Lockheed Martin via author).*

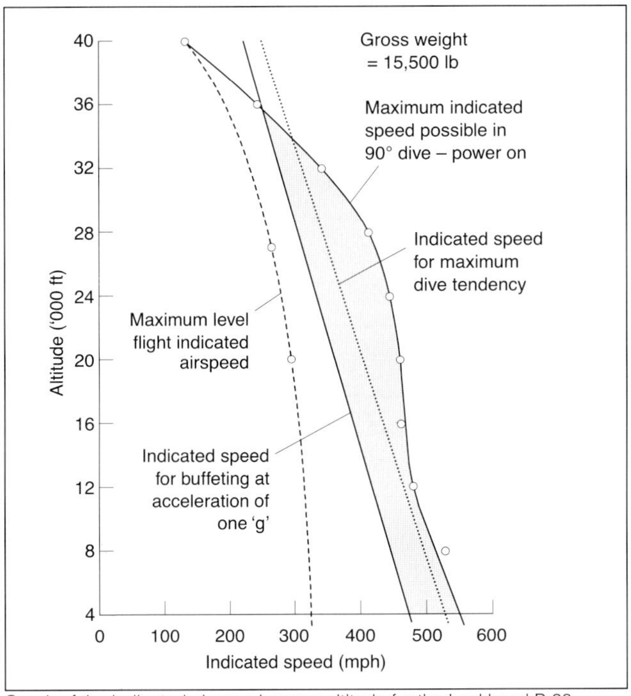

Graph of the indicated airspeed versus altitude for the Lockheed P-38 Lightning derived from its early dive test programme. The chart shows regions of difficulty for the aircraft. The originator of this 1942 chart was Clarence L (Kelly) Johnson, designer of Lockheed's P-38, P-80, F-104, U-2 and SR-71.

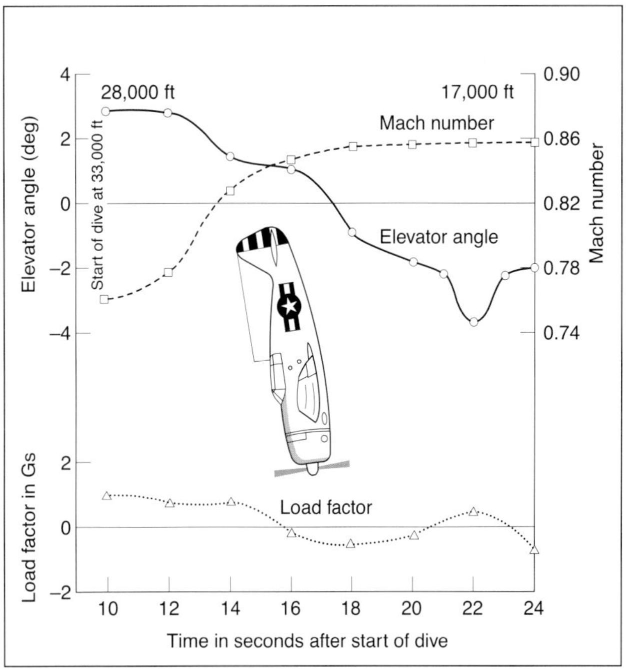

Time history of a vertical dive test of a Republic P-47C Thunderbolt. None of Republic's test pilots were anxious to do the tests and would only have agreed for very high fees. Major Perry Ritchie of the US Army Air Corps at Wright Field agreed to perform the dives as part of his regular duty and carried out 30 of them, so earning an Air Medal.

the local speed of sound rose and the aircraft Mach number started to reduce. It was first believed that the buffeting problem was caused by tail flutter, a rather natural conclusion since this was a familiar situation, and a likely source of trouble. As for the 'tuck-under' problem, as it was also called, no one had an explanation.

It was eventually realised that the aircraft, with their thick cambered aerofoils and straight wings, were encountering very profound changes in lift curve slope, downwash, shift in zero lift angle, loss in elevator effectiveness and rearward movement of the lift centre at these high Mach numbers. As they entered the transonic regime the aircraft were becoming more stable, while their control effectiveness was reducing. The culprit was found to be shock-induced flow separation on the wing that drastically reduced the lift. This in turn reduced the downwash at the tailplane location, made the tail much more effective and increased the aircraft's nose-down pitching moment, thereby steepening the dives.

Flight control systems

The simplest form of flight control system was the one used by the Wright brothers, namely a direct mechanical linkage between the pilot's controls via cables and pulleys to the control surfaces. With time, the inherent flexibility in the early linkages gave way to rods and levers. The pilots' control forces were directly related to the aerodynamic loads on the control surfaces but could be influenced by bobweights and springs in the circuits.

The compressibility problems encountered in the 1940s included very abrupt changes in aircraft trim. The characteristics of the manual elevators with which aircraft were equipped varied rapidly with small changes of speed, thus making precise control difficult. It also proved increasingly difficult to ensure freedom from flutter as aircraft speeds moved further into the transonic region. At the end of World War Two there was very little information available regarding control surface hinge moments at transonic speed. In addition, the control forces to operate such manual controls were becoming unacceptably high. Flight experience had shown that aileron flutter caused the control column to be wrenched from the pilot's grasp. This led to the suggestion that any tendency for the ailerons to flutter could be resisted by a powered irreversible control system in which the pilots' sense of 'feel' would be largely removed.

The first jet fighters such as the Gloster Meteor, de Havilland Vampire and Lockheed F-80 Shooting Star had fixed tailplanes with manually operated elevators, and experienced the same compressibility problems as their piston-engined forebears. To alleviate these difficulties, the fully powered 'all-flying tail' was developed and first used on the North American F-86E Sabre of 1952. This is one in which the moving tailplane and elevators were linked for better pitch control when immersed in shockwaves. Thus, the move was made to fully-powered flying controls, often employing full duplication of the system (rather than manual reversion, which gives very heavy controls) to deal with the failure case. In consequence, artificial feel had to be provided for the pilot. This was derived essentially from bellows subjected to the ambient dynamic pressure, thus leading to a relation between control force and movement similar to that obtained with aerodynamically balanced controls. Powered controls also provided a better insurance against flutter.

With the advent of the first truly supersonic fighters, the North American F-100 Super Sabre and MiG-19, came the almost universal adoption of the fully-powered all-moving tailplane, known also as a slab tailplane. There were three pressing reasons for the adoption of the all-moving tailplane: the effect of a flap-type surface (i.e. elevator) in supersonic flow is limited to the area downstream of the hinge, compared with the whole-surface effect obtained in subsonic flow; the rearward shift of the centre of lift, as an aircraft accelerates through the transonic region, produces excessive longitudinal stability, and only a large pitch control surface could produce the forces required for adequate manoeuvrability; and the aeroelastic distortion of a relatively high-aspect-ratio elevator further reduced effectiveness and lowered the flutter speed.

The Republic F-84F Thunderstreak was one of only two American swept-wing fighters derived directly from an existing straight-wing design. The F-84F was evolved during a period of reduced defence spending. The original idea was to marry 40° swept-wing and tail surfaces to the fuselage of the existing F-84 Thunderjet to increase its limiting Mach number. It first flew in 1950 but, because of numerous problems, not least of which was unsatisfactory high speed longitudinal and lateral control, did not enter service for 30 months after the aircraft was given an all-flying tail. With a touchdown speed of around 135kts squadron pilots, more used to lighter and slower-landing aircraft found it demanding. The F-84F did not achieve IOC until 1954, nevertheless, it remained in Air National Guard use until 1971 and sold in great numbers around the world. *(via author).*

The first production Soviet fighter to exceed Mach 1 in level flight, the twin-engine MiG-19 of 1954 was found to have inadequate control effectiveness at supersonic speed and required an all-moving tail. To provide limited adverse-weather capability the MiG-19P (one of the prototypes of which is shown) was equipped with the *Izumrud* AI radar mounted in the extended nose on the intake splitter. Characteristic of many contemporary Soviet designs are the large wing fences to resist tip stalling. *(Alexei Mikheyev via author).*

The Jet Age

Tailplane position

Many early swept-wing aircraft, with tails mounted low on the fin, had difficulty in countering pitch-up caused by tip stalling leading to a forward shift of the centre of lift on the wing and increased downwash at the tail. Various palliatives were, and are still, being used to delay, or at least make predictable, the root cause: flow separation. The most common among these are wing fences, saw-tooth leading edges and vortex generators. During the 1950s both high and low tailplanes had their proponents among designers seeking to avoid loss of pitch stability. High tails were used on the McDonnell F-101 Voodoo and Lockheed F-104 Starfighter, and low tails on nearly every other type. The trend, during the 1950s and 1960s, was decisively towards low fuselage-mounted tailplanes. The aim was to keep the tailplane as far away as possible from the region of intense downwash/wake behind the wing, particularly when aspect-ratio is low, as it is for all fighters. The two

The area of the McDonnell F-101 Voodoo's tiny swept wing was little more than half that of the F-102A, while maximum take-off weight approached 52,000lb, giving it the very high wing loading of 141lb/ft^2. Stalling speed was frighteningly high, 184kts at sea level, rising to 472kts at 50,000ft where it was restricted to 1.2g. Although roll rate was adequate, the Voodoo was a complete turkey in any situation that required a moderate amount of manoeuvrability, and this was not helped by the high set tail, which was blanketed by the wings at quite low AoA, leading to pitch-up. A pitch-up inhibitor was installed at an early stage, but like so many systems of that era, it didn't always work. Left to right: F-101B, F-101A and RF-101A in 1957. *(Boeing via author).*

Designed to Specification F.4/48, the Gloster Javelin's delta wing of 10% thickness/chord ratio, gave it very low wing loading of 34lb/ft², far less than even the Douglas F4D Skyray. Because of its high span loading (W/b) deceleration during turns was dramatic. Due to its high T-tail the Javelin suffered deep stall problems during its early development which was marred by a series of fatal crashes. Only later was it found that to recover from a deep stall it was necessary to enter a spin, then recover from that. Shown is an FAW 9 with provision for air-to-air refuelling. *(Flight Refuelling via author).*

critical flight conditions are when shock-induced flow separation occurs on the wings at high AoA and high subsonic speed.

Roll controls

Fighters had historically used ailerons for roll control, but the deflection of an aileron causes aeroelastic distortion. This is particularly severe on thin, swept wings at high dynamic pressure. The problem was predicted for the relatively short-span, highly tapered wing of the North American F-100 and the aircraft was fitted with inboard ailerons. An alternative, adopted in the 1950s for the Republic F-105 Thunderchief, is to use spoilers, which operate asymmetrically and dump the lift on a wing. They have the additional benefits of reducing the adverse yaw (e.g. nose right when rolling left) associated with ailerons and allowing more wing trailing edge to be used for flaps. From the 1960s, with the advent of all-moving tail surfaces, the differential tail or taileron has been used to further augment roll performance. They are particularly useful at high speed (high dynamic pressure), where ailerons are prone to control reversal whereas the differential tail retains reasonable effectiveness.

In the early 1980s, the McDonnell Douglas F-18A

First of the USAF's 'Century Series' fighters, the North American F-100A, had insufficient vertical stabiliser area that led to several accidents due to inertial coupling. During the pull-out from a dive, in which slightly more than 8g was reached in October 1954, an F-100A suddenly yawed nose right to 15°, then disintegrated. The aircraft had been built to survive 8° yaw at maximum dynamic pressure but in this case the aircraft had made a supersonic sideslip well beyond the design limit of the airframe. The solution was to increase fin and wing area back to the original design. *(NASA via author).*

Hornet was found, during flight testing, to suffer from a marked loss of roll rate in transonic flight, due to its relatively flexible wing. This was in spite of having a differential tailplane and inset ailerons for roll control. However, because of the aircraft's digital flight control system, it was relatively straightforward to rewrite the flight computers' software to phase in the leading- and trailing-edge flaps to give extra roll control power. Differential tailplanes are all-moving, and operate either symmetrically for pitch control or differentially for roll control. They retain their effectiveness to very high AoA and do not have the adverse yaw effects associated with ailerons at low AoA. In addition they are effective at high Mach numbers, hardly affected by wing sweep angle, and have become universal on high-performance, conventionally tailed fighters. For close-coupled canard configurations differential deflection of the foreplane produces negligible rolling moment due to opposing aerodynamic interference of the wing. It is therefore necessary to resort to wing trailing-edge flaperons to provide rolling power.

Inertia coupling

In 1947 the Bell X-1 flew supersonic and, along with other experimental aircraft, investigated the problems lying in wait on the other side of the 'sound barrier'. The most costly lesson learned from these research aircraft was their loss of directional stability with increasing supersonic Mach number. Fin effectiveness, which increases slightly transonically, diminishes beyond about Mach 1.4 to below its subsonic value. Both the X-1A and X-2 went out of control at their highest Mach numbers, when rolling into quite gentle turns. This phenomenon, referred to as inertia coupling, has played a significant part in the evolution of supersonic fighters. When an aircraft rolls, the distributed masses representing the wings, fuselage and tail unit are all put into motion about some axis and then brought to rest again at the end of the roll. If the aircraft is of a slender configuration with small span, it has most of its mass concentrated in the fuselage (e.g. Northrop F-5). Such aircraft are usually capable of high rates of roll resulting in a tendency to inertia coupling problems. A slender aircraft, on rolling, will rotate about its own principal axis, the line of least resistance, rather than the flight path. If the aircraft rolls sufficiently quickly, the destabilising moment from the fuselage inertia forces will overcome the stabilising aerodynamic moment of the fin. The aircraft will become directionally unstable, with the yaw angle diverging to the point where the aircraft's structural limit is exceeded, leading to break-up.

Rapid rolling is not the only source of inertia coupling. The case of the F-100A dramatically drew the industry's attention to the phenomenon. One particular accident prompted a very expensive programme of modifications. In October 1954 George Welch, North American's senior engineering test pilot, made the last of the series of test flights to take the F-100A into one corner of its flight envelope. This involved a dive from 45,000ft (13,700m) to maximum permissible speed (strictly maximum dynamic pressure), followed by a 7.5g pull-out. During the pull-out, in which slightly more than 8g was reached, the aircraft suddenly yawed nose right to 15°, then disintegrated. The F-100A had been built to survive 8° yaw under maximum dynamic pressure, but in this case the aircraft had made a supersonic sideslip well beyond the design limit of the airframe. Apart from F-100s engaged in testing, all others were grounded until the cause of the problem was found. After an exhaustive accident investigation, it was found that the aircraft's inherent directional stability was insufficient to keep the aircraft pointing the way it was going when pulling out of a high-speed dive. Welch was doomed before he left the ground, because the test he was to perform was bound to destroy his F-100A, just as it would any aircraft with its mass distributed mostly in the fuselage and lacking directional stability. A modification plan, including increased wing span and more fin area, was agreed, so that by the middle of 1955, F-100As already in service, awaiting delivery or on the production line were brought up to the new standard. The modifications postponed yaw divergence in rolls from about 75°/s to about 180°/s, still not an exceptional rate but acceptable. On current aircraft the reduction of directional stability at supersonic speed is countered by the use of the automatic flight control system (AFCS) which produces rudder deflection as a function of lateral acceleration.

Twin fins

The deterioration in directional stability with increased Mach number and/or increased angle of attack meant that 1960s combat aircraft needed much larger fins than their older and slower predecessors. Fins were usually sized to meet stability requirements at low and moderate angles of attack (up to 15° typically). Most of that generation of combat aircraft (e.g. McDonnell Douglas F-4 Phantom, Ling-Temco-Vought A-7 Corsair II) cannot be flown safely up to the wing stall boundary because of lateral/directional control deficiencies. On many of these aircraft, vortices shed from the near-circular-section forebody and/or the wing-root junction increase the destabilising effect of the forward fuselage and reduce the stabilising effect of the fin. In addition, as angle of attack is increased, the fin is borne deeper and deeper into the wake shed by the fuselage, so that fin effectiveness is further reduced.

A tall swept fin loses a large proportion of its effectiveness at high speed as a result of aerodynamic loads twisting the tip. Typically, 20 to 25% of fin effectiveness can be lost from rudder distortion and spanwise twist resulting from bending. The use of twin fins can overcome the problem of tall flexible single fins provided they are far enough apart laterally to overcome mutual biplane interference. The main argument in favour of twin fins is that there is always one vertical surface in relatively clean flow to provide directional stability regardless of the AoA/sideslip combination. Twin fins come into their own at high supersonic speed, when mutual interference disappears as the Mach lines from each fin pass behind its neighbour. However, for low Mach numbers, twin fins are not necessarily the most efficient way of improving directional stability. With the increasing requirement for stealth as originally exemplified by the Lockheed SR-71, twin fins, if they are canted, are useful in reducing radar cross-section, as more recently illustrated by the Lockheed Martin F-117A and F-22A.

Ventral fins

There have been many instances of an initial lack of fin effectiveness having to be rectified by the use of ventral fins. A fin set below the fuselage becomes more effective at

The MiG-25 'Foxbat' was designed as a counter to the CIA's Mach 3 Lockheed A-12. Powered by two Tumansky R-15B turbojets of 22,500lb afterburning thrust, 'Foxbat' could reputedly reach Mach 2.8 and could operate at up to 78,000ft. In extremis it could be wound up to Mach 3.2 although this was the engine overspeed limit. With the reduction of directional stability at such high speeds the 'Foxbat' needed twin vertical stabilisers and ventral fins. *(Alexei Mikheyev via author).*

The Chance Vought F8U-3 Crusader III flew off against the McDonnell F4H-1 Phantom for major US Navy orders in 1957. Like its F8U-1 forebear, it had a very thin two-position wing plus flap blowing. Unusual were the forward-raked air intake and large ventral fins that folded down from horizontal with undercarriage extended, to near vertical when the landing gear was raised. Powered by a Pratt & Whitney J75-P-5A of 29,500lb in reheat, one of the three aircraft was flown to Mach 2.6 and believed capable of Mach 2.9. *(AFFTC/HO via author).*

high AoA, since it will often be working in clean air. Ventral strakes are sometimes more efficient than additional upper fin area because of the favourable interference effect created with the rear fuselage and their comparatively high structural stiffness. Ventral strakes have appeared on 80% of fighter-type aircraft entering service since 1960, but with the need for more stealthy designs and the abilities of the latest flight control systems, they will not be used on future combat aircraft.

Autostabilisation

Early jet aircraft were relatively hollow with long tail moment arms which gave them adequate pitch damping. The need for autostabilisation arose due to the increasingly dense aircraft being designed in the 1950s. These contained a lot of rotational energy once disturbed in pitch, and because they had much shorter tail arms, their damping was reduced. In the case of rolling motion, caused by, say, flight in turbulent air, swept-wing aircraft can suffer from severe Dutch roll, to the extent that the aircraft may not settle down before it is hit by the next gust. This can make precision flight difficult and accurate weapon aiming impossible. Simple autostabilisation – that is, artificially aiding the damping – required the motion to be sensed, followed by application of a control to oppose the motion. The simple (low-authority) systems used in the 1950s required that, without autostabilisation, the aircraft had to have handling qualities sufficient for recovery from any manoeuvre to be acceptable for service. These placed significant limitations on aircraft due to safety considerations. If large authority autostabilisation was required, some degree of redundancy was needed in the event of failure. This meant more complexity, cost, volume and weight. However, with enormous improvements in the reliability, weight and volume requirements of electronics, the acceptance of the increased complexity steadily flourished. This, to some extent, was forced by the increasing demands made by airframe designers. Having arrived at a basic configuration whose handling qualities were often judged to be inadequate, the aircraft needed the control system designers to remedy stability and control deficiencies.

By the mid-1960s, experiments with electrical signalling had indicated that significant benefits could be achieved by breaking the mechanical link between the pilot's control column and the power control actuator. This led to the concept of the Control Configured Vehicle (CCV) in which the flight control system is a major driving force in shaping the aircraft, rather than an afterthought. The concept relies on active controls that constantly monitor and react to disturbances, much like autopilots had done for the previous two decades. However, such Active Control Technology (ACT) implied acceptance of feedback control going far beyond previous applications. Not only could it modify aircraft stability and control, but it could be used to improve aircraft performance in the broadest sense, including size and cost. These improvements result from the use of the control surfaces to redistribute, alleviate or enhance the aerodynamic forces acting on the aircraft. ACT systems were seen to offer such high reliability at acceptable weight and cost that an aircraft could be designed to be inherently unstable, relying on ACT for its stability. But with the penalties of failure so severe, the many design teams working to exploit ACT in the 1960s had to be attracted by some very significant benefits. ACT offered a remarkable range of hitherto unattainable facilities, one of which was relaxed static stability (RSS). This called for a relaxation of the demand for inherent airframe stability, while keeping the aircraft controllable by electronic means. Up to the late 1960s the requirement for natural longitudinal static stability was treated as an inviolate design constraint. It was considered difficult if not impossible to fly a modern combat aircraft which was inherently unstable in pitch. General Dynamics boldly put its money on the CCV concept contender in the USAF Light Weight Fighter programme, which, as the YF-16, first flew in 1974. The aircraft featured RSS and envelope-limiting under the command of a full-time, fly-by-wire (FBW) flight control system. Since then all high-performance fighters have been designed as CCVs.

Stall/spin problems

Most early jet fighters were limited to subsonic speeds and were designed primarily for air-to-air combat. This meant that they had relatively low wing loadings, aerofoils designed for manoeuvring flight, and moderate sweepback, with most of the lift carried by the wings. For this type of aircraft, the wing generally stalled before the AoA at which directional (or weathercock) stability was lost. Stall was characterised by the classic 'g' break with the nose falling through. To get these types of aircraft into a spin you had to

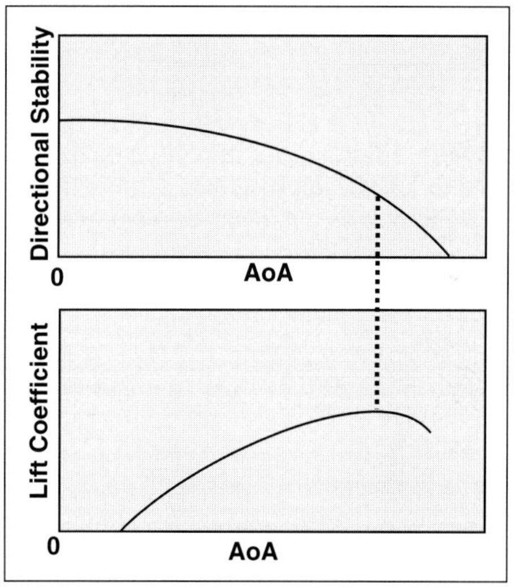

Stall and directional stability characteristics of early jet fighters, where the wing stalled before directional stability was lost. *(Pete West/AIR International).*

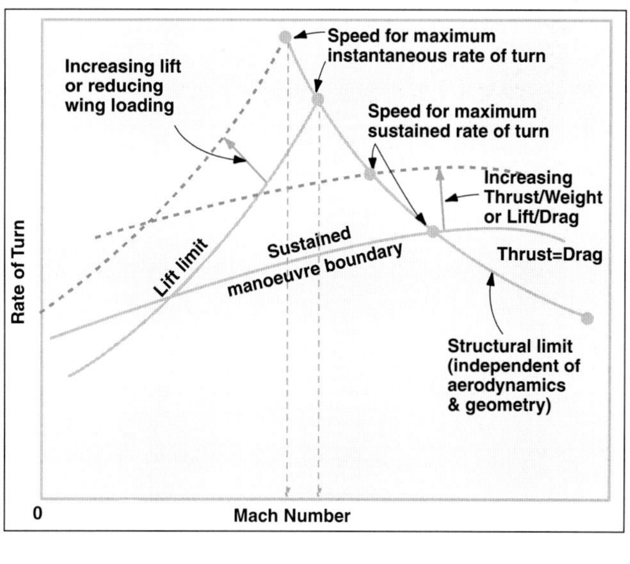

Sustained and instantaneous turn rates versus Mach number. *(Pete West/AIR International).*

really want to.

In the late 1950s, the emphasis in air-to-air combat switched from close-in gun firing to the supersonic interception of high-flying bombers, requiring longer-range missile launching. With this driving philosophy, fighter designers aimed at obtaining high manoeuvrability, low drag and large speed envelopes. This led to thin, highly swept wings with sharp leading edges, large flat fuselages that continued to lift after the wing had stalled, and large horizontal tailplanes to obtain the desired supersonic manoeuvrability. The rapid strides made in fighter performance, with speeds increasing from 500kts/Mach 0.8 in the late 1940s to 1,300kts/Mach 2 within a decade, had left little time for research into their stalling and spinning behaviour. The traditional fighter design approach was to go for performance and manoeuvrability at the expense of high AoA handling qualities, conduct stall/spin testing, install a warning device in the aircraft and provide appropriate precautions in the flight handbook. As a result of this and the lack of specifications, analytic methods and criteria for stall/spin prevention, most high-performance fighters designed in the 1950s exhibited less than desirable stall/spin characteristics. All modern fighters are extremely dense and obtain a lot of rotational energy in a spin, and at low airspeeds the controls are slow to take effect, causing a great loss of altitude. An additional hazard in spins is compressor stall, due to separated flow passing down the inlet duct, leading to engine surge or flameout.

In the late 1960s, fighter-to-fighter combat began to assume greater importance in operational training and in the requirements for new aircraft. Although an air-to-air engagement might start with a supersonic first pass, the speed quickly unwinds to subsonic and, if continued, stabilises at the value for maximum sustained turning. If the sustained (thrust-limited) manoeuvre boundary is exceeded, either altitude is lost or the speed rapidly decreases to the stall (lift-limited) boundary. This sets up the ideal conditions for a departure from controlled flight into a spin, particularly when the pilot is trying to track a more manoeuvrable target. With the advent of high-speed configurations, aircraft could be manoeuvred to increasingly higher AoA in the subsonic flight arena since more than adequate tailplane power was available. A new type of high AoA limitation was encountered, namely yaw divergence. It came very suddenly and usually resulted in loss of control before the pilot was aware of it.

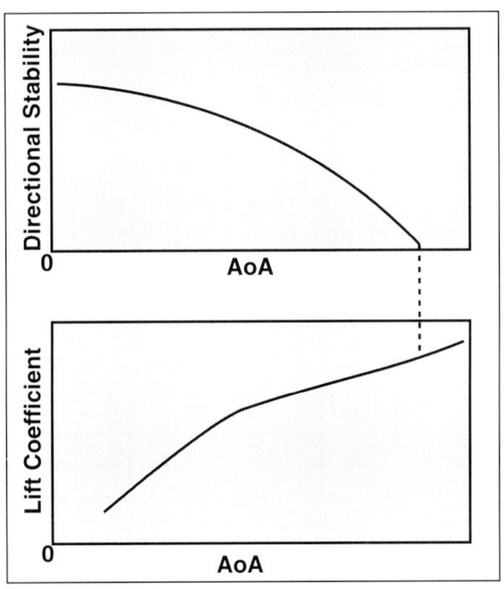

Lift and directional stability of slender fighters, whose directional stability is lost before the wing stalls. *(Pete West/AIR International).*

Because, for these aircraft configurations, directional (weathercock) stability is already zero at the time of stall, the aircraft were very prone to enter a spin. This is indicated by the fact that between 1966 and 1970 the US alone lost approximately 225 military aircraft due to stall/spin problems. To be limited in usable lift by lateral/directional instability or deterioration in controllability that often led to departure from controlled flight into a spin became unacceptable. At best, a spin takes the aircraft out of the fight, and at worst, into an expensive hole in the ground.

This is why, in the 1970s, the emphasis changed from spin recovery to spin avoidance. When building in spin avoidance, but without restricting the lifting capability of the wing, designers had to ensure positive dynamic directional stability. This is the combination of weathercock stability and dihedral effect along the flight path. The AoA at which it becomes negative was found to correlate well with spin entry. In addition, if adverse yaw due to aileron deflection is present, further application of aileron to control bank when sideslip occurs can override a low level of directional stability and lead to yaw divergence. Consequently, an adequate level of directional stability must be incorporated.

Spin prevention

True spin prevention could be achieved only by artificially restricting the AoA by building in progressively heavier stick forces as the limit was approached. At the same time, lateral stability had be maintained and the pilot's lateral control authority also needed to be progressively restricted as the AoA limit was approached. Although spin prevention could theoretically be implemented by a large-authority stability augmentation system as an adjunct to a mechanical control system, in practice it can be much more effectively engineered with a full-time, full-authority, FBW control system. This could prevent spins and over-stressing and, with effective aerodynamic controls on all three axes, allow carefree manoeuvring over the full flight envelope. Thus, although the aircraft itself may appear to be restricted, pilots of previous aircraft had implemented a no-go zone anyway, being quite aware of their aircraft's limitations. Carefree manoeuvring enables the full lift limit of the airframe (potentially a 40% increase in maximum lift was thought attainable) to be exploited in air combat without risk of departure. In addition to the benefits to combat manoeuvrability and consequent combat success, there is a direct economic benefit in aircraft saved due to the elimination of spinning accidents.

The spin prevention AoA limit can be scheduled with dynamic pressure, in effect airspeed, to provide an over-stressing limit, thereby limiting the product of normal acceleration and mass of the aircraft. The limit can additionally be scheduled with altitude to provide separate subsonic and supersonic 'g' limits. Similarly, the roll rate limits can be scheduled with AoA and airspeed to prevent inertia cross-coupling and lateral over-stressing. The pilot then has full freedom to manoeuvre in pitch and roll and the flight control system will restrict his ability to apply such vigorous disturbances as would cause the aircraft to spin or result in structural overload. With such protection the pilot can then devote all his attention to the fight rather than to controlling a frisky beast, knowing that he will eventually come up against a safety barrier.

Boeing F/A-18F-1 Super Hornet being prepared for its first catapult launch from USS *John C. Stennis* in 1997. The E/F is a structural upgrade of the F-18, designed to increase range and warload, and has a stretched fuselage, bigger wing, increased fuel capacity and uprated General Electric F414 engines. The aircraft's control surfaces have been enlarged, their deflections increased, and actuator power and rates stepped up. Noteworthy in this photo are the very large full-span flaps/flaperons, tail deflection and the toed-in rudders to give extra nose-up pitching moment. *(Boeing via author).*

Supermanoeuvrability

The provision on fighters of large pitch authority, which they need for supersonic manoeuvrability, is partly behind the dramatic 'Pougachev Cobra', first demonstrated by the Sukhoi Su-27 'Flanker' in 1989. This low-speed, very high AoA manoeuvre is one in which the 'Flanker' is able to achieve a pitch angle change of 90° to 100° in about 2.5 seconds. The really significant part of the manoeuvre is that there is no tendency to roll off or depart from controlled flight. Three design characteristics are required for an aircraft to successfully execute it: firstly, a high nose-up pitch control power; secondly, strong lateral/directional stability and control (both static and dynamic); and thirdly, a robust pitch-down must be available.

Nose-up pitch control power is, from a design standpoint, the easiest feature to add to an aircraft. Contemporary fighters have manoeuvre trim requirements at supersonic speeds that typically result in excess nose-up control power at subsonic speeds. Furthermore, there has been a trend towards more negative static margins at subsonic speeds to reduce the supersonic trim drag of aircraft, which further results in excess nose-up pitch control power. The third feature that influences nose-up pitch control power is that many contemporary fighters (such as the F-16, F/A-18, MiG-29 'Fulcrum' and Su-27) use vortex lift to enhance manoeuvrability. This results in increased forebody lift at high AoA. However, as noted below, a high pitch rate is essential for successful high AoA excursions of the Cobra type. This requires, on the Su-27, that the AoA limiter of the FBW system be disengaged to allow extremely rapid tailplane deflections to be attained at the start of the pitch-up. The resulting motion is highly dynamic with the acquired kinetic energy of rotation allowing the aircraft to overshoot its trim AoA of around 60°. The stick is held back until maximum AoA (90° to 100°) is achieved when the FBW system is re-engaged with the pilot pushing the stick to the trim condition to allow a recovery.

Fighter evolution over the last fifty years has led to slender configurations that have a substantial portion of the aircraft ahead of its CG. The provision of high lateral/directional stability at high AoA for such shapes is a

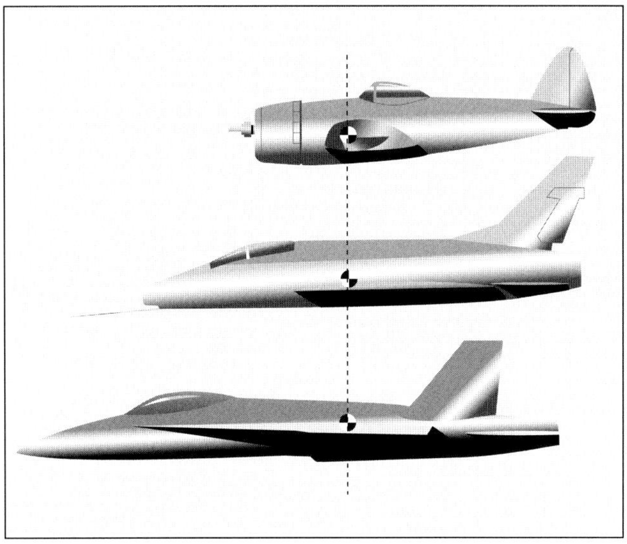

Fighters have become more slender with a greater side area of the aircraft ahead of the centre of gravity. *(Pete West/AIR International).*

very elusive design feature. This aspect of fighter design has often been referred to as a black art, since to achieve robust lateral/directional stability at high AoA and good resistance to loss of control requires a very careful balance between static and dynamic stability. This arises because at high AoA, the behaviour of slender fighters is dominated by the aerodynamics of the vortices on the forebody (whose shape is heavily influenced by radar performance) and the wing/body vortex strake. Appreciable instabilities of the aerodynamic characteristics are caused by flow separation at AoAs in the 30 to 40° range. However, even cursory examination of the Su-27's radome indicates that its radar requirements were not compromised in order to have good post-stall lateral/directional stability. The way to reduce lateral disturbances was found to be an increased rate of attaining the highest AoA, to give reduced time in the 30° to 40° range before passing into the region of fully separated flow beyond 60°. The short duration of the manoeuvre prevents the development of lateral motion despite the presence of aerodynamic disturbances. In other words, the pilot passes through the unfavourable region as quickly as possible. The gyroscopic moment due to the engines' rotation is another major lateral effect, especially when pitch rates are high.

Nose-down pitch at high AoA was thought to be the most difficult attribute to achieve without heavily compromising other design and performance aspects. Analysis of the 'Flanker's' Cobra manoeuvre showed that its pitch angle was reduced from its maximum value back to wings level in about three seconds. One of the characteristics of modern fighters is the appreciable nose-down moment at AoAs beyond about 60°. This is due to the 'lift' centre of the wing moving back, under the influence of the complete flow separation occurring on the wing, from about quarter chord at 20° AoA to about half chord (the centre of area of the wing) as the AoA goes to 80°. With the 'lift' centre now behind the CG, a restoring pitch-down results, the pitch motion is arrested and a recovery made. The reason for placing the term 'lift' in quotation marks is that at 90° AoA the force that is holding the 'Flanker' up against gravity (albeit momentarily) is the engine thrust.

In addition to the aerodynamic methods of conferring supermanoeuvrability, multi-axis engine thrust vectoring (MATV), as demonstrated by the MATV F-16 and X-31, may become the definitive enabling technology. At present the only operational fighters that have thrust vectoring are the Harrier and Su-37 'Flanker', though both the F-22 and JSF will enter service with it, though only in the pitch direction. 'Speed', it has been said, 'is life. A slow-moving target is what you want to see, not what you want to be.' So the tactical value of post-stall manoeuvring remains a matter of controversy, though pilot support for it has grown in recent years.

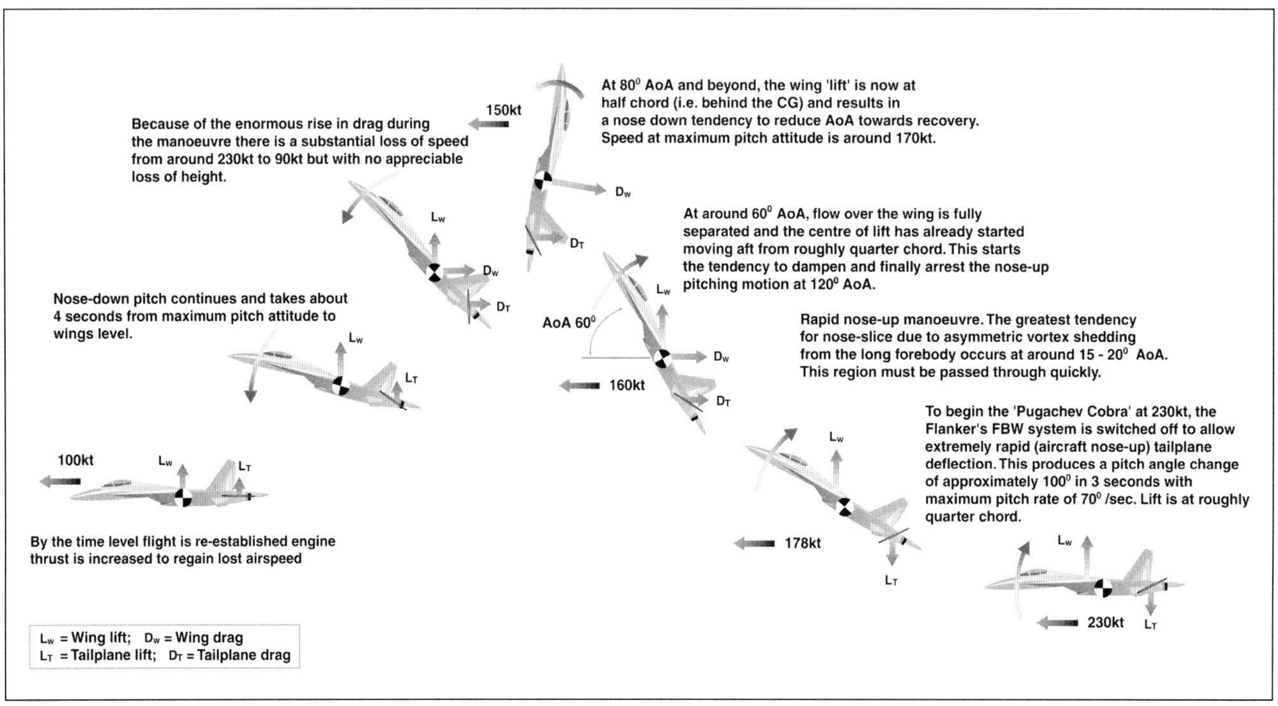

By its ability to demonstrate the 'Pougachev Cobra' manoeuvre, the 'Flanker' shows its tailplane's high effectiveness, resistance to lateral/directional instability at high AoA and robust nose-down moment as the wing's centre of 'lift' moves aft of the CG. *(Pete West/AIR International).*

Chapter 4

Propulsion

Flight of heavier-than-air aerial vehicles marked the triumph of man over gravity, while the struggle of power versus weight was a feature of technical achievement. It leads from the early work by the Wright brothers to the introduction of jet engines during World War Two. However great the progress in aerodynamic and structural efficiency has been over the years, the advances in light, compact and efficient powerplants have been paramount for the growth of fighter performance. From the start, lightness was a determining factor. It remains so, but is now allied to ensuring that 'carefree handling' of the engine harmonises with that of the airframe.

Glossary

Bypass ratio (BPR): The ratio of external mass flow of the cold jet from the fan to the mass flow passing through the engine core.
Power loading: Total aircraft weight/total installed power.
Pressure ratio: The ratio of the pressure of the air leaving the last stage of the compressor to the pressure of the air entering the first stage.
Propeller efficiency: Power delivered to the air by the propeller/power delivered to the propeller by the engine.
Propulsive efficiency: Useful propulsive power/engine power $= \dfrac{2}{1 + U/V}$ (where V is aircraft velocity and U is the engine's jet velocity)
Solidity: Ratio of total area of propeller to disc area.
Specific fuel consumption (sfc): The rate at which fuel is burned for a given power or thrust output. For gas turbines a rough rule of thumb, which no one endorses but which seems to work, is:

$$sfc \propto \sqrt{\frac{1}{1 + \mathrm{BPR}}}$$

Specific thrust: Engine thrust/air mass flow through the engine.
Specific weight: Engine weight/power output (lb/hp).
Thermal efficiency: Power output/rate of fuel energy consumed. In its simplest form this is expressed as the

$$\text{Carnot (ideal cycle) efficiency} = 1 - \frac{T_1}{T_2}$$

where T_1 is the lowest temperature in the cycle (typically at inlet) and T_2 is the maximum temperature (TET).
Turbine entry temperature (TET): Temperature of the gas leaving the combustor and entering the turbine.
T/W (thrust/weight ratio): The engine thrust divided by engine weight.

Fighters such as the Airco DH.2, RAF FE.8, Sopwith 1½ Strutter, Pup and Camel and Moranes had Gnôme or le Rhône or Clerget rotaries: the original Nieuport 28 (seen here in replica form) being powered by a 160hp Gnôme Monosoupape rotary. The le Rhône or Clerget were sufficiently different from the Gnôme to circumvent legal actions over patent rights. *(Cliff Knox).*

Pre-World War One

The Wright brothers calculated that an engine delivering a minimum of 8hp (6kW) would be required to maintain flight of their first aircraft. Simplicity of construction, few parts, ease of manufacture and, above all, lightness were the governing factors in their design. When assembled and tested in February 1903, the home-built, four-cylinder, in-line, water-cooled engine developed 13hp for a power loading of 11.5lb/hp (9.5kW at 7kg/kW) and this was the output on which they flew. They understood the importance of the proper relationship between the rotational tip speed of the propeller and the forward speed of the aircraft. By running their propellers at one-third engine rpm they achieved a propeller efficiency of around 70%, though bearing in mind the state of the art it would have probably been as efficient if they had turned at the engine speed.

The period from 1903 to the outbreak of war was one of great variety in engines. The earliest intensive development of aero-engines had taken place in France, and the best of the early engines had been air-cooled of all three types, rotary, radial and in-line. By 1914, France had introduced the greatest number of designs: twenty-two of the fifty engines of the time. There were about a hundred new designs introduced; most were built in both air-cooled and liquid-cooled forms. In addition to engines with stationary cylinders, the air-cooled rotary was greatly developed in France. In Germany, the six-cylinder, in-line, liquid-cooled engine had come to the fore, particularly exemplified by the 'Mercedes'. Wooden propellers of laminated construction, first introduced in Germany, had become more or less standard. The emphasis was placed on getting a greater power/weight ratio and improved reliability.

World War One

When war broke out in 1914, twenty-five different aircraft and six engine types had been designed and built by the Royal Aircraft Factory (RAF) at Farnborough which, up to 1917, was largely responsible for engine design in the UK. At the outbreak, however, the RFC and Royal Naval Air Service (RNAS) did not have a single aircraft powered by a British engine; indeed, only about a hundred aero engines had been built in the UK. The only unit available was the RAF 1, and this was a close copy of a Renault 80hp (60kW) air-cooled V-8. Fighters like the Airco DH.2, RAF FE.8, Sopwith 1½ Strutter, Pup and Camel, Nieuports and Moranes had Gnôme, le Rhône or Clerget rotaries. Indeed, the Allies saddled themselves with a part supply problem by fielding small quantities of many different types of engine. This was not really a matter of policy, for there was simply no standardised large-scale production of engines to uniformly equip the aircraft. In Germany the Benz and Mercedes companies were pre-eminent.

The rotary engine was so-called because it was a type in which the crankshaft remained stationary, while the engine with propeller attached spun around it. In order to feed fuel and oil to the engine, the crankshaft was hollow. Over a tortuous route was pumped a rich mixture of gasoline to run the engine and castor oil to lubricate it. Unorthodox though it seems, the rotary engine was a great success in fighters during World War One. Properly maintained and repaired, with attention given to spark-plugs, the Gnôme was good for twelve to fourteen hours between overhauls. The most distinctive feature of the engine was not its spinning cylinders, but the fact that it was the first successful air-cooled engine. It eliminated the excess weight of the radiator and cooling liquid (water at the time). The problem with air-cooled, in-line engines was invariably local hot spots. In comparison, the rotary engine, with all its cylinders equally exposed to the wind and the spinning motion, had very effective cooling. It was not until the early 1920s that aluminium (good heat conduction per density) cylinder heads brought about effective cooling without having the engine itself whirling around.

There were, however two principal disadvantages with the rotary engine. One came from the castor oil fumes and spray that flew back into the pilot's face. Old photographs show pilots wearing scarves, giving them a dashing appearance, but the scarves were useful in providing a means to wipe the oil off the pilot's goggles. The second drawback was the gyroscopic action set up by its whirling motion, which tended to give the aircraft a mind of its own. Moreover, carburation was tricky since there was no carburettor or throttle in the conventional sense. The pilot on approach to landing set the fuel control for a near idling speed, then 'blipped' the engine by cutting the ignition with a thumb switch on the control column.

Liquid-cooled engines

During the war air-cooled engines of sizes up to about 14lt (850in^3) displacement and about 150hp (110kW), mostly of the rotary type, had been used with great success. The early builders found it difficult, however, to make air-cooled engines much larger than this and thus they turned to the water-cooled type. Between 1912 and 1914 the Daimler Motor Company in Germany had developed the 'Mercedes' type, a vertical six-cylinder, water-cooled unit of 160 to 180hp (120 to 135kW) with welded steel cylinders. This type of cylinder construction was the basis of not only the standardised American wartime engine the Liberty, but also of the Rolls-Royce aero-engines (Hawk, Falcon and Eagle) and a number of others. In 1915 Hispano-Suiza in Spain designed a 150hp (112kW) V-8 water-cooled engine with cylinders cast *en bloc* and brought it to France where it was initially manufactured. It was subsequently built by twenty-three licensees including the Wolseley Motor Car Company in the UK and Wright-Martin in the USA, which was formed in 1916 to build the engine. Various attempts were made to fit propeller reduction gears but these proved unreliable because gears were not ground at that time.

Beginning of the radial engine

By the middle of World War One the most powerful engines were water-cooled, though before the end of the war the development had begun of competitive simpler, less vulnerable and lighter air-cooled fixed radial engines. The British Admiralty had been encouraged about the prospects of the air-cooled radial engine in preference to the hitherto popular rotary. It was believed that more power could be derived for the same size and weight, even before deducting the 25% or so power that was lost in the windage of the rotating cylinders. This could benefit taking-off from aircraft carriers and alleviate the operational difficulties of those times with water-cooled engines. The plumbing and radiator leaks from which all water-cooled engines suffered were particularly serious for the Navy because of frequent forced landings in overwater service. In addition to leaks, water-cooled engines in those days were very slow and

A seven-cylinder 50hp Gnôme rotary engine in a Blackburn Monoplane. *(Philip Jarrett)*.

troublesome to put into operation. They required as much as fifteen minutes to warm up, and in winter, since there was no satisfactory antifreeze in use, they had to be drained as soon as they were stopped and then refilled with hot water when needed again. There were periods of several weeks when, because of frost, no aircraft with water-cooled engines could fly at all. Thus, in April 1917, the Air Board issued a specification for an air-cooled radial to develop not less than 300hp and to be not over 42in in diameter.

Altitude effects on power

One of the first lessons to be learned during the war was that aeroplanes suffered a loss of power as they climbed to altitude. Since performance was of prime importance in military operations, the attention of engine designers became focused on improving power output at altitude. Three different approaches were taken to solve the problem. One was to over-size the engine to give it sufficient piston travel to generate the necessary power. However, such engines were often of too light construction to be used at full throttle at sea level. Mercedes followed this practice. The second approach was to design an engine with a very high compression ratio and to depend on throttling to prevent damage at low altitude. The third approach was to provide an external device for compressing the air before it entered the engine, namely a supercharger. This did not play any role in fighter aircraft during World War One; indeed it was not until the 1930s that supercharging was widely applied, but considerable research was carried out in the UK, France and the USA.

While the engine developments of World War One were certainly crude by later standards, a good deal of progress was made through constant attention to reliability, reduced fuel consumption and increased power. Hence, the end of the war saw the weight of liquid-cooled engines down to a little over 2lb/hp (0.7kg/kW) and that of air-cooled engines slightly lower. By considering the thermal efficiency, namely the portion of the energy in the fuel that is finally converted into thrust energy, engines at the end of the war were achieving over 20% (twice that obtained by the Wrights).

Between the Wars

US air-cooled engines

The development of high-powered air-cooled engines in America began with a design competition held by the USAAC in 1919. Well aware of progress being made in Britain with air-cooled engines of over 300hp (Armstrong-Siddeley Jaguar and Bristol Jupiter), an American engine of the same general size and type was needed. A specification was issued calling for an engine to develop 350hp, or only 50hp less than the Liberty, which was then the largest American one of any type. In 1923 the USAAC issued a specification for a new engine designated the R-1454 (R: radial engine of 1,454in^3 (24lt)). Curtiss delivered the first R-1454 in 1924, nine months after contract signing, and on test it delivered up to 405hp at 1,650rpm. Before the end of 1925 the engine had passed a fifty-hour type test consisting of forty-five hours at 90% of its rated 390hp and five hours at fully rated power for a weight of 830lb. However, by 1926, when flight tests with the R-1454 were being made, a much better engine appeared, the Pratt & Whitney R-1340 Wasp, that not only delivered slightly more power (410hp) but weighed only 650lb and was a good deal cleaner and more reliable.

Whereas the British paid very little attention to the specific design of naval aircraft, the US Navy had begun to get an important share of total aircraft appropriations. At the time all such USN aircraft were British, the best being the Sopwith Camel with a 130hp Clerget air-cooled rotary and the SE5 with a water-cooled 180hp Hispano-Suiza. For reasons it shared with the Royal Navy, the US Navy had an announced policy from 1927 of using only air-cooled engines for ship-based aircraft.

The USAAC remained unconvinced over the use of air-cooled engines for fighters until 1928, when at the National Air Races the Navy entered the first US fighter designed around a radial engine, the Wasp-powered Boeing XF4B. While there was no direct competition between it and the Army's most advanced liquid-cooled fighter, the Curtiss P-1B, the XF4B achieved 149.6kts versus 128.3kts of the P-1B with the heavier V-1150 (D-12) engine developing about the same power. Even more telling was a race to 10,000ft and back which the XF4B achieved in 5.92 minutes, while the P-1B took 7.08 minutes. General Fechet, Chief of the Air Corps, was so impressed by this performance that he placed a verbal order on the spot. This order for the Boeing P-12 fighter with the Wasp engine marked the establishment of the air-cooled engine in American military aviation. Air-cooled types (Jaguar and Jupiter) had become standard for British fighters about five years previously.

Engine cooling

The cooling of engines was initially achieved without much thought for the resulting aerodynamic drag. Air-cooled engines were placed in the nose of the aircraft with their cylinders projecting into the airstream, while for water-cooled engines a radiator of suitable size was slung under the aircraft or behind the propeller. Typically, fighters had blunt-nose nacelles, prefaced either by a naked radial engine or a flat radiator.

Despite the lead given by Lanchester before 1900 it was not until the late 1920s that the basic principles to minimise the drag of either type of engine were finally understood. These were that the drag directly due to cooling was minimised if the heat-radiating surfaces were made as large as possible and the velocity of the cooling air over them was reduced to a minimum; the stream of cooling air should be made to rejoin the general airstream in such a way as to cause as little turbulence as possible; and the air not used for cooling should be led around the engine and the radiator (if any) as smoothly as possible.

Cooling radial engines

Efforts to reduce the drag of the air-cooled engine by enclosing it with a cowl started in 1918, but the need to provide adequate cooling and accessibility had made designers cautious of such features. For a time it was difficult to see clearly any future for the air-cooled engine, for which 80% of the drag was due to turbulent airflow around the exposed cylinder heads and 20% was due to the skin friction, necessary to provide the cooling effect. It seemed probable that aircraft with such engines would be slower than similar types with water-cooling. However, any loss of speed might be outweighed by the better rate of climb, greater cruising range, smaller size and short take-

A seven-cylinder 64hp Le Rhône rotary engine with gravity feed from cylindrical tank. *(Philip Jarrett).*

off, all due to the lower specific weight of the air-cooled engine. Moreover, without the need for water pumps and plumbing, maintenance was easier on the exposed cylinders of the radial engine than on in-line water-cooled types, where a whole bank had to be disassembled to get at a single cylinder.

During the late 1920s there was an accelerated development of cooling and cowling of engines. Firstly, in 1927, from the National Physical Laboratory (NPL) in England came the Townend Ring which was a narrow chord wing wrapped in a circle around the engine. Its downwash forced air onto the engine and reduced drag by preventing flow breakaway downstream of the engine cylinders. It was very widely used and had the advantage of not interfering much with accessibility to the engine since it was not closed down in front and did not have the long skirt of the NACA cowl that was introduced shortly afterwards.

The NACA cowl took the form of a fairing completely enclosing the engine and extending back to the fuselage. A comparatively narrow annular slit between the cowling and the fuselage was for the escape of the cooling air. It made a definite separation between the streams of cooling and non-cooling air and made the former rejoin the latter smoothly. Its aerodynamic effectiveness was excellent but it reduced the flow of cooling air so much that in some cases the power that could be taken from engines larger than the 200hp Wright Whirlwind was seriously limited by overheating, with the result that the top speed was scarcely increased by use of the cowl. Nevertheless, in preparation for the 1929 National Air Races, baffling which greatly improved the cooling under the cowl was developed at Wright Field and a NACA cowl incorporating it was installed on the USAAC's Curtiss XP-3A. The speed attained with this combination was far superior to that with the Townend Ring.

Hence, whatever the problems it created for cooling, the NACA cowl was realised to be essential if the desired speeds were to be attained. However, the cooling problems became more serious because by now two-row engines were becoming common. These were even harder to cool. Better baffles were designed so that, rather than simply deflecting the airflow around the rear of the cylinders, the baffling forced the air through the cooling fin passages and cut off all other airflow. Moreover, it was realised that the best way of reducing the flow of cooling air when less than maximum was required was to vary the size of the exit from the cowl. The first adjustable cowl flaps of the type that became standard were first tested in 1934.

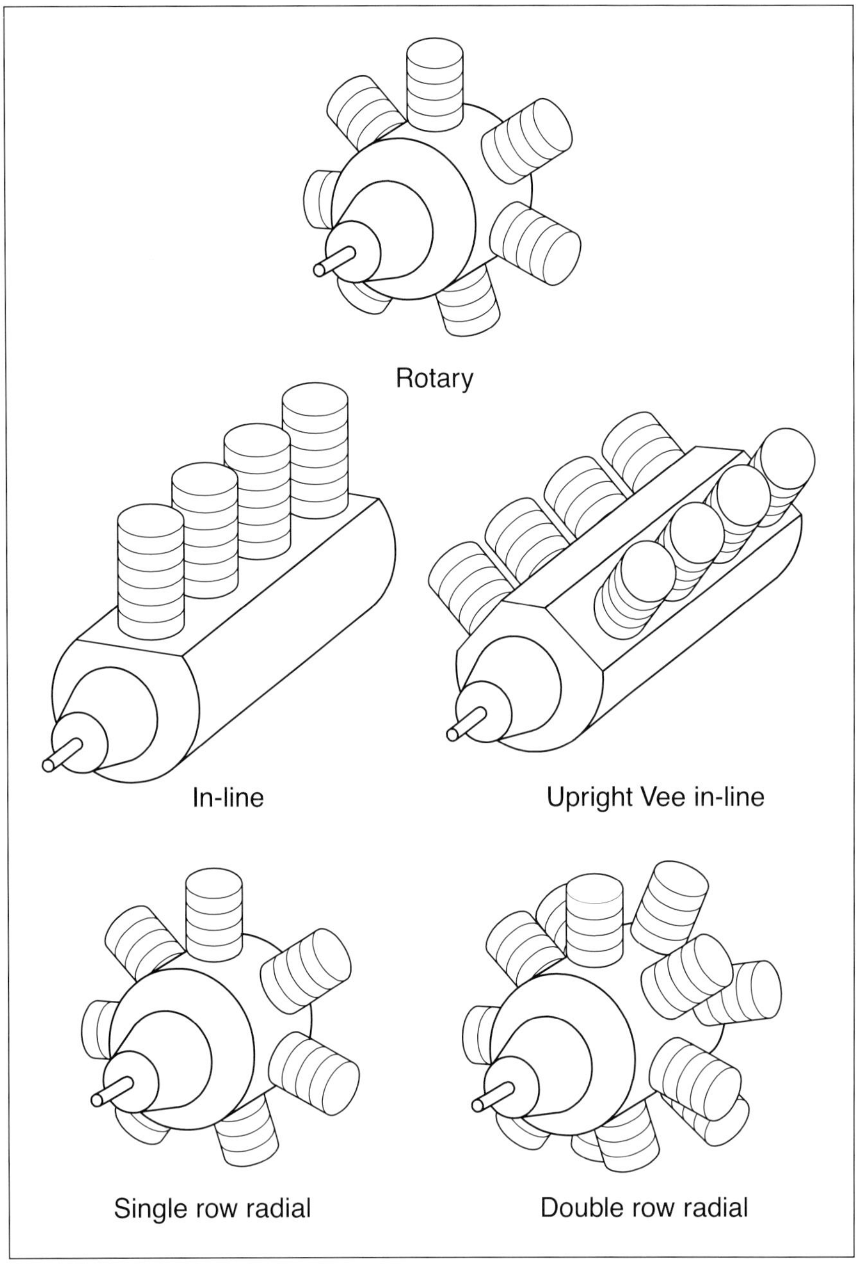

Basic piston engine configurations. It was not until the early 1920s when the development of aluminium cylinders brought about effective air cooling without having the engine itself whirling around (the rotary engine), that the fixed radial engine became practical. Cooling of 1930s double row radials became problematic though.

Despite the NACA cowl's superiority, use of the Townend Ring continued for several years on service fighters (Boeing P-12C and P-26, Bristol Bulldog IIIA of 1932) chiefly because pilots insisted on being able to see forward between the cylinders. They held the lack of visibility due to the NACA cowl responsible for several accidents. By the late 1930s, however, pilots had become used to the cowl and few, if any, complaints were heard. With further development the cowling gradually became a highly developed metering device, incorporating baffles and shutters to direct airflow as required for the cooling air, while minimising drag.

Cooling in-line engines

While the NACA cowl remained largely unused, the cooling drag of liquid-cooled engines was greatly reduced by the use of high-temperature cooling with ethylene glycol (commonly known as Prestone) instead of water. As early as 1915 it had been suggested at Farnborough that a coolant with a boiling point higher than that of water would be advantageous, because if the temperature of the coolant and radiator was increased, heat could be dissipated at the same rate by a smaller and lighter radiator containing a lesser weight of coolant. This development made possible a great reduction in drag since the cooling surface needed to be only one-third the size. Although this improvement also remained unused in service for about ten years until 1935, owing to the lack of a production engine that operated successfully with glycol, it gave substance to the argument that the total drag of liquid-cooled engines could be kept below that of air-cooled types. It was argued throughout the 1930s that with sufficient development, the liquid-cooled engine with the cooling system and coolant included could be made nearly if not just as light as the air-cooled engine. The US Army converted some Curtiss D-12s and Conquerors to glycol cooling and demonstrated that with the glycol at 150°C the weight of radiator and coolant could be reduced to half of what it was with water. For the D-12 the figures were 70% less radiator and 27% less coolant by weight.

In 1926 it was realised that airflow through the radiators on liquid-cooled engines could, if properly ducted, eliminate the cooling drag and even produce a little thrust at speeds above 260kts (483km/h). However, at this speed the air would be moving too fast for efficient cooling. To achieve this, the air had to enter some form of cowl in which it slowed down before it met the radiator, but this slowing down needed to be accomplished with minimum dissipation of energy. Having achieved a slow-speed airstream through the radiator, so that effective cooling could be performed, the heated air had to be speeded up so as to rejoin the external flow at a speed not less than the flight speed, to avoid extra drag. Thus the air at high pressure, caused by both the slowing down and the heat input, could be accelerated via a simple convergence to a speed higher than the flight speed and produce some thrust to offset the drag caused in the slowing-down process. It was a mild form of jet propulsion, and as the speed of flight increased, the propulsive effect, like all jet propulsion, became more efficient. It could even outweigh the drag due to skin friction in the air passages, and cooling could therefore be attained at a negative cost in drag. This contrasts with the 6 to 10% of the engine power lost in cooling drag of previous aircraft.

Meanwhile, the surface radiator had been developed for the water-cooled engine, particularly in connection with the Schneider Trophy races of the 1920s. In this radiator a portion of the wing had a double skin, and water was circulated in a thin layer between the two surfaces. On the Supermarine S.5B, nearly the whole wing surface was used in this way. It was a highly powered flying radiator; even the surface of the vertical fin was used as an oil cooler. The wing radiator seemed to solve the problem of cooling without additional drag, but it had several disadvantages that prevented its adoption for other than racing aircraft. It was difficult to construct and maintain, it added complexity to the already difficult problem of metal-covered wing construction, and, for military aircraft, was far too vulnerable.

The importance of low frontal area

Curtiss brought out its D-12 water-cooled engine in 1922. It was a V-12 engine of 1,145in^3 (18.7lt) displacement developing 400hp (300kW) that quickly proved to be the best engine in the world at that time for high-speed aircraft. In a racing aircraft also built by Curtiss it set a new world speed record of 194.5kts, and in 1923 the sleek Curtiss CR-3 biplane won the 1923 Schneider Trophy Race at 231.9kts. The performance of these Curtiss racers was so far ahead of any British aircraft that the engine naturally attracted a good deal of attention, particularly that of Sir Richard Fairey. His appreciation of the importance of the streamlined engine nacelle and the influence of this on the engine shape and type was eventually to lead to the Rolls-Royce Merlin. He obtained rights to manufacture the D-12 engine, the Reed metal propeller, wing surface radiators and other aids to streamlining. The Reed propeller, of twisted aluminium plate, had a small hub and pointed spinner that blended well into the improved nacelle lines made possible by the slim engine, and largely set the

Surface radiators had been developed for the water-cooled engine in connection with the Schneider Trophy races of the 1920s. In this radiator a portion of the wing had a double skin and the water was circulated between the two surfaces. On the Supermarine S.6B, nearly the whole wing surface was used in this way and even the surface of the fin was used as an oil cooler. *(via author).*

pattern for advanced propulsion. He designed around it the Fox two-seat day bomber, which in 1925 astonished both British government authorities and aircraft industry by having a speed and general performance at least equal to any of the single-seat fighters of the day. The newest and best of these, Fairey Flycatcher and Gloster Gamecock, were respectively powered by Jaguar and Jupiter air-cooled radial engines of about the same power as the D-12. The opinion had been spreading that the air-cooled engine was the only one suitable for fighters, the water-cooled type being used chiefly for long-range types on account of its lower fuel consumption. The government was sufficiently persuaded of the merits of the D-12 that it obtained several examples for testing by RAE Farnborough, from which it concluded that it was essential that Britain should have an engine of the same type.

Rolls-Royce was persuaded to interest itself in an engine of small frontal area and stripped down one of the D-12 engines. In 1925 it began the design of an engine later to be called the Kestrel, a factor in the decision being the knowledge that the larger part of new aircraft procurement by the government would be fighters, the type that could benefit from a clean compact engine. The Type F, a 1,295in^3 (21.2lt) V-12 engine of 525hp (390kW), allowed even better streamlining and was put into production before the end of 1927 with a normal rating of 480hp at 3,000ft for a weight of 865lb.

Henceforth Rolls-Royce planned its V-12 engines with frontal area in mind. The company took its Buzzard engine of 850hp (630kW) basic power, geared and supercharged it, for the first time supercharging at sea level, as distinct from maintaining power at altitude. This was the R engine that, in 1929, achieved 1,900hp at a power loading of 0.8lb/hp (1,420kW at 0.5kg/kW) at 2,900rpm. It was the powerplant of the Supermarine S.6 that won the 1929 Schneider Trophy race at 286kts (530km/h) and was subsequently developed to 2,780hp (2,000kW), achieving over 350kts (645km/h) in the S.6B in 1931.

The Schneider Trophy aircraft were dedicated racing machines and every other consideration was subordinated to the attainment of the highest possible speed. They carried no payload, and 'witch's brew' fuel sufficient only for the race. The engines were run under conditions that would have given impossibly short lives for ordinary flight purposes. Most importantly, though, the wing loading was, for the time, very high at around 200kg/m^2 (40lb/ft^2), giving landing speeds of about 85kts (160km/h). After some fatal accidents involving their Schneider Trophy designs, the Italian Air Ministry decided to abandon the development of liquid-cooled engines for military aircraft. Consequently, in 1940, when Italian industry again needed high-powered liquid-cooled engines, it had to build German ones under licence.

Liquid-cooled versus air-cooled engines

As for the attitude towards liquid-cooled engines in general, the races contributed a very great deal towards establishing two convictions in the UK. The first was that clean, liquid-cooled engines could be used in the fastest types of aircraft, fighters. Even before 1929 the aircraft

Liquid-cooled engines of the late 1920s provided aerodynamically much cleaner installations than their contemporary air-cooled competitors. A radiator was needed of course and on the Rolls-Royce Kestrel of this Hawker Fury it is mounted in the middle of the undercarriage. *(BAe via author).*

industry had been forced by the Fairey Fox to alter its belief that only air-cooled engines were suitable for fighters, but the government continued to buy fighters with air-cooled engines, notably the Bristol Bulldog with the Bristol Jupiter, and it was not until 1931 that the first squadron was equipped with the Hawker Fury powered by the Kestrel. Once the Fury had been accepted, however, few squadrons were equipped with new fighters with air-cooled engines. The second conviction was that far greater power could be obtained from a liquid-cooled engine than from an air-cooled engine. It was not until the 1940s that an air-cooled engine anywhere in the world equalled the power of the Rolls-Royce R engine of 1931. These two convictions formed the basis of the monoplane fighters of the mid-1930s.

If the liquid-cooled engine proved ultimately to give superior performance, an air force without a good one would be at a serious disadvantage in time of war. Every major country apart from America consistently promoted the development of both engine types, no matter which seemed the better at a given moment. Indeed, the two most successful countries, UK and Germany, even promoted competition by having two major firms working continually on each type. By 1932 the USAAC had become convinced that Curtiss-Wright, whose Conqueror was the only American liquid-cooled engine still in production, could no longer be looked to as a source for a really modern type. The Conqueror-powered P-6 fighter was inferior to the contemporary Wasp-powered P-12 and P-26 at a time when their performance at altitude was inferior to that of the Hawker Fury with its liquid-cooled Kestrel. Further development of the Conqueror was not seen to be feasible and support of it stopped.

Detonation

Another important advance was understanding the destructive effects of detonation and setting limits in engine design and in the operating practices, to avoid its occurrence so that the compression ratio could be increased to give more power. Detonation is uncontrolled ignition of the fuel/air mixture as it is compressed and heated by the piston. Supercharging makes detonation more likely to occur and more difficult to avoid because the temperature and pressure of the charge are increased by compression in the supercharger before it enters the cylinder. There are three principal methods of preventing detonation. First is the use of fuels that are naturally less liable to detonate. Secondly, the compressed charge can be cooled in an intercooler before it enters the cylinders. However, since intercoolers add so much weight and drag to an aircraft they were never used in service engines with single-stage superchargers. Thirdly, detonation can be prevented by elimination of 'hot spots' in the engine itself, usually the spark-plugs and exhaust valves. In liquid-cooled engines such regions could be largely avoided by the flow of coolant. In air-cooled engines the design could to some extent be arranged to increase the flow of air past potential hot spots, but was much less effective.

Fuels

Work by Ricardo into the tendency for some fuels to detonate led to a research programme by the Shell Company at the end of World War One at a time when octane ratings of around 50 to 65 were common, though at that time there was no method of measuring such properties. The high boost of the Rolls-Royce R engines for the Schneider Trophy races could only be tolerated because of the exotic fuel cocktail used. Such a fuel was totally impractical for service use because of the high freezing point of the benzol used and the poor heat-to-weight ratio of the alcohol employed as a charge-cooler. However, in 1932, 87-octane gasoline (the first standard British fuel to contain lead) was introduced into UK from the US and all engine development began to be based on it. Until 87-octane fuel became operationally available in 1934, the standard fuel had been between 73 and 77 octane rating. The introduction of anti-knock additives such as tetra-ethyl lead to produce high-octane leaded fuels and other developments in fuel chemistry opened the way to extensive widening of the detonation-free limits.

Supercharging

The power output of piston engines is proportional to the density of the air used for burning, so that at 20,000ft (6,100m) the shaft power will have dropped to almost half that at sea level. To maintain engine power at altitude, and increase power at sea level, for a given size engine, required supercharging to compress the air before it entered the engine.

Superchargers were used at sea level to supply the engine with air more dense than atmospheric in order to increase power, and at altitude, despite the decreasing density, to supply air at least as dense as the engine could use and thus prevent any falling off of power. Supercharging at sea level creates serious difficulties with detonation, and, in the case of air-cooled engines, with cooling as well. There is a difficulty with detonation at altitude also, because despite the air being colder, compressing it to sea-level density results in the delivery temperatures and pressures being substantially above those at sea level. This was true even with highly efficient superchargers, and much more so with the inefficient ones of the 1930s. As for cooling, difficulties arise with air-cooled engines at altitude because the density of the air decreases proportionally *more* than the temperature difference between the ambient and the cylinder fins increases. Thus cooling fin area that is able to dissipate the heat generated at sea level is not able to dissipate the same heat from the same power at altitude. Here the liquid-cooled engine had a double advantage: not only was the size of the radiator free from any inherent limitation, but up to very high altitude there was no need to increase it, because even with glycol cooling, the radiator is so cool compared with the fins of an air-cooled engine that the proportional increase in the differential between radiator and atmospheric temperature fully compensates for the decrease in air density.

The Kestrel was Rolls-Royce's first engine with an integral gear-driven supercharger, a feature strongly desired by the government, which was becoming convinced that this was a better bet than the turbosupercharger and was pressing Bristol Aero Engines and Armstrong-Siddeley to develop supercharged versions of their engines too. However, there was a lack, which was common to all engine firms at this time, of personnel with experience in compressor development, and particularly its aerodynamic aspects which were completely different from the problems

A Kestrel on a Rolls-Royce test rig. During the 1930s full supercharging, automatic boost control, 87-octane fuel and sodium-cooled valves resulted in improved Kestrels, the final versions delivering 745hp at 14,500ft at 3,000rpm, with a total of 4,750 being delivered. *(Rolls-Royce via author).*

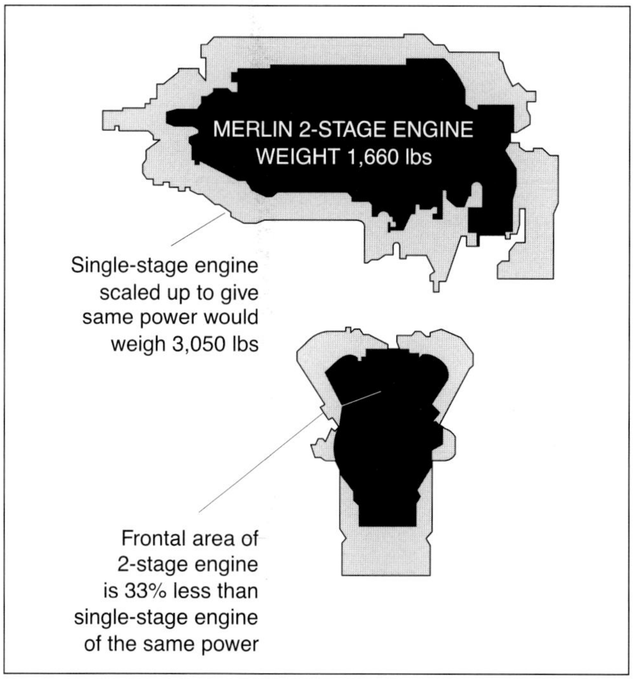

The saving in size and weight by using a two-stage supercharger and aftercooler on the Rolls-Royce Merlin (both engines have the same power at 30,000ft).

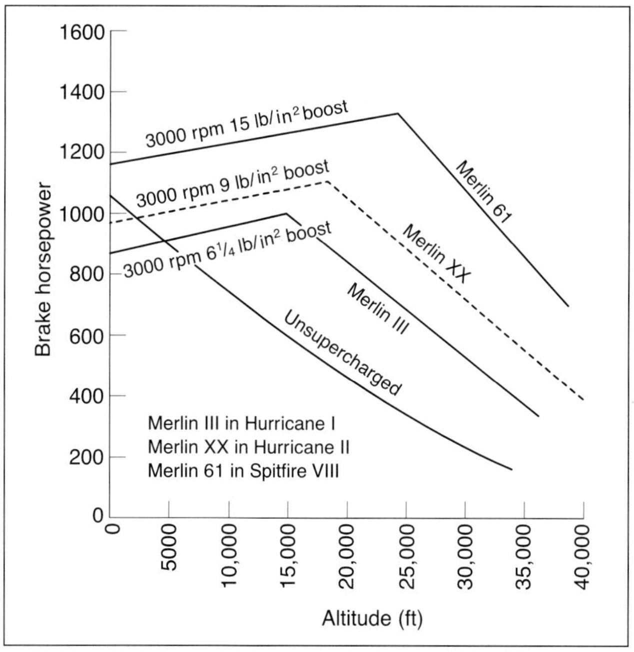

This chart shows the enormous effect that supercharging has on power at high altitude. A normally-aspirated Rolls-Royce Merlin would produce about 370hp at 25,000ft. The Merlin III gave about 700hp at that altitude, whereas the Merlin 61 with a two-stage supercharger produced 1,350hp. Note: Boost pressure is the difference between ambient atmospheric pressure and that of the supercharger delivery.
(Pete West/AIR International).

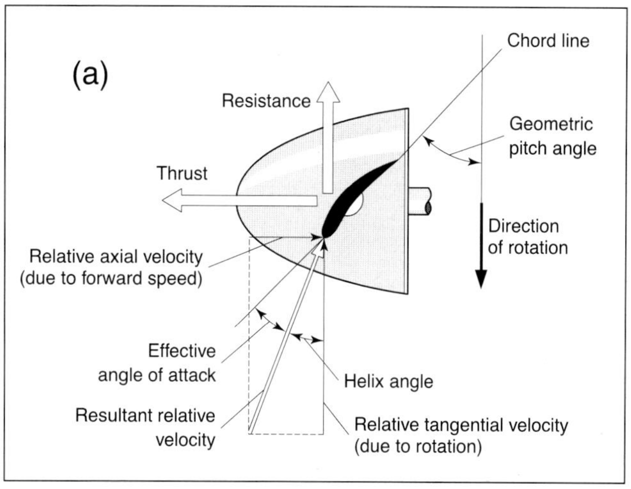

(a) A propeller blade section, having a profile similar to a wing, will produce forces analogous to lift and drag. These can be more conveniently resolved into forward thrust and tangential resistance. The relative velocity at each station is made up of two components, one along the axis of rotation (e.g. the flight speed) and the other due to rotation (which varies with the radius) at 90° in the plane of the propeller disc. Thus the blade needs to have twist, in that at the root where the relative wind is predominantly the forward velocity of the propeller, the blade pitch (its setting angle) is high ('coarse') whereas at the tip the pitch is low ('fine').

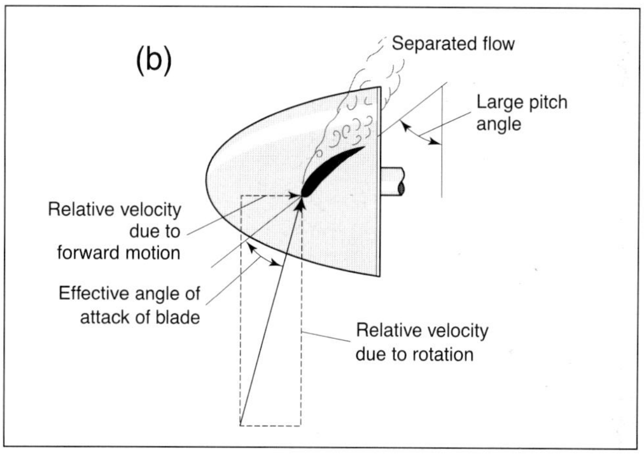

(b) If the engine is running at high rpm and the blade pitch angle is set large for efficient cruising, then at low forward speed, say on take-off, the blade angle of attack may be so high that, as shown here, portions of the blade will stall.

encountered in the rest of the development of reciprocating engines. Without the advances in detonation prevention, improved fuels and efficient superchargers, the power of World War Two engines would have been halved.

Variable-pitch propellers

A development in propulsion that was given considerable emphasis by World War One was the variable-pitch (vp) propeller. This device was first thought of as a means of giving the propeller sections appropriate angles so that they could operate throughout the aircraft speed range with a high propeller efficiency. It was soon realised, when use of a suitable device was available, that it was equally important as a means of controlling the speed of the engine under varying flight speeds. The power output of the engine could be maintained from take-off to maximum speed. For the aeroplane speeds and engines current at the time, fixed-pitch propellers were fitted so that the engine turned at its rated rpm at the aircraft's top speed. This restricted the engine on take-off, to around 85% of its maximum rpm. Thus an improvement of 15% in engine power could be had by the use of the variable-pitch propeller, while at the same giving a small gain in propeller efficiency. However, the benefits of variable-pitch to World War One aircraft, which had modest top speeds, would have been very limited. Wooden blades were not well adapted for fitting into the vp

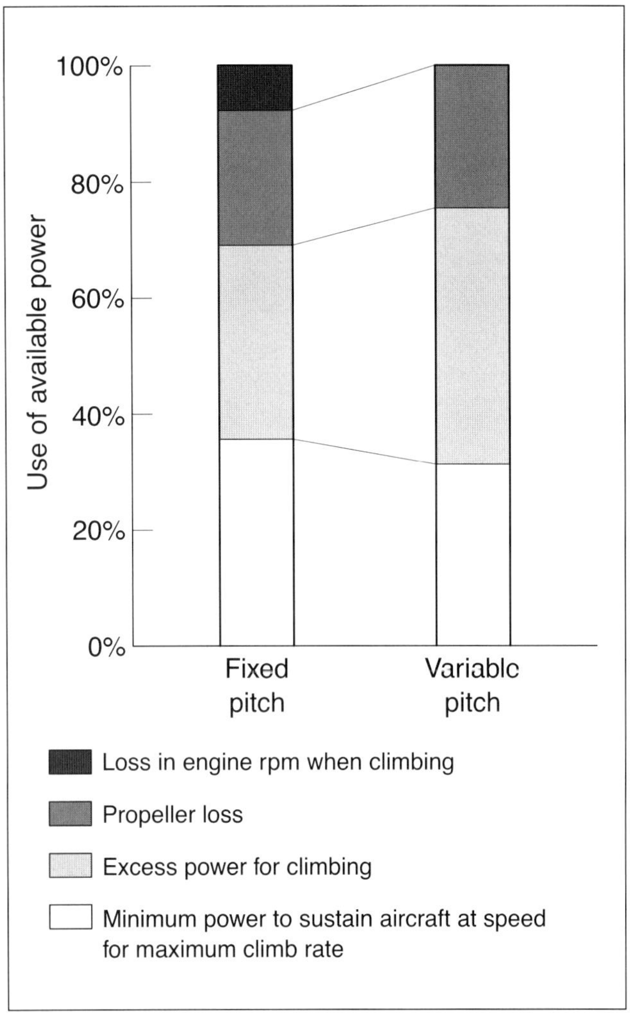

For a single-engine biplane with a vee air-cooled engine the use of a variable-pitch propeller could provide an extra 13% of power to be used for climbing.

hub, since they required a clumsy and heavy hub structure, and variable-pitch propellers were at least twice as heavy as fixed ones. Also, in Britain propeller reduction gear had been developed to high efficiency and reliability, and together with low wing loadings of around 10lb/ft^2, work was pushed ahead on the metal propeller blade.

Metal blades were adapted to a simple type of hub whose pitch could be altered on the ground but not changed in flight. This type of propeller was readily accepted in the US, and by the 1930s it almost completely displaced wooden propellers for large engines. The critical need for the variable-pitch propeller came not from the military but from the introduction of airliners like the Boeing 247 and Douglas DC-2 and DC-3. These aircraft with wing loadings of around 25lb/ft^2 cruised at a sufficiently high speed (140 to 165kts) that the blade angles of the fixed-pitch propellers were high enough to cause stalling of portions of the blades during take-off. In 1933, Caldwell in the USA provided a two-position hydraulic propeller: low (fine) pitch for take-off and high (coarse) pitch for cruising. On the Boeing 247, this reduced the take-off run by 20%, increased rate of climb by 22%, cruising speed by 5% and ceiling by 4,000ft (1,220m). This was quickly superseded by the constant-speed propeller, in which a speed governor produced the necessary signal to alter the pitch of the propeller blades, to automatically keep the engine rpm at a predetermined value. Despite this development, the Hawker Hurricane, for example, entered service with a two-bladed, fixed-pitch wooden propeller and only received a vp metal type in 1939.

Air-cooled versus liquid-cooled engines

Throughout the late 1920s and 1930s the furious controversy between the proponents of air-cooled and liquid-cooled engines continued to rage. Advantages of the air-cooled engine were seen as: less weight, easier to maintain, less vulnerability, unaffected by frosty conditions and quicker warm-up. Advantages of liquid-cooled engines were: more uniform cooling and lower frontal area, although the radiator had to be placed normal to the airstream. For fighters the importance of minimum frontal area could not be denied, so the only response available for the radial air-cooled engine was to increase its power/weight ratio.

The US industry's concentration on the air-cooled engine rested on its generally accepted conviction that sales of that type (especially for civil use) would far exceed that of liquid-cooled engines. This was especially true after about 1932 when it was accepted that US liquid-cooled engines were by no means as highly developed as air-cooled ones. Moreover, whereas the RAF had always strongly emphasised the importance of single-seat fighters, in the USA it was never so important in relation to other types, and in the mid-1930s the USAAC seriously considered abolition of single-seaters.

The importance of weight

Of all the figures measuring the performance of an aero engine, one of the most important was its specific weight (lb/hp). For the Jaguar and Jupiter and the first successful American high-power air-cooled engine (R-1340 Wasp), the chief advantage of the air-cooled engine over its liquid-cooled competitor was a definitely lower specific weight. At the end of the 1920s the specific weight of air-cooled engines was little more than that of in-line engines without their radiator and coolant (1.6 to 1.7 versus 1.5lb/hp

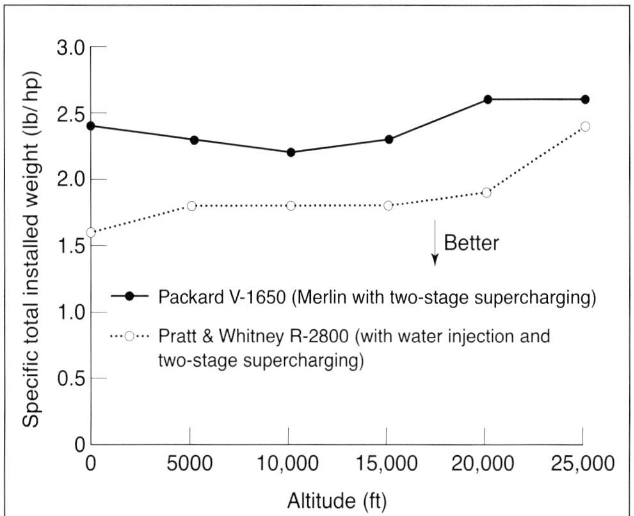

Comparison of specific total installed weight. Although the Pratt & Whitney R-2800 provided the benefit of a significantly lower specific installed weight (lb/hp) at low altitude than the Packard V-1650 (Merlin), the difference became almost negligible at 25,000ft.

respectively); with water used as the coolant this added about 0.5lb/hp or about one-third of the 'dry' weight of the engine itself. Though high-temperature cooling would cut the weight of the cooling system, it could not be completely eliminated, so the specific weight of the liquid-cooled engine itself had to be reduced. This had to be done in spite of the fact that even without its cooling system, the in-line engine weighed about 10% more per cylinder volume than the radial type. The difference was primarily due to the in-line engine's greater length and hence much greater weight of crankcase and crankshaft per cylinder.

By 1930 the air-cooled engine's weight advantage was so persuasive as to be decisive so far as comparison between American engines actually in production was concerned. As for what might be achieved by further development, however, the situation was less clear. Was it possible to increase the power of the liquid-cooled engine much more than the air-cooled type? The power of an engine varies with the product of two factors, the rpm and the mean pressure within the cylinder, which depends primarily on the degree of supercharging. The liquid-cooled engine had a considerable advantage in rpm, the Conqueror, for example, being rated at 2,450rpm versus 1,900rpm for the R-1340 Wasp. The push-rod valve gear of the radial is less suited to high speeds than the overhead camshaft of the in-line, whose cylinders were normally smaller and their pistons and rods correspondingly lighter than those of the radial. The great obstacle to higher mean pressure was the problem of detonation.

Ultimately, both the weight-saving anticipated from high-temperature cooling and the high-power output of 1hp/lb were fully realised in the Rolls-Royce Merlin and Daimler-Benz DB 601. The realisation, however, took far longer than enthusiasts had anticipated in 1931, and in the meantime the air-cooled engine had also made great advances. Probably the most balanced view was the air-cooled engine would continue to have the advantage at low altitude. On the other hand it was only late in the 1930s, after much improved fuels had become available and methods had been developed for producing air-cooled engine cylinders with much closer finning, that air-cooled engines could equal the liquid-cooled engine at altitude.

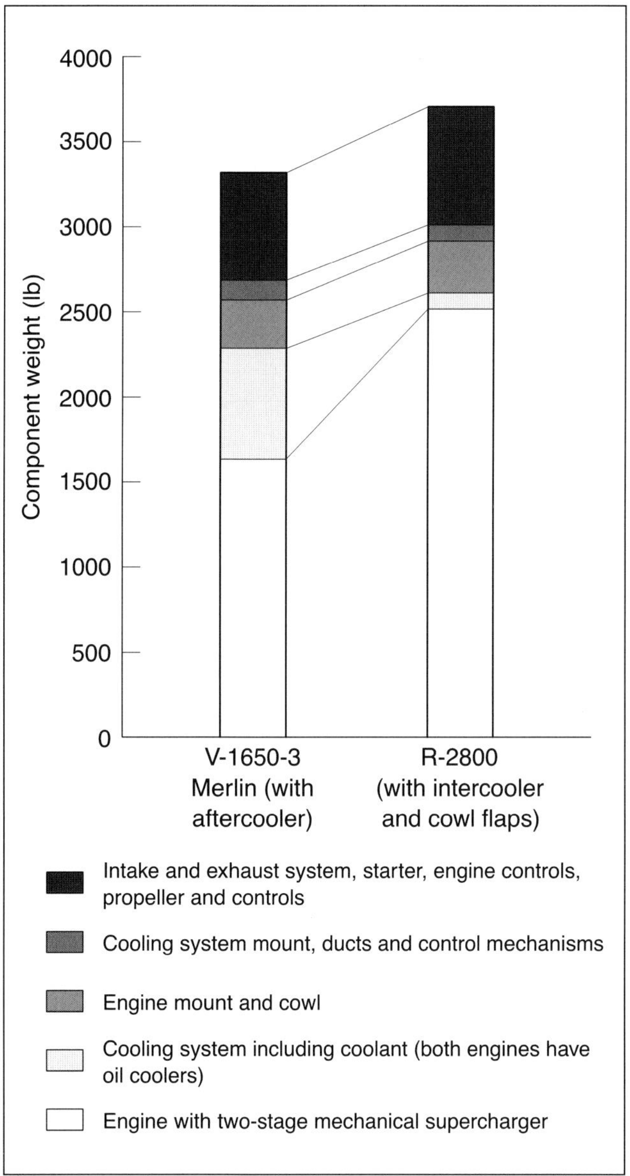

Comparison of liquid-cooled and air-cooled engines. Installed weight breakdown of the Pratt & Whitney R-2800 (Republic P-47 Thunderbolt) and Packard V-1650 Merlin (North American P-51 Mustang).

Capacity for overload

An argument frequently used in favour of liquid-cooled engines during the 1930s was that they could sustain maximum instantaneous power for longer than could air-cooled ones. This was because the reservoir of coolant acted as a buffer that took time to overheat whereas air-cooled cylinders overheated almost instantly. In the 1920s and early 1930s this was perhaps a real advantage. The cooling systems of the time normally operated at about 80°C, and, taking the Rolls-Royce Kestrel as an example, the use of 20% more power than the cooling system was designed for would have meant about five to six minutes before boiling occurred. Air-cooled engines of the early 1930s, on the other hand were barely able to cool sufficiently for their rated power and there was hardly any margin between overload and failure.

The situation of later liquid-cooled engines (with glycol) was quite different owing to the much smaller amount of coolant available. For example, taking 20% excess power from a V-1650-3 (Packard-built Merlin) of the P-51B of 1943 would have raised the temperature of the radiator and coolant far quicker. Contemporary air-cooled engines, however, could then stand an appreciably higher cylinder temperature than they had been designed for, and while this undoubtedly shortened the life of the engine it did not induce immediate failure.

Vulnerability

For use in ground support, with the risk of being hit by small-arms fire, the most important advantage of the air-cooled engine was its lower vulnerability. A single puncture in the cooling system of a liquid-cooled engine would put it out of action within a couple of minutes, whereas air-cooled engines continued to survive for hours even with bullet holes in a cylinder. Apparently this was the principal reason for the use of the BMW 801A air-cooled engine in the Focke-Wulf Fw 190. The advantage was not decisive in air combat, since if one fighter was fairly hit by a burst from the heavy armament of another it was likely to be fatal in any case.

Two-row radials

By 1930 it was generally agreed that a single-row radial had reached its limit of displacement at about 30lt (1,800in³). No successful radial had ever had more than nine cylinders in a single row, nor cylinders of greater individual displacement than 3.4lt (207in³). There was never any doubt that a good liquid-cooled engine of the classic V-12 form could be built with a displacement at least this large, with considerably greater output and of smaller frontal area than any single-row radial. A fighter had to have a fuselage just large enough in cross-section for the cockpit, but as such was not wide enough for a radial engine which were all between 1.2m and 1.4m (4 to 4.5ft) in diameter.

To meet this challenge Pratt & Whitney decided early in 1931 to develop a two-row radial engine of 1,830in³ (30lt) displacement as its first of this type. The Navy was the first US customer to be seriously interested in two-row engines, wanting to use one not to obtain higher power but to obtain a smaller diameter with the same power that could be got from a big single-row. Even the Pratt & Whitney R-1830 was intended to produce only 15 to 20% more power than the big single-rows, but by 1936 it was delivering 1,200hp (895kW).

Probably the greatest weakness of the air-cooled engine at this time was its inability to use a high degree of supercharging. This meant that although the specific weight at low altitude was lower for air-cooled engines in the first half of the 1930s, the difference became very much smaller above 12,000ft. In the US the comparison was at this time more potential than actual, since the only superchargers capable of maintaining full engine power to these altitudes were turbosuperchargers, which were experimental and had so many problems that they were far from ready for general service.

By 1940 much more powerful air-cooled engines were under development, the Wright R-2600 and Pratt & Whitney R-2800 of 1937, the Bristol Centaurus (53.6lt/3,270in³) of 1938 and the BMW 801 (41.8lt/2,550in³) of 1940. These were much larger than the most prominent liquid-cooled engines, the Merlin (27lt/1,649in³), DB 601A (34lt/2,070in³) and Allison V-1710. However, liquid-cooled engines fitted beautifully into the sleek fighters being schemed in the mid-1930s. Indeed, none of the vee-type engines used in World War Two were more than 0.75m (2.5ft) wide and a little over 0.9m (3ft) high. In these single-engine aircraft the radiator could be and was located wherever it was most convenient. The advantage of the liquid-cooled engine is shown by the fact that the substitution of the V-1710 for the R-1830 in the Curtiss XP-40 Hawk reduced its total drag by 22%. Whereas the XP-40 with a V-1710 rated at 1,090hp reached 318kts at 15,000ft, the P-36 with an R-1830 rated at 900hp managed only 260kts at 10,000ft.

Turbosuperchargers

Parallel development occurred of gear-driven superchargers, both single- and two-stage, especially in the UK and Germany, and the exhaust-driven turbine-supercharger (turbosupercharger) in the USA. The latter used the hot exhaust gases, which contain a very large

The Grumman F6F-5 Hellcat first flew in 1944 and was powered by a Pratt & Whitney R-2800 Double Wasp of 2,000hp that could be boosted by 10% with water injection. These are two F6F-5s (in post-war service with US Navy Reserve at NAS Floyd Bennett, NY) of the 7,870 built when production ended at the end of 1945. *(via author).*

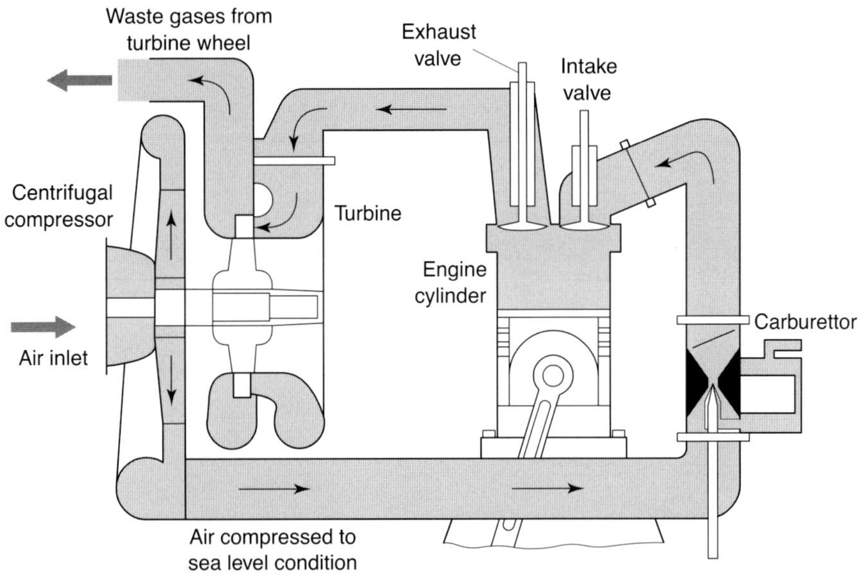

Diagrammatic illustration of an early model (circa 1924) turbosupercharger which permitted piston engine aircraft to reach otherwise impossible heights. At high altitude, the thin air was compressed by a centrifugal compressor and directed through the carburettor, where fuel was added, and then through an intake valve into the cylinder. Following combustion, the exhaust gases were chanelled to drive the turbine which in turn drove the compressor. *(Pete West/AIR International).*

amount of energy that might otherwise be wasted, to drive a turbine, which in turn drove a compressor. However, severe metallurgical difficulties were encountered due to the high temperatures, and it took Wright Field and General Electric about twenty years to produce a relatively reliable unit.

The superior performance shown in Table 5 of the XP-37 (a Curtiss P-36 fitted with a turbocharged V-1710) in 1937 had an immediate impact on the USAAC. Contracts were let for experimental prototypes of two new fighters based on the V-1710 with a General Electric turbocharger, the twin-engine Lockheed XP-38 Lightning and the single-engine Bell XP-39 Airacobra. However, there were concerns about the turbosupercharger and its controls before it could reach the reliability necessary for general service use. In addition, problems still existed with propellers and ignition at the 20,000ft reached with the turbocharger. (There was still a lot of trouble with the controls of the turbocharger in the P-38 as late as 1942.) During 1939, about a year after delivery of the turbocharged XP-38, XP-39 and YP-37, the USAAC became convinced that it was unwise to risk everything on the turbocharger and that a medium-altitude fighter should be developed. As a result of this the Curtiss P-40 (a reworked P-36 with the V-1710) won the Army competition held in the spring of 1939.

Table 5 Performance benefits of re-engining with a liquid-cooled engine

Aircraft	*Engine*	*Sea-level power*	*Speed @ altitude*
Curtiss P-36	R-1830-25	1,100hp (820kW)	270kts @ 10,000ft
Curtiss XP-37	V-1710-11 (turbocharged)	1,150hp (860kW)	286kts @ 20,000ft

The V-1710 was the only American liquid-cooled engine to see service in World War Two, and from 1937 to 1939 it looked as though it might compete with the Merlin. The top speed attained by the XP-37 in 1937 was the same as that achieved by the Merlin-powered Spitfire prototype in 1936, and the 20,000ft altitude at which the XP-37 reached its top speed was 4,000ft higher, due to the turbocharger, than that for the Spitfire. In 1939 the Bell XP-39 Airacobra reached 350kts and the Lockheed XP-38 registered 360kts, both at 20,000ft. However, continuing uncertainty about the possibility of developing a fully satisfactory turbocharger led the USAAC to drop it from the P-40 Hawk and the P-39 Airacobra in 1939, leaving them with inadequately developed single-stage mechanical superchargers.

Alternative use of exhaust gas

By the late 1930s, about 50% of the energy released in burning the fuel was wastefully dumped overboard via the exhaust as hot gas, while at best only 30% was delivered to the propeller. An important innovation was the exploitation of thrust available from the engine exhaust. In some cases this added 20% to the thrust of the entire propulsive system, an example being the ejector exhausts on the Spitfire's Merlin. For highly supercharged engines with stub exhaust pipes, the exit velocity of the exhaust gases was about 600m/s (2,000ft/s). It was not possible, however, to deflect the gases backwards at these velocities without considerable increase in back pressure that would have merely reduced the power output from the engine. However, with careful design, the power equivalent of the exhaust gas momentum was made large compared with the engine power reduction due to the increase in back pressure.

World War Two

100-octane fuel

At the beginning of World War Two the frontline British fighters Hurricane I, and Spitfire I, both with Merlin engines, were inferior to the Bf 109E with its DB 601A in

Design of a single-seat fighter (later to be called the Mustang) around the unsupercharged Allison V-1710 engine to meet British needs was started by North American Aviation early in 1940 and as the NA-73X, the first example flew 117 days later in October of the same year. When Allison did produce a two-stage supercharger for the V-1710 in 1942, it proved to be of such a form that it could not be installed in the Mustang. A version of the Mustang was fitted with the Rolls-Royce Merlin 61 during 1942, it proved such an outstanding success that henceforth all P-51s were powered by the Packard-built Merlin (V-1650). This is a P-51D with triple-cluster rocket launcher. (via author).

climb at all altitudes. Though the DB 601A weighed very little more than the Merlin (660kg versus 625kg) it had 25% greater displacement and was installed in a considerably lighter aircraft. Rolls-Royce was obliged to develop the Merlin to tolerate much higher manifold pressure, in order to obtain equal performance at the low altitudes, and to develop superior supercharging, so as to compete at the higher altitudes. Toleration of higher manifold pressure was the simpler of the two problems since the Merlin was inherently stronger than the DB 601A, weighing 23kg/lt versus 19.6kg/lt. An increase in permissible boost for the Merlin depended almost entirely

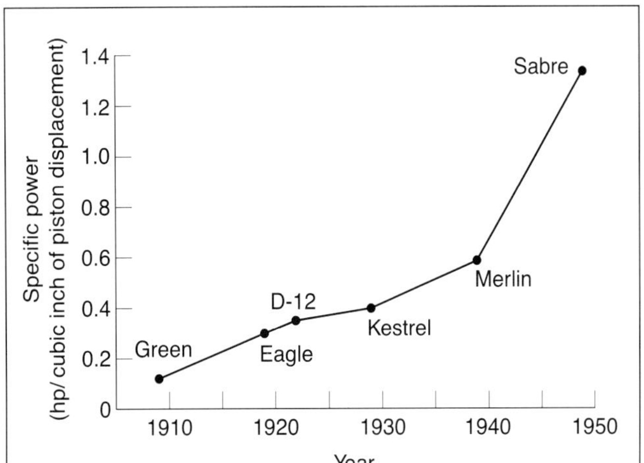

The power extracted from a cylinder of given size increased by a factor of three between the beginning of World War One and the start of World War Two. The rapid increase during World War Two was due mainly to higher intake manifold boost pressure, via more efficient supercharging, and the availability of 100-octane fuel.

on the appearance of a service fuel that would tolerate it without detonation. Fortunately, 100-octane fuel became available (albeit in limited quantities) to the RAF in September 1939. Its value can be seen from the fact that the Merlin II, which with a manifold pressure of 1.4bar on 87-octane fuel had a combat rating of 1,030hp at 16,250ft at the outbreak of war, was authorised before the end of 1939 to use 1.6bar on 100-octane fuel to give a combat rating of 1,160hp at 13,500ft with no increase in rpm. By the middle of 1940 the combat rating was 1,310hp at 9,000ft again with no increase in rpm but with higher boost. The engine development for this rating consisted simply in an alteration in the automatic boost control.

UK and US engines at the outbreak of war

It has often been asserted that whereas British air-cooled engines were designed and built with military service (i.e. the highest possible maximum power for the lowest possible weight) exclusively in mind, American types were heavy because their design and development was aimed at securing reliability in commercial service. The most likely explanation is simply that Wright and Pratt & Whitney were, rightly or wrongly, more conservative than Bristol, believing it was better to go for a larger engine than to try to push a smaller one too far. In this respect Germany followed a policy much like the American one, although they were thinking in terms of military use exclusively.

In the early part of World War Two American air-cooled engines did have higher specific weight than Bristol ones (1.3 to 1.4 compared with 1.2lb/hp). The true inferiority of US engines in this respect was probably not quite so great as the ratings seem to imply, since American ratings were guaranteed by the manufacturer whereas British ratings were not. Despite this tendency to rate US engines lower

Increases in power from the Supermarine Spitfire Mk I's 1,030hp Rolls-Royce Merlin II or III (above) to over 1,600hp in the Spitfire IX's Merlin 66 (below), necessitated greater propeller solidity — from the Mk I's two-bladed wooden propeller to the later Griffon-powered five-bladed Rotol units. *(Philip Jarrett).*

The Spitfire Mk IX of 1943 with its 1,600hp Merlin 61 (designed for bombers) housed in an extended nose, came with ejector exhausts and a four-bladed Rotol propeller that was able to absorb the extra power at high altitude. The supercharger cut in automatically at about 19,000ft (6,000m) which gave it an extra 10,000ft to a ceiling of 43,000ft and a 70mph increase in top speed of 410mph (660km/hr). It was probably the best fighter of its day and had the tactical advantage that it looked rather like the inferior Mk V. However, the Spitfire's narrow undercarriage track caused difficulties using rough airfields and led to many landing accidents. Shown is a Spitfire IX in the markings of the RAF's Polish 316 Sqdn. *(Ryszard Malachowski via author).*

than British engines, it was not until combat experience had shown the necessity of more power that American engines were developed of specific weight fully equal to the Bristol engines. On various occasions military necessity led US authorities to issue technical orders permitting operation of engines at outputs higher than those guaranteed by the manufacturer.

In addition to being heavier than British engines of the same power output, it has often been said that American engines showed their semi-civilian character in various specific features of their design. Probably the leading example is the absence on American engines of the variable-datum automatic boost control present on all British ones in 1939. This was a regulator that enabled the pilot to safely use maximum power from the engine under any conditions by simply opening the throttle as far as it would go; he had no need to pay attention to the various factors such as altitude, which set the limit to safe power, and then be careful not to exceed this limit. Because it was not needed in airline operations, its lack led to the ruining of many American engines by inexperienced military pilots.

Late in 1941 the Fw 190A powered by an air-cooled radial engine (the BMW 801C of 1,600 to 1,700hp at sea level) appeared with performance decisively superior to the Spitfire V. Like the DB 601 in the Bf 109, this engine had only a single-stage supercharger, but its displacement was so much larger than the Merlin's that it was hopeless to compete at altitude unless the British engine had improved supercharging. Even though Rolls-Royce had originally wanted to avoid the difficulty, inherent in a two-stage compressor, of having the faults of the first stage magnified in the second, it introduced a two-stage supercharger. Incorporated was an aftercooler, in which the cooling was carried out after the completion of the entire supercharging process. This was to limit the temperature of the charge entering the cylinders to increase its density and to avoid detonation. As a result the Merlin 61 (1,300hp at sea level) entered production in March 1942, giving the Spitfire IX an increase of 10,000ft (3,000m) in fighting altitude and 60kts (113km/h) in top speed.

Absorbing the power

The problems of absorbing the increased power and converting it into thrust by means of a propeller at high altitude received much attention. A propeller of normal solidity could do this efficiently, provided its diameter and/or its rpm were large enough. The diameter of fighter propellers was limited by ground clearance and weight. Compressibility due to excessive speed could cause such tip losses that neither increased propeller diameter nor speed was practical. The only remaining alternative was to increase the propeller solidity, which could be done in two ways. The first was to increase the blade width, keeping the number of blades constant. However wide, paddle-type blades (e.g. Republic P-47 Thunderbolt and Vought F4U Corsair) produced a high centrifugal twisting moment which had to be resisted by heavier variable-pitch mechanisms. The second method was to increase the number of blades. Most fighters at the start of the war were fitted with two-bladed propellers (many of them wooden), but three-, four- and five-bladed propellers were introduced later. The possibility of using contra-rotating propellers in tandem was considered, but apart from some trials no production fighter was so equipped during World War Two.

The development of gas turbines

The period from 1920 onward saw a great improvement in the efficiency and performance of compact centrifugal compressors. The principles of turbine design and the art of fabricating suitable high-temperature materials were greatly advanced as a result of turbosupercharger development. This furnished the background for turbojet development, and was available to engineers when the necessary advance in aircraft performance made jet propulsion timely.

A. A. Griffith (mathematician and aerodynamicist) of the RAE at Farnborough had, in 1929, put forward the proposition of an axial-flow gas turbine, taking power out as a drive shaft (i.e. a turboprop). Such an engine could be smaller, lighter and more fuel-efficient than contemporary piston engines. He had concentrated on improvements in axial compressor and turbine design to try to achieve high efficiencies that would allow the gas turbine to compete with the piston engine on grounds of thermal efficiency. This was also Whittle's problem. Until he demonstrated that he could improve component efficiencies, and have a reasonable amount of power left over, few people were prepared to accept gas turbines as a practical powerplant for aircraft. The turbojet as conceived by Whittle in 1929 was aimed at a means of propelling aircraft at speeds of 435kts (800km/h) at high altitude where air temperature is low. This was where both the propulsive and thermal efficiencies of the engine would be high and where such speeds would be feasible. Where Whittle's patent differed from Griffith's was in the use of the residual gas energy as a propulsive jet. Whittle's first engine ran in 1937, after much dogged perseverance and work on a shoestring due to lack of official backing. Meanwhile, von Ohain in Germany independently had a similar idea in 1933 and work progressed in that country at a faster pace. German manufacturers were able to provide greater funding and resources than Whittle could muster, and it took von Ohain and Heinkel a little over three years to develop an engine reliable enough to fly in an aircraft, the Heinkel He 178 in 1939.

Propeller versus jet propulsion

By the early 1940s, piston engine power loading was heading to a trough at 1.5lb/hp (0.9kg/kW). At the time, few piston engines were capable of producing 3,000hp (2,200kW) and none had delivered 4,000hp (3,000kW). There were major problems in design and development of piston engines of such power. Thermal efficiency could only be improved by higher compression ratios within narrow limits. This was because of limitations on piston diameter due to detonation and heat flow into the pistons, the number of pistons, and the increased complexity of such things as valve gears, bearings, ignition. On the other hand, propulsive efficiency had improved significantly by means of reduction gearing and the vp propeller. However, airscrew propulsion, having been the practice to the mid-1940s, was the generally accepted means of propulsion without examining the price to be paid at aircraft speeds exceeding 350kts (650km/h). A contemporary fighter requiring 1,000hp (750kW) to do 260kts (480km/h) at sea level needed 5,000hp (3,700kW) at 430kts (800km/h), but 7,600shp (5,600kW) from the shaft was required to be delivered to the propeller in order to obtain this. The

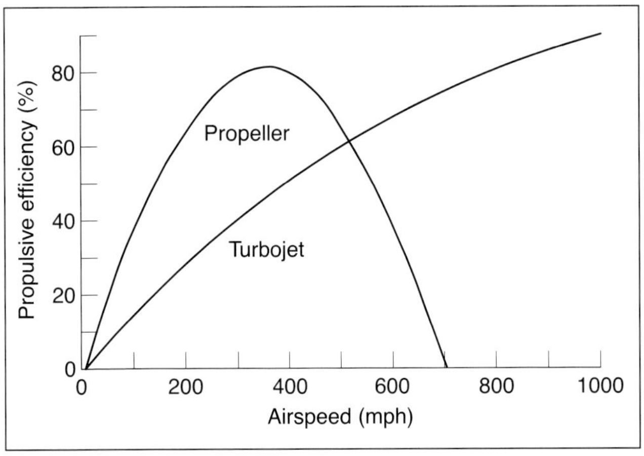

Graph showing the comparative propulsive efficiencies of propeller and turbojet. At speeds below approximately 430kts the turbojet is less efficient because its propulsive efficiency depends largely on its forward speed. The turbojet is most suitable for high speeds, whereas a propeller's efficiency drops off due to the compressibility effects of the high Mach number at its tips. *(Pete West/AIR International)*

wasted 2,600hp was about equally divided between normal rotational losses (torque and drag) on the propeller blades and compressibility effects on the blade tips. At 260kts (480km/h), the propulsive efficiency of a propeller was approximately double that of a jet, largely accounting for the fact that the specific fuel consumption of the jet was more than double that of the piston engine. At about 450kts (830km/h), the jet and propeller are roughly equal in propulsive efficiency, and since the jet is much simpler and lighter, it followed that jet propulsion had to be better at this speed. However, a major criticism was the rate at which turbojets burned fuel, though this overlooked the power being generated.

One of the other advantages which Whittle foresaw for the gas turbine was the ability to use less volatile and therefore safer fuels than gasoline. Furthermore, the use of a jet engine meant that pilots had merely one power lever rather than the controls on the piston engine that included throttle, mixture, propeller, radiator, supercharger, air intake, and slow running.

The first jet engines

The flights of the Heinkel He 178, though successful, were recognised as a misapplication of the turbojet principle due partly to the low speed of the aircraft. Consequently they failed to enlist enthusiastic support on a broad scale. With the outbreak of war, all technical development in Germany had come under direct government control. Another factor that inhibited the German designers was their shortage of nickel, forcing them to use materials that did not have the high-temperature qualities of those available to the British engines. Indeed, the engines used on the first flight of the He 178 were virtually burned out in the ten-minute duration of the flight.

The achievement of high maximum cycle temperature has always been the main objective in aero-engine technology since it is the key to high thermal efficiency. The first Whittle engine ran red hot at a maximum TET of about 900K. Turbojet development began with very low pressure ratios and with a tight limit on the TET due to the tendency of the turbine blades to become plastic and stretch at high temperature. Towards the end of the war, however, the reduction in manhours and materials required, compared with piston engine and propeller, coupled with the increased power output and lower weight, caused an increase in activity. However, it was not until the Junkers 004 engine had made a number of flights in the Messerschmitt Me 262 at over 450kts (800km/h) that jet-propelled aircraft were accepted in Germany as a useful tactical weapon. When hostilities ceased, the Junkers Jumo 004 had accumulated more than 25,000 hours of bench testing and over 5,000 units had been built.

In Britain, work on the turbojet had also been hampered by government policy, and it was not until Rolls-Royce started work on turbojets in 1941 that rapid advances were made. On instructions from the Air Ministry, the company undertook the development and manufacture of the Whittle engine. The only British design to enter production before the war's end was the Welland of 1,700lb (7.5kN) thrust, Rolls-Royce's development of the Whittle W.2B. Problems encountered in the combustion system and turbine blades were overcome and both mass flow and performance were increased. At a thrust of 2,200lb (9.8kN) it evolved as the Derwent, and became the standard powerplant of the Gloster Meteor.

News of the work in Britain and Germany filtered across the Atlantic in 1940. The Whittle W.1X (a test-bed engine), and its design drawings, was sent to the US in 1941. General Electric (having extensive experience with turbochargers and their high-temperature alloys) was given the task of producing a batch of the engines to power a prototype aircraft, the Bell P-59 Comet. The USAAC also made arrangements to have de Havilland deliver one of its H1 Goblin engines of 3,000lb (13.3kN) thrust to Allis-Chalmers, which was to develop it as the J36 for the Lockheed P-80. However, this did not happen, and General Electric's I-40 (J33) producing 4,000lb (17.8kN) became the powerplant for production P-80 Shooting Stars.

The established aircraft engine manufacturers in the USA, Pratt & Whitney and Wright, were kept out of the turbojet programme during the war because the US government felt that they should concentrate on their piston engines. The effect of this decision was to limit the supply of turbojets in America for five years after the war, predominantly to those of General Electric and Westinghouse. Pratt & Whitney started to build their J42 and J48 versions of the Rolls-Royce Nene in 1948.

The Jet Age

All but one of the British turbojets made in World War Two used the centrifugal compressor, the type which Whittle and von Ohain had chosen for its simplicity and ease of development. However, most German engines used the axial-flow compressor that subsequently became standard for fighters. The relatively simple wartime turbojets' performance was little affected by the type of compressor used, though the choice of the axial-flow type gave Germany more development problems. The only British engine with an axial-flow compressor built during the war was the Metrovic F.2, test-flown but not ready for production in 1945.

Centrifugal versus axial-flow compressors

The centrifugal compressor uses an impeller (of similar principle to that in older hair dryers) which when rotated at

Designed specifically for the purpose of testing von Ohain's new jet engine, the He 178 was an experimental aircraft and flew in August 1939 which was almost two years before the first British jet. With its 1,000lb thrust Heinkel S3B turbojet weighing around 800lb and specific fuel consumption of 1.6lb/hr/lb the He 178 reached 378kts. The Gloster E.28/39 used to test Whittle's engine, achieved 294kts on roughly the same thrust. *(Philip Jarrett collection).*

high speed by its turbine thereby induces air into the centre of the impeller. Centrifugal action of the impeller's vanes accelerates the air outwards (and raises its pressure), whereupon it passes into a diffuser formed by divergent passages. The air's kinetic energy is then converted into pressure energy, so that the pressure rise is about equally divided between impeller and diffuser. The biggest advantage of the centrifugal compressor is its robustness. It is much less susceptible to ingestion problems due to the inherent strength of its impeller. It was also easier to develop and manufacture both in terms of cost/lb and manhours/lb. The single-stage centrifugal compressor gave much more flexibility in operation, it maintained its efficiency over a wide range of flight conditions, and was more stall-resistant compared to the axial-flow type. Furthermore, it was easier to match compressor and turbine and there was a wealth of experience on which to draw.

Axial-flow compressors have a very large number of small individual blades mounted on discs arranged in stages on a shaft along the length of compressor. A stage consists of a row of rotating blades followed by a row of stationary blades (stators). From the front to the rear of the compressor, there is a reduction of the annulus area through which the air flows. The shaft is driven by the turbine at high speed, so that air is continuously induced into the compressor. In simple terms, the rotating blades scoop air in, then throw it out, slightly compressed at the following stator blades, which slows the air down further, to give another rise in pressure. The blades are both costly to produce and vulnerable in service. Axial-flow compressors are also relatively heavy. Despite these factors the axial type had great potential, since it could have its capacity more easily increased by adding more compressor stages, whereas the centrifugal compressor needed its diameter increased. The efficiency of the axial type was about 7% higher over a wider pressure range, due to the centrifugal's high losses associated with its tortuous flow path and higher tip speeds. The axial compressor's main virtue, though, was its much larger mass flow rate (and hence thrust) per frontal area. For the same airflow the diameter of the centrifugal might be 80% larger than its axial counterpart. This gave the axial engine a huge advantage for high-speed aircraft where a small cross-sectional area is vital to keep transonic and supersonic drag low.

The axial turbojet began to exert its supremacy, both on the score of small frontal area and lower specific fuel consumption (due to its higher pressure ratio and compressor efficiency). The transition to axial-flow compressors was inevitable, though not without pain, since there were stalwart supporters of both types.

Afterburners and water injection

In the late 1940s, thoughts were given to the possibility of reheat (afterburning) to augment the basic thrust of the engine to improve acceleration for take-off, climb and combat performance. Since only about 25% of the air entering the combustor was used in burning, there remained an abundance of oxygen in the gas stream for further combustion. Introducing and burning fuel between the turbine and the propelling nozzle results in a rise in the temperature of the exhaust gas. This can be converted to increased velocity of the gas leaving the propelling nozzle, thereby increasing thrust. The potential thrust augmentation depends on the square root of the ratio of the absolute temperatures after and before the extra fuel is burned.

Because of the noise and pyrotechnics, afterburning was

TOP: Aircraft designers faced with the task of creating the first jet aircraft, kept as close as they could to piston-engine aircraft practice, adopting a single-engine configuration, with conventionally straight wings. Since they were building small experimental aircraft, the main concern was to give emphasis on the new features associated with the installation of the new turbojet engine. The Gloster E.28/39 first flew on 15 May, 1941, at Cranwell, piloted by Flt Lt P.E.G. Sayer. *(Philip Jarrett).*

ABOVE: Problems with long air inlets and jet pipes, and the desire to get more thrust than one of the early engines could provide, led most designers to select two engines for the fighters they planned in 1939–40. The He 280 was a typical example, sticking to conventional design practice with the engines mounted immediately below the wings (uncowled here). *(Philip Jarrett collection).*

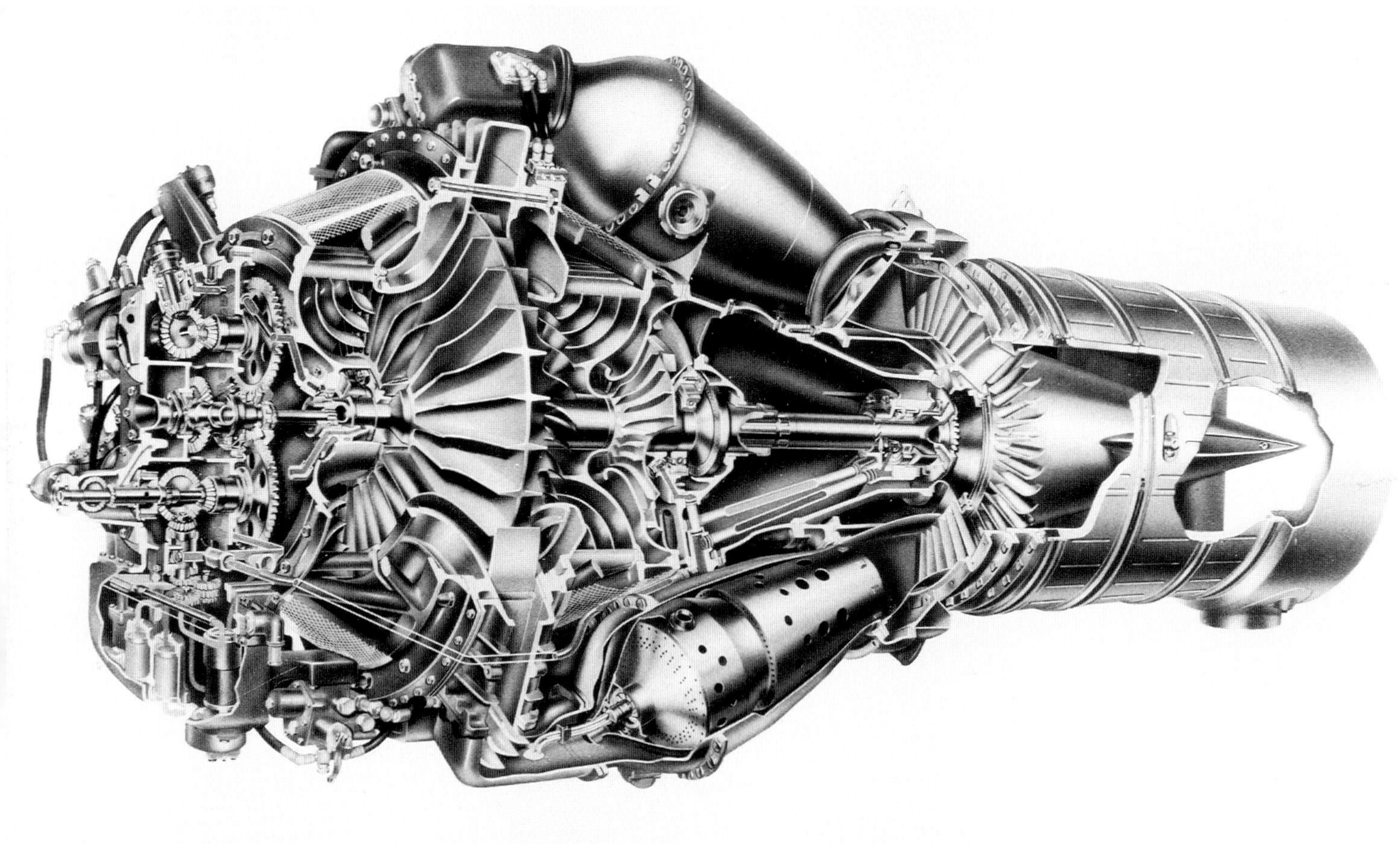

The Rolls-Royce RB41 (Nene) first ran in October 1944 giving a thrust of 22kN (5,000lb) making it the most powerful turbojet at the time. It was made under licence in Canada, Australia, United States and France. Over 39,000 unlicensed versions were made in the Soviet Union as the VK-1 of 26kN (5,950lb) thrust for the MiG-15 among other aircraft. It was the higher pressure ratio of the centrifugal-flow Nene compared with that of the F-86's heavier axial-flow J47, that was, in part, responsible for the higher operational altitude of the MiG-15. *(Rolls-Royce via author).*

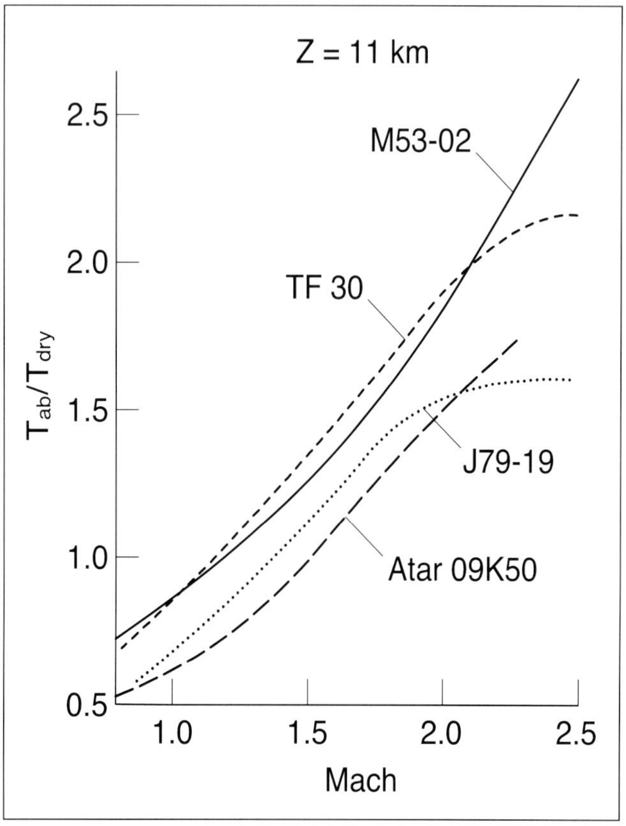

A turbofan has a high afterburning thrust potential and its percentage thrust augmentation is much greater than for a turbojet because of the large quantity of uncombined oxygen its bypass duct provides. A turbofan of BPR 1.0 at 36,000ft and Mach 2 has only about one third of its sea level static thrust, but by using its afterburner its thrust can be increased by a factor of four. The figure shows a comparison of T_{ab}/T_{dry} for two turbofans, the M53, TF30, and two turbojets, J79 and Atar 09. *(Pete West/AIR International).*

not regarded favourably, especially by test pilots. However, the performance improvements were dramatic. The de Havilland Vampire, for example, whose thrust augmentation was limited to only 13% due to its fixed nozzle, had its time to 40,000ft (12,000m) reduced by 30% with a modest increase in total fuel consumption, even though the specific fuel consumption was high (perhaps three times as much). Maximum speed was increased, and pilots were quickly converted to this form of thrust augmentation. Afterburning of typical single-spool axial engines, like the Rolls-Royce Avon, Snecma Atar and Armstrong-Siddeley Sapphire, gave an increase in thrust of up to 45% for take-off and combat. It became the preferred route to higher engine thrust.

Water injection proved a useful method of obtaining a modest thrust boost of up to 10%. The mass flow was increased and the compressor outlet temperature dropped, due to the cooling effect. This permitted more fuel to be burned, and the temperature drop through the turbine for a given power output was reduced and gave higher residual jetpipe temperature and exhaust velocities. Additional fuel

WB188 was the first prototype Hawker Hunter but was converted to the Mk 3 version shown here for the successful attempt on the World Airspeed Record in 1953. For this the aircraft had a Rolls-Royce Avon RA.7R (reheated, simply on/off) of 42kN (9,600lb), a two-position nozzle, pointed nose and, unusual for a Hunter, side-mounted airbrakes. *(BAe via author).*

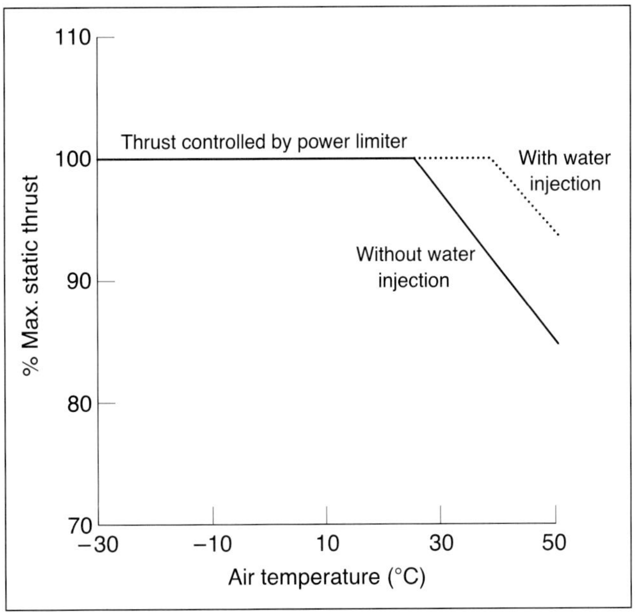

The maximum power output depends to a large extent upon the mass flow passing through the engine. There is a reduction due to the density decrease with altitude and/or temperature increase. Under these conditions the power output can be restored or boosted for take-off by cooling the airflow with water or water/methanol.

was required to restore engine speed, this being done by injecting a water/methanol mixture, the methanol burning in the combustor. Though used on several engines, water injection never became general, due to the extra volume, weight, cost, and maintenance. One of the engines on which water/methanol was used extensively was the General Electric J47 for the North American F-86 and Boeing B-47.

Development trends

If a complicated mechanism is to be developed successfully, it is essential to break it down into its individual components and examine each of these in detail. The reciprocating engine did not lend itself to this process, because most of the development had to be done on the complete engine. Moreover, because of the intermittent nature of the processes, a theoretical approach was difficult, if not impossible, and progress had to be made on the basis of empirical knowledge gained from testing. The gas turbines scored on all counts. The cycle was continuous, each stage being accomplished by an individual component which could be separated from the engine and tested as a unit. Furthermore, a theoretical approach was possible, so theory and practice could progress simultaneously. The problem lay in the air mass flows and powers needed for testing. In 1945, 7.5MW (10,000hp) was needed to test a typical compressor; by 1975 around ten times that power would have been needed, had testing been done at full pressure.

Five years after the end of World War Two, the pattern of fighter engine evolution had become reasonably clear. The piston engine had ceased to play any further part in fighter development. For carrier-based aircraft, though, flight path control is critical. They need to approach at the lowest possible speed, on the back side of the drag curve where increased thrust is required with decreasing airspeed. In contrast to jet engines that are relatively slow to respond to throttle, thrust modulation of turboprops comes from propeller pitch changes, with very short time lags. This kept alive Navy interest in turboprops and led to the Westland Wyvern (powered by the Armstrong-Siddeley Python driving contraprops), which entered service in 1953 and remained until 1958. Thus, after a brief spell in the limelight, the

The first prototype Grumman F9F Panther flew in November 1947 powered by a Rolls-Royce Nene of 5,000lb thrust, while the second and third had Pratt & Whitney J42 (licence-built Nene) and Allison J33 respectively. The F9F-5 shown was powered by a Pratt & Whitney J48-P-6 of 6,250lb boosted 12% by water injection. The wing fold of this shipboard fighter was just outboard of the air intakes. *(via author).*

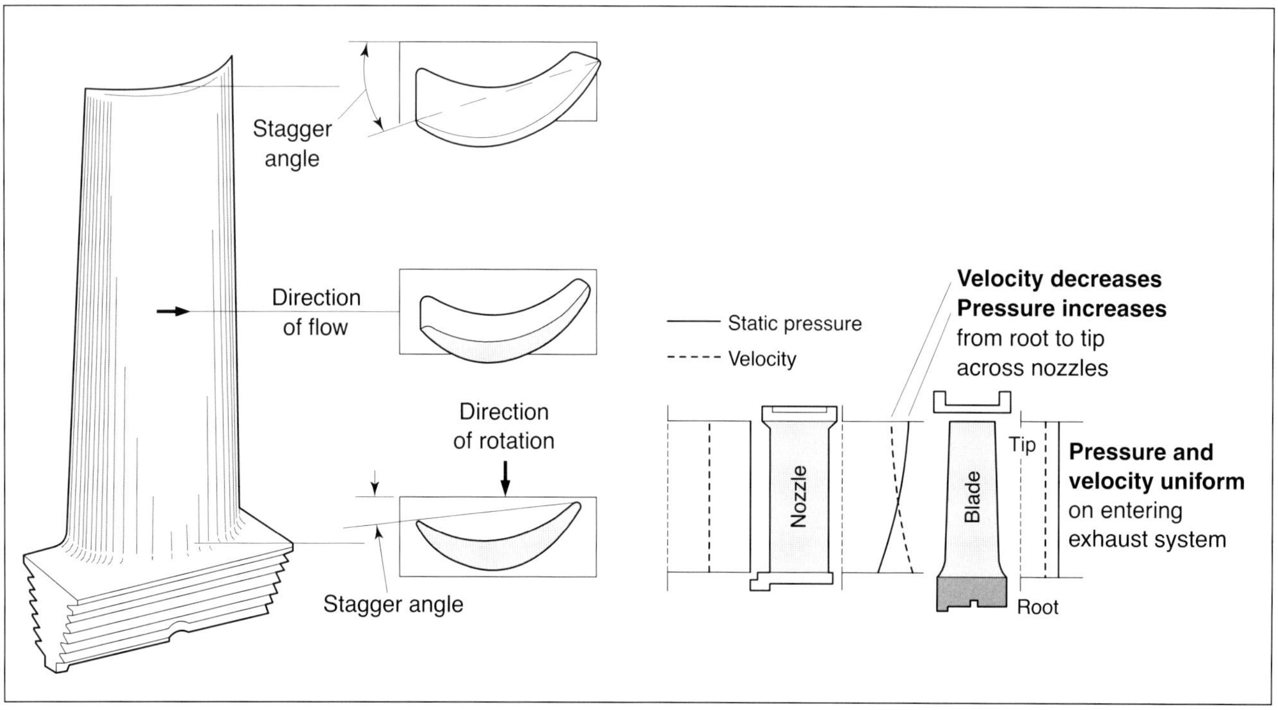

Typical turbine blade showing twisted contour. The reason for the twist is to make the gas, flowing from the combustor, do equal work at all positions along the length of the blade and to ensure that the flow enters the exhaust system with a uniform axial velocity. Compressor blades have a similar twist. This is to produce a pressure gradient along the blade such that the tip has a higher pressure than the root, this being designed to balance out the centrifugal action of the rotor on the airstream. Note the fir tree type root fixing. This is being superseded by the use of blisks (integrally-bladed disks) on the latest fighter engines, to reduce their weight.

Proposed for the air-defence of the UK, the mixed-powerplant Saunders-Roe SR53 was powered by a de Havilland Spectre rocket engine operating on hydrogen peroxide and kerosene, controllable within the range 4.4–35.0kN. Endurance was provided by a 44kN thrust Armstrong-Siddeley Viper lightweight turbojet. This and its SR177 descendant fell victim to the 1957 Defence White Paper. *(Philip Jarrett).*

The broad design of the Westland Wyvern had been settled by the end of 1944 and included provision for a turboprop engine. However, the Wyvern first flew with the underdeveloped Rolls-Royce 24-cylinder Eagle, the firm's most powerful (3,500hp) and last piston engine. Pioneering work was also required of the eight-blade contra-rotating propeller to be used. In the event the Wyvern TF Mk 2 (VW867 shown which was used in the flight test programme for deck landing trials on HMS *Illustrious*) flew in 1950 with an Armstrong-Siddeley Python. The later and definitive Wyvern S 4 entered service in 1953 and saw service during the Suez campaign of 1956 but was withdrawn from front-line service in 1958. *(GKN Westland via author).*

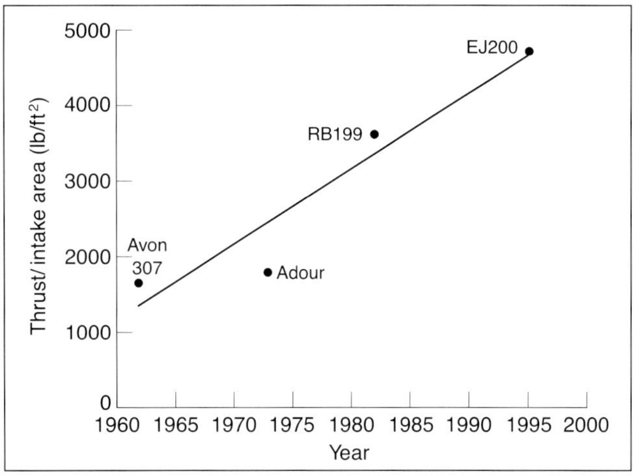

Increasing thrust per intake area, so important to minimise aircraft frontal area and drag.

turboprop also slipped from the fighter scene, yielding the field to the pure jet engine with an axial-flow compressor.

The turbojet of single-spool design was limited in its pressure ratio and hence thermal efficiency. It also had low propulsive efficiency because of the high jet velocity. What could be done if the turbine entry temperature (TET) was raised by use of improved materials, or by cooling the turbine blades and other parts? If more fuel was burned without increasing the pressure ratio, the velocity of the jet stream would increase without a compensating improvement in the thermal cycle. If both pressure ratio and turbine entry temperature were increased, the thermal efficiency could be improved. But to do this, considerably more fuel per pound of air must be burned, in order to overcome the increased internal losses associated with the higher pressure ratio. This had to lead to a higher jet velocity if the increased fuel burn was to be justified, but the higher velocity results in increased momentum losses. A balancing of the losses soon leads to a new optimum. The net result of these considerations was that in the early 1950s, no marked improvement in specific fuel consumption of the simple turbojet was in prospect. In addition, under certain conditions, the forward stages of the single-spool compressor became overloaded while rear stages were operating at less than optimum efficiency.

Table 6 Reduction in number of aerofoils

Rolls-Royce engines/aircraft	*Aerofoils*	*Engine T/W*
Spey (TF41)/F-4 (A-7D)	3,420	5
RB 199/Tornado	2,700	7
EJ200/Typhoon	1,800	9+

As the number of blades has decreased their chord length has increased. However, due to the small bearing surface at the blade root fixing, stress concentrations are extremely high. This has led to the development of blisks, in which the compressor blades and their supporting disc are made from a single forging as one component. Blisks, which offer at least 20% weight saving, maybe more, are already in the EJ200 and General Electric F412. Blings, made up as an integral series of blisks, are being ground-tested.

Multi-spool engines

With increased component efficiency and the load per stage, it was possible to limit the number of compressor and turbine stages. For example, the single-spool Rolls-Royce Avon 207 in the Hawker Hunter F.6 of 1956 had fifteen stages to give it a pressure ratio of 7.5 (i.e. 0.5/stage), whereas forty years later stage performance has more than doubled. Table 6 shows the trend in reduction of parts for a few representative Rolls-Royce engines.

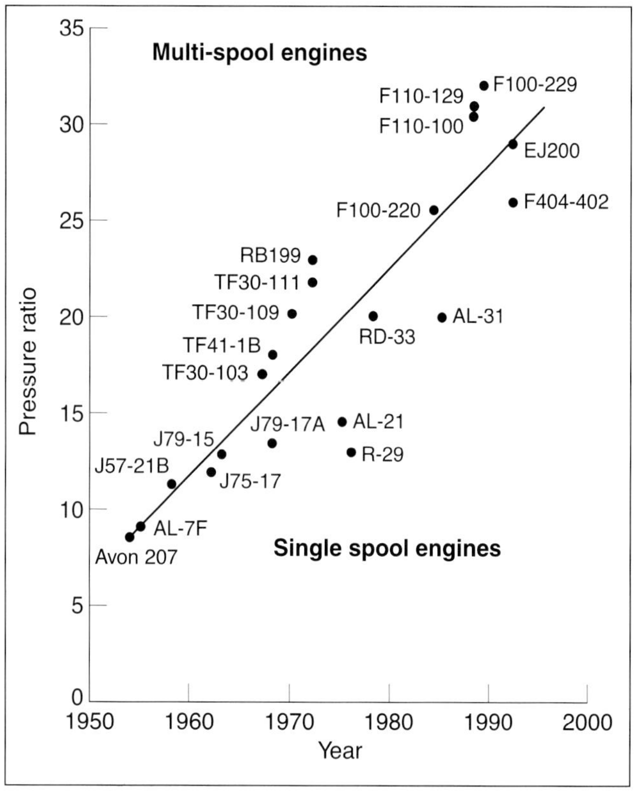

Pressure ratio with time.

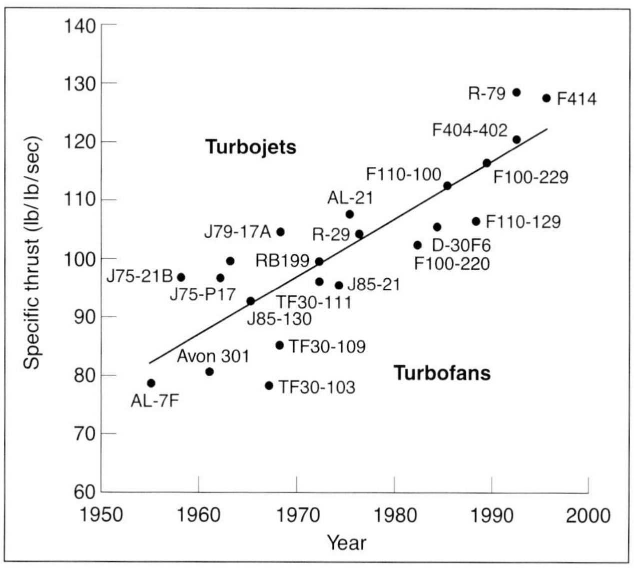

Specific thrust with time.

To efficiently achieve higher pressure ratios a twin-spool engine was needed. The compressor was divided into two: a low-pressure (low-speed) compressor and a high-pressure (high-speed) compressor, each driven at different 'best' speeds by independent turbines through concentric shafts. The first production engines of the type were the Bristol Olympus and Pratt & Whitney JT3 (military J57). The J57 of 1953 produced 44.5kN (10,000lb) thrust and thus had a big advantage over the Rolls-Royce Avon, then giving 32kN (7,200lb). By 1957 the J57 was giving 58kN (13,000lb), while still retaining the same thrust difference. As pressure ratios began to rise, causing a higher combustor pressure, trouble began to be experienced with carbon in the exhaust gases, especially on US engines that used a different combustor design. From a performance standpoint this had little significance and, although it accelerates turbine blade erosion, smoky exhausts were tolerated. For military aircraft, however, smoke is important since it enables an aircraft to be seen more easily, and carbon particles are some of the things preyed upon by the seeker heads of infra-red missiles. Combustor design remains something of a black art compared with that of other components, but engineers continue to strive for increased efficiency and reduced emissions.

Turbofans

Following the successful introduction, in the early 1960s, of the Rolls-Royce Conway and Shvetsov D-20P bypass engines for civil aircraft, all major manufacturers adopted the turbofan. The US approach was to attach fans to the front of their existing engines. The fan was combined with an intermediate compressor to give an even higher pressure ratio.

However, the bypass ratio (BPR) of the Conway (used in the Handley Page Victor Mk 2) was only around 0.3 (and did little more than cool the engine casing), so that the specific thrust (or jet velocity) was still high. Pratt & Whitney's JT3D had a BPR of 1.5 with its reduced fuel burn for the Boeing 707 and DC-8. Turbofans with a BPR greater than 1 are not the best engines for supersonic flight though, because of their low jet velocity (specific thrust). They also have a lower altitude limit than turbojets because their bypass flow is cooler and less compressed than the flow exhaust of a turbojet at similar TET and overall pressure ratio, and this reduces their stability limit. Weighed against this is the lower fuel burn of the turbofan. The Pratt & Whitney F100 turbofan was designed for the F-15, whose mission demanded long range, and its engine required a BPR of approximately 0.6 to give low specific fuel consumption. This was not so important for the F-16 air combat fighter, where greater tolerance to flow distortion and a wider stall/surge margin of a lower bypass ratio engine would have been more appropriate. One of the problems with afterburning turbofans is that good mixing of the bypass air and core gas flows is essential. Because the combustion process is so sensitive to temperature, it is less easy to burn fuel in the cold and less compressed bypass air than in the hot core stream. This was one of the problems afflicting the F100 engine. It was found in service that when afterburner was selected at high altitude, a rough light caused pressure pulses to travel upstream in the engine's bypass duct. This led to stagnation stalls of the fan, followed by the compressor stalling and engine flameout. Although re-lighting was not difficult, it was a serious threat for the single-engined F-16 and required a redesigned, and longer, mixer between the bypass and core flows. To minimise duct length and weight, most modern designs use separate core and bypass burners, with mixing occurring downstream of the combustor.

General Electric's F404 (the engine in current Northrop-Grumman F-18s) has a BPR of about 0.4 and was designed about five years later than the F100. It is regarded as a very good engine for a fighter requiring high manoeuvrability. Subsequent European engines have similar bypass ratios (the Snecma M88 has a BPR of about 0.25, and the EJ200 has a BPR of about 0.4).

The thrust/weight (T/W) ratio of the dogfighter engine is of paramount importance, and the prize for increasing this ratio is great indeed. Thus there is a premium for squeezing the maximum air flow possible through an engine of given weight and size, and for using the maximum turbine entry temperature allowable. This was increased significantly by the introduction of air-cooled turbine blades in the mid-1960s.

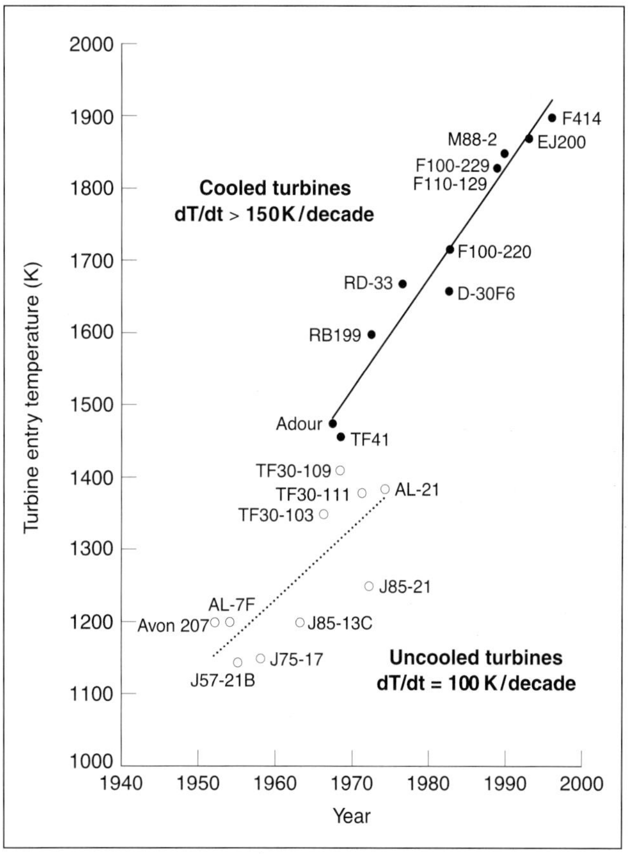

Turbine entry temperature with time.

At a constant pressure ratio, increasing turbine entry temperature causes an increase in specific fuel consumption due to reduced propulsive efficiency arising from the higher jet velocities. However, the gain in specific thrust is more important than the increase in fuel burn, especially at supersonic speed where a small engine is vital to reduce weight and drag. That said, specific fuel consumption (sfc) is neglected at peril, since at maximum dry ratings it dominates combat persistence which may ultimately be as important as the T/W. Moreover, maximising long-range combat air patrol endurance and minimising the frequency for inflight refuelling require low partial-power sfc. All of these contribute to overall fighter force effectiveness.

The thrust/weight of turbojets rose from about 3 in 1950

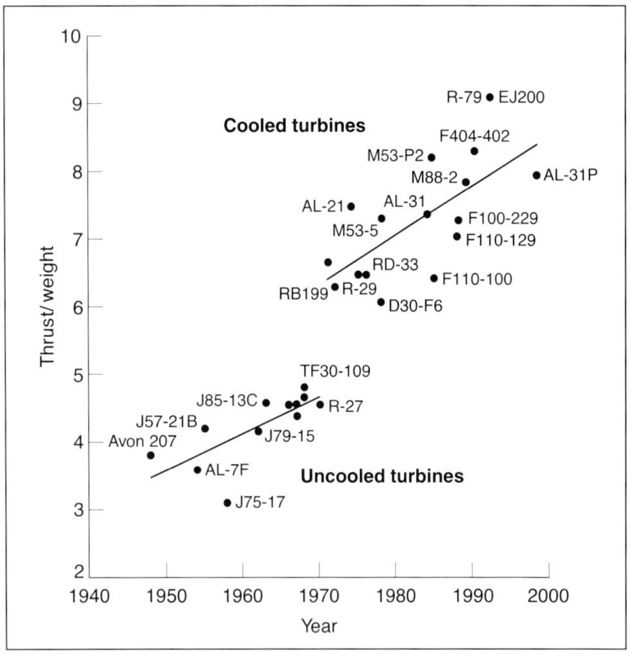

Thrust/weight with time

to almost 5 by 1970. Within the same period the thrust per unit frontal area increased by 40%. The time between overhaul for military engines in 1950 was around a hundred hours. The J47, for example, had a time between overhaul (TBO) of fifty hours in 1952, though by 1955 this had been increased to 650 hours. This was far better than the large radial piston engines in use at the time. By 1970 TBOs had risen to 1,000 hours. After twelve years' service with Tornados, the turbine blades of the RB 199 have lasted more than 1,000 hours (three to four years). The life requirement for the carcass of Eurojet's EJ200 is twenty-four years, compressor blades more than twelve years (unless affected by foreign object damage), hot gas parts about five years and hot nozzle parts two to three years, with repair as necessary.

Efficiency improvements, namely better compressor blade design (sharp leading edges and much lower thickness ratios that allowed tip Mach numbers to reach 1.5 with only weak shockwaves being formed) and advances in the properties of high-temperature materials, gave an increase in load per stage. It was possible, therefore, to limit the number of compressor and turbine stages as mentioned. However, to reach pressure ratios of 25:1, multiple shafts were required, as mentioned. The Tornado's RB 199, for example, is a three-spool turbofan, with separate fan, intermediate compressor and high-pressure compressor, each designed for optimum pressure ratio and efficiency. Some of today's fighter turbofans deliver pressure ratios of over 30.

Airframe/engine integration

It was some years before the need became evident to develop the jet engine as a powerplant integral with intake and tailpipe. To the aircraft designer, intakes were discontinuities on his fuselage or wing, tolerated but kept as small as possible. Designers of aircraft with built-in wing-root inlets accepted that provision had to be made to keep the fuselage boundary layer out of the intake. However, the effects of duct loss and poor velocity distribution on engine performance, coupled with the cyclic effect of the asymmetric distribution arising from the bifurcated inlet (e.g. Hunter), were not understood. Other effects such as throttling of intakes, flight at high AoA and the need for adequate leading-edge radii, the shock and ignition gases from gun and missile firing, and possible masking of the bifurcated intakes by the fuselage during manoeuvres, had to be uncovered in flight. The need for careful airframe/engine integration was highlighted in the 1960s by the General Dynamics F-111 which ran into very serious difficulties. That said, with the much greater understanding and appreciation of the problems involved, intakes and their integration now receive as much attention as the rest of the powerplant.

For subsonic or low-supersonic flight the simple pitot inlet is the optimum. It is fairly tolerant of AoA, has uniform distribution, and pressure losses are low. While the rounded-lip intake is good for subsonic and manoeuvring

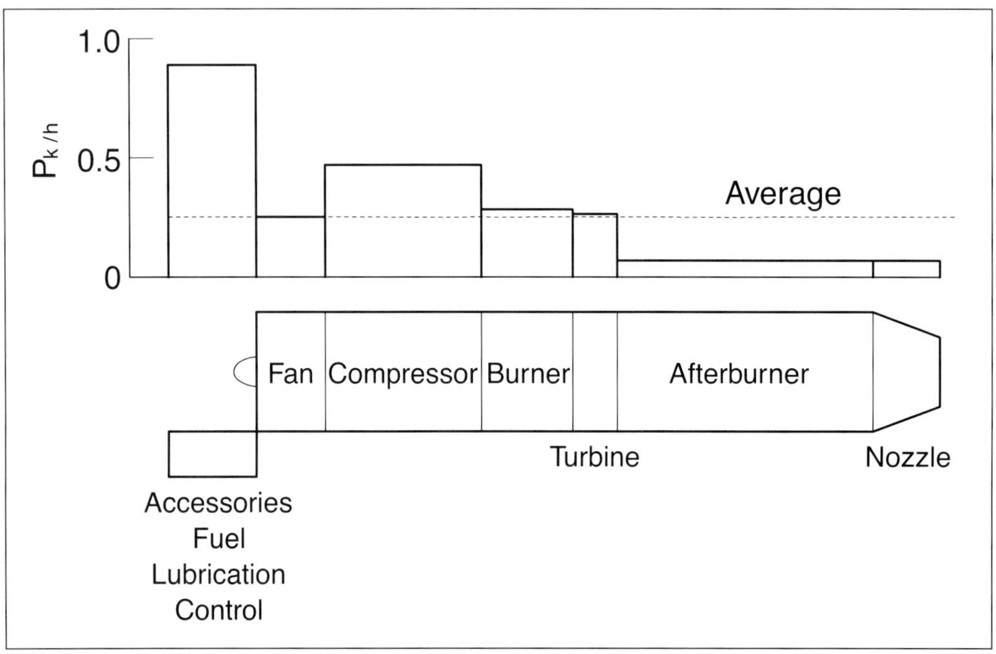

Engine turbomachinery damage from projectiles and warhead fragments has always been a major concern. This figure shows the relative vulnerability of engine components of a typical afterburning turbofan. $P_{k/h}$ is the probability of a kill given a hit. *(Pete West/AIR International).*

The fixed air inlets for the SNECMA M88 turbofans on the Dassault Rafale are highly integrated into the fuselage. This photo of the first two-seat B version clearly shows the splitter plate designed to prevent the fuselage boundary layer entering the intake and the V-shaped forward fuselage. Mounted on the inlet duct is the starboard forward-oblique ECM antenna, a component of the Spectra threat warning and self-defence suite along with the RWR antenna shown just forward of the foreplane. *(Francois Robineau via author).*

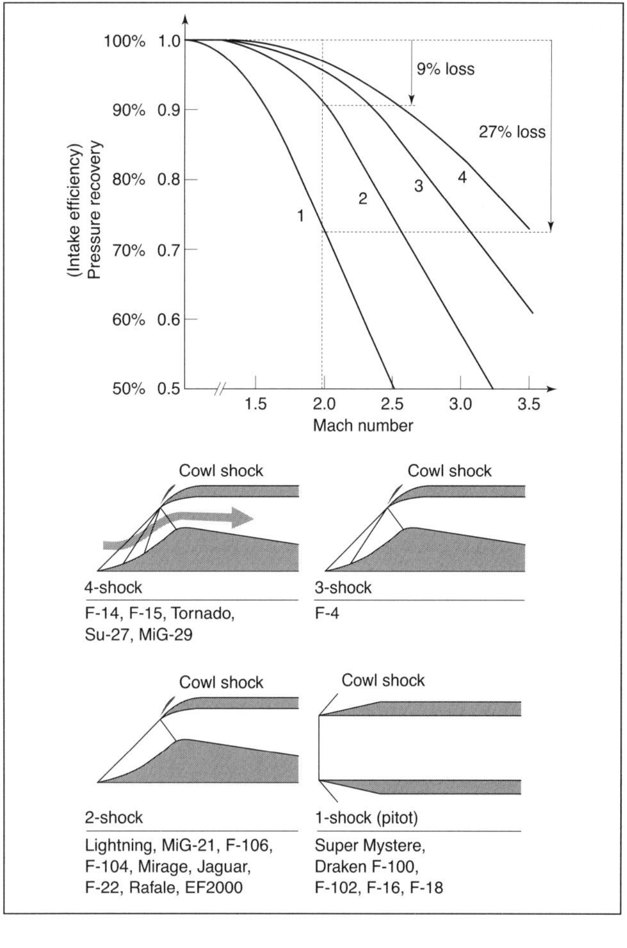

RIGHT: Various types of intake geometry and their ideal efficiencies, at each Mach number, in terms of pressure recovery. Without any pre-compression surface (e.g. a spike or wedge) a normal shock wave forms, across which there is an abrupt and large loss of total pressure. The figure shows that the crippling 27% loss at Mach 2 for the pitot inlet is reduced to 9% when a suitable wedge is added. Bear in mind that a 9% loss will reduce net thrust by 15% and increase specific fuel consumption by 6%. Total pressure losses can be further reduced by using more sloping surfaces, though at the expense of extra complexity, cost, weight and subsonic drag. *(Pete West/AIR International).*

flight, the intense shockwave that forms just ahead of the inlet lips at supersonic speed results in a loss of total pressure. Bear in mind that a 9% loss will reduce net thrust by 15% and increase sfc by 6%. The earliest supersonic aircraft, typified by the 1950s North American F-100 Super Sabre, used sharp lips on its pitot intake, an oval opening with a very long subsonic duct. As speeds moved further into the supersonic regime, intake design became more important, and at Mach 2, for example, the pressure rise in the inlet will be greater than in the engine. The basic problem was still the same, namely to reduce the intake air flow to a Mach number of 0.4 to 0.5 at the compressor face, at the same time diffusing it to the maximum static pressure with minimum loss of total pressure.

The only means of decelerating a supersonic airstream is by means of a shockwave, which causes a loss of total pressure. The higher the Mach number ahead of the shock and the stronger the shock, the lower the Mach number behind it and the greater the loss. The pitot inlet, being without a precompression spike or wedge, causes a single normal shock, with a very large loss of total pressure at high supersonic speed. This loss is reduced by generating a number of weaker shocks (via precompression surfaces) rather than a single strong one. For example, the 27% total pressure loss at Mach 2.0 for the pitot intake may be reduced to 9% when a single wedge is introduced.

Supersonic variable-geometry intakes

The engine airflow requirements of a combat aircraft vary greatly across the flight spectrum. The intake must be capable of adequately handling the full range of air flows. Whether this demands variable intake area depends largely on the maximum design speed. Broadly speaking, variable geometry is unlikely to pay off unless the design Mach number exceeds about 1.6. The F-16 has a fixed pitot intake, and is capable of Mach 2. The Dassault Mirage 2000 of comparable size, however, is faster with much less installed thrust, mainly because it does have a variable-geometry, two-shock inlet.

Means of altering the areas of intakes were, until the 1970s, provided on Mach 2+ aircraft to accommodate the large variation of engine air flow and to minimise the pressure loss across the supersonic region of the flight envelope. Then came the demand from fighters such as the F-14 and F-15 for enlarged flight manoeuvre envelopes. Formidable difficulties in integrating propulsion system and airframe were surmounted to produce the very wide AoA ranges of today's highly manoeuvrable fighters. Such intakes, at high AoA, have to supply air of relatively low flow distortion to avoid encroachment on the surge margin of the engine. The weight, complexity and life-cycle cost of variable-geometry intakes are such that they can be justified only when there is a clearly identifiable and worthwhile performance improvement. For example, the stealthy (and therefore expensive) F-22 has a fixed inlet type.

Evolution of the variable convergent-divergent nozzle

Owing to the high pressures developed in the jetpipe, efficient supersonic expansion of the jet gases is necessary. The need for variability comes from the quite disparate (dry versus reheat) temperatures that have to be handled by the exhaust nozzle. As temperature doubles (e.g. 1,000 to 2,000K), the area required at the nozzle throat also roughly doubles. The disadvantages of the simple convergent nozzles fitted to the engines of early low-supersonic aircraft (e.g. North American F-100) were not too severe, provided

Unquestionably the most significant and successful fighter of the 1960s, the McDonnell F-4 Phantom of 1958 started life as a fleet defence fighter but saw far wider usage as a land-based multi-role aircraft. Variable geometry for two-dimensional intakes was first used on the F-4, which features a vertical fixed ramp, followed by a movable ramp (with boundary layer suction) to create a variable throat area and engine bypass slot. This is a dramatic view of an F-4M Phantom with Rolls-Royce afterburning Speys of 56 Sqdn carrying four AIM-9L Sidewinders and being filled to the gills with venting from wing trailing-edge fuel pipes. *(MoD via author).*

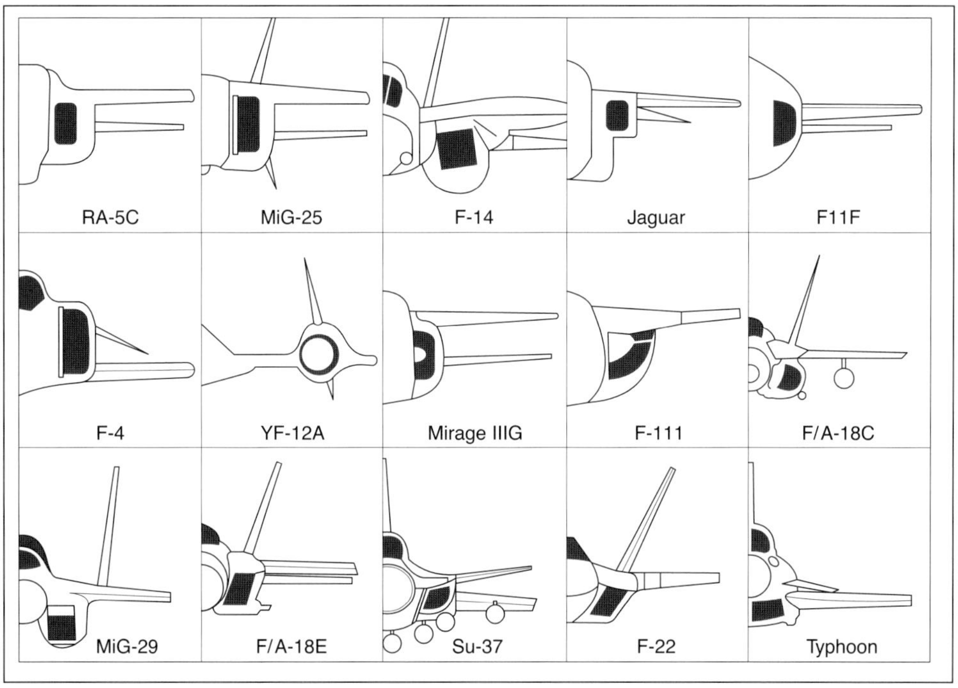

Evolution of air inlets showing the wide variety of configurations adopted over the years. *(Pete West/AIR International).*

This photo of British Aerospace Tornado F.3 shows the rectangular air intakes typical of 1970s supersonic fighters, contained within which are two movable, horizontally mounted compression ramps to control intake air mass flow and position the shock waves which form across the intake face. Also shown are the two blow-in doors which open at low speed to allow extra air to enter the subsonic portion of the inlet duct. *(via author).*

Soviet fighters often show features designed to enable the aircraft to operate from austere airfields. The air inlets of the MiG-29 have intake ramps that form doors to close over the intake face during ground running to block most of the airflow and prevent debris being vacuumed into the engine. They are clearly evident on this MiG-29SMT. Louvres on the top of the leading-edge strake open to provide intake air until the aircraft reaches nosewheel lift speed when the main doors open. *(Alexei Mikheyev via author).*

some form of exit (throat, in this case) area-variability was embodied. With afterburner lit, the nozzle opens to give a larger area for the greatly increased volume of the gas stream. However, for flight at altitude at Mach 2 and beyond, nozzle pressure ratios become so large that the potential thrust loss from a convergent-only nozzle becomes unacceptable. It comes down to balancing the extra thrust at high Mach numbers (above Mach 1.5) against the extra weight and cost of a convergent-divergent (con-di) nozzle.

In the 1950s, ways to get the exhaust gases to expand efficiently without prejudicing nozzle kinematics were studied exhaustively. Designers found themselves for some time seeking in vain for the true variable con-di nozzle. The mechanical complexity of the moving parts, losses due to imperfect shape and sealing at high speed, and high weight and cost appeared to preclude its use. A simpler solution was therefore sought. In the US it had long been the practice to cool jetpipes with relatively large amounts of air bled from the intake. In the search for a way of controlling the expanding flow from the nozzle, use of this cooling air to govern the effective area ratio of a con-di nozzle was explored. And so the ejector nozzle was born. The earliest interim types tended to be heavy and required a large amount of engine cooling air and under-performed to such an extent (e.g. on early General Dynamics F-111s) that risk reduction became an important consideration.

The first true variable con-di nozzle appeared in 1967. The iris type was used on the Grumman F-14 and its motion, applied to the carefully shaped flaps, produces a convergent shape for the internal contours in the rearward non-afterburning (closed) position. This gives a smooth external boat-tail shape. In the forward position the realignment of the movable flaps causes a con-di nozzle shape to be formed internally, yielding a very small boat-tail angle. The variations in throat size and divergence are mechanically coupled. The 'conventional' hinged and coupled petals of the con-di nozzles on the McDonnell Douglas F-15 and F-18, Lockheed Martin F-16, Saab JAS 39 and Eurofighter Typhoon lead to the nozzle length remaining almost constant, tending to reduce slightly as they close from reheat to dry. For the latter aircraft there was lengthy debate regarding whether it should be fitted with a con-di nozzle, since originally the UK's Ministry of Defence didn't want one.

In low-BPR turbofans the same fan air that cools the afterburner is used to cool the nozzle. This air discharges near the nozzle throat to form an internal film over the short length of flap exposed to the hot exhaust gases. Because pure turbojets have no fan, and because compressor bleed air is so costly, the above form of cooling was impractical. The ability of turbofans to self-cool (keeping their engine bays cooler) not only reduces weight but produces a drag reduction in the internal air passages.

Thrust vectoring

In the early 1950s, it began to be realised that long runways were vulnerable to attack, and aircraft that could operate independently of them were seen as a valuable asset. The design concept of the Hawker P.1127, which was developed into the Harrier, arose in 1957. It was aimed at evolving a lightweight V/STOL strike/reconnaissance aircraft using the Bristol BE.53 (Pegasus) vectored thrust turbofan. Apart from the unsuccessful and now discarded Yak-38 'Forger',

This photo gives a limited indication of the degree of area variability of the convergent-divergent nozzles of the Pratt & Whitney F100-220 turbofan on this F-15C of 52nd Fighter Wing, Spangdahlem AB, Germany. In addition, the control rods and support structure previously hidden by the nozzle petals, typical of earlier F-15A F100-100 engines, are exposed. It was found through experience that the lack of petals produced an insignificant increase in back-end drag. *(via author).*

British Aerospace Harrier GR.7 sitting on the thrust of its Rolls-Royce Pegasus contra-rotating turbofan, clearly showing the extended length zero-scarf forward nozzle and blow-in doors open. Control in the hover is achieved by reaction controls at the extremities of the wing, nose and tail that use air ducted from the engine's compressor. *(BAe via author).*

Photo shows the ±20° range of up/down thrust vectoring capability of the Pratt & Whitney F119 engine nozzle used on the Lockheed Martin F-22 Raptor. Apart from conferring high AoA capability to the F-22, small amounts of thrust vectoring can also be used to reduce cruise trim drag by avoiding use of an aircraft's tailplane. *(Pratt & Whitney via author).*

Pratt & Whitney F100 engine with pitch/yaw balance beam nozzle (vectored to port) being installed in McDonnell Douglas ACTIVE F-15 in 1995 to provide multi-axis TVC. Vectoring was achieved to 20° about the aircraft centre line to produce 4,000lb normal force, with a vectoring rate of 120°/sec being demonstrated. *(Pratt & Whitney via author).*

the Harrier is the only V/STOL fighter to see operational service. That the use of the four-poster nozzle system has worked extremely well is evident from the fact that highly developed Harriers are still in production thirty-five years after the P.1127 first flew in 1961.

Thrust vectoring will be used on the Lockheed Martin F-22 primarily to assist in permitting trimmed flight at extreme AoA rather than to confer STOL attributes. Though the F-22 has had to sustain a not insignificant weight penalty from its stealthy, two-dimensional thrust-vectoring nozzle, General Electric's AVEN nozzle on the multi-axis thrust-vectoring (MATV) F-16 has shown a much lighter solution to thrust-vectoring. A stealthy version is under development. With the experience of super-manoeuvrability gained from MATV F-16, DASA/Rockwell X-31 and Sukhoi Su-37, it is possible that existing fighters may be retrofitted with thrust-vectoring nozzles, and could be an option for Eurofighter's Typhoon. As with other aspects of super-manoeuvrability, such as the operational wisdom of low-energy flight in combat, the costs have to be weighed against the benefits. The benefits may include use of thrust vector control to recover aircraft from out of control situations, and so pay for itself quite quickly.

Variable-cycle engines (VCE)

For multi-role aircraft good performance at both subsonic and supersonic speeds is essential. A high BPR turbofan is best for the former, whereas a straight turbojet is best for

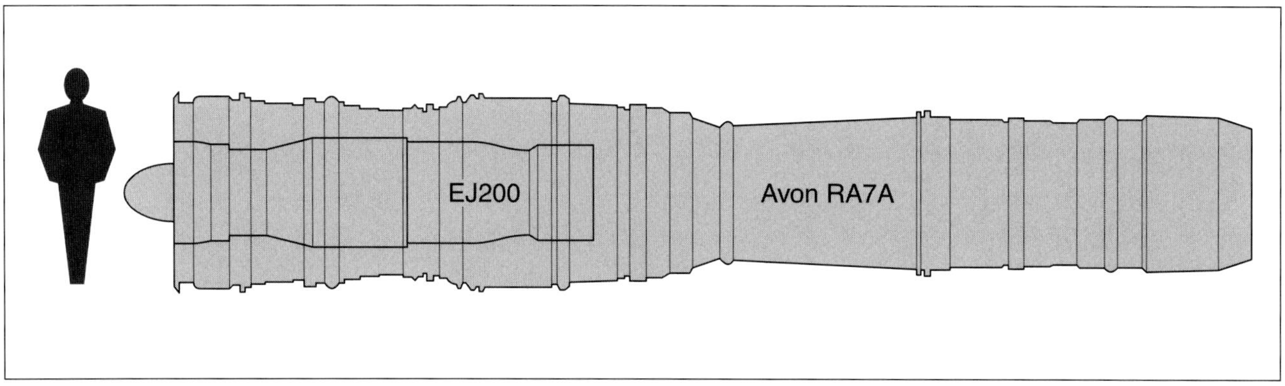

What 40 years has achieved: size comparison of the Rolls-Royce Avon RA7R (100 series) of 1952 rated at 33kN (7,500lb) thrust dry and 42kN (9,500lb) with reheat (used in the record-breaking Hunter F3) with Eurojet's EJ200. The Avon's enormously long jetpipe was governed by its installation in the Hunter. Thrust augmentation of the Avon was limited by the variability of the two-position nozzle. Note: both engines (at a similar stage of development) are scaled to the same sea level static reheat thrust. *(Pete West/AIR International).*

the latter. For air superiority fighters an engine somewhere between the two is required. The engines required for the Advanced Tactical Fighter (ATF) being designed in the early 1980s had to provide the aircraft with supercruise capability (cruising supersonically without afterburning) and with a range significantly longer than the F-15. This demanded a moderate-pressure ratio core with a high-pressure ratio fan together with high TET to give high dry thrust. The ATF engine had to be about the size of the F-100 but of lower weight and with half the number of parts. General Electric chose to pursue variable-cycle engine technology because of the dry thrust requirement. On its YF120 General Electric is understood to have used a variable-area bypass injector to switch the flow between the core and bypass duct to vary the bypass ratio. The dry thrust at supercruise comes from using the turbojet mode (BPR roughly zero) with a small bypass flow for cooling, without paying the penalty in terms of fuel consumption at subsonic speed, where the turbofan mode is used. The highest BPR of General Electric's YF120 was around 0.3, again to maximise dry thrust, which at a round 125kN (28,000lb) is almost double that of the early F100s. The other contender for the ATF engine, Pratt & Whitney, had designed the still-classified J58 VCE in the 1950s for the SR-71, so P&W knew the strengths and weaknesses of its rival's VCE and went for a more straightforward design in its YF119, believing that it could meet the ATF requirement without using a variable-cycle design. After the ATF engine decision General Electric acknowledged that its YF120 may have been ahead of its time and presented higher development risks than the YF119. However, the concept will probably be the preferred route for future fighter engines to combine stealth and agility with low fuel burn.

The Future

Despite the staggering advances that have been made over the last fifty years, the field of fighter propulsion is offering an acceleration of progress. The engines being installed in today's fighters are much smaller, lighter and more powerful than those regarded as outstanding performers only twenty years ago. Demonstration engines already achieve thrust/weight ratios over 12. A continued drop in the number of parts as more efficient blisks and blings, for example, find increasing use, will lead to production engines with T/W ratios of around 15 within a decade.

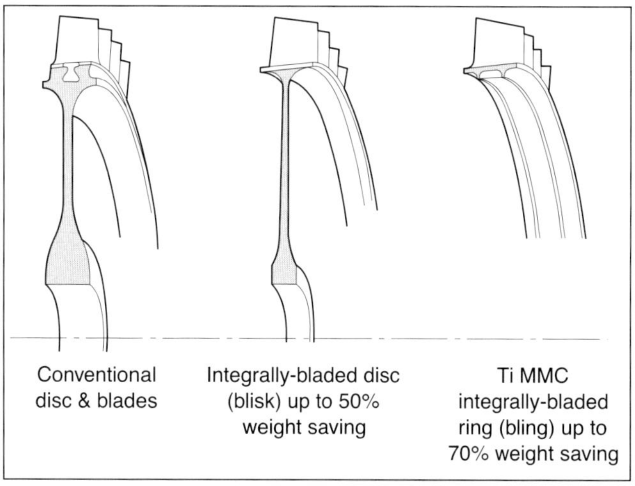

Historically the thrust/weight ratio of fighter engines has doubled in 40 years. Thus the next generation of engines will require thrust/weight ratios of between 15:1 and 20:1 with at least half of this improvement being sought from stronger and lighter materials. The use of integrally-bladed discs avoids traditional blade attachment, resulting in significant weight and cost savings. Titanium metal matrix composites (TiMMC) offer about 50% more strength with twice the stiffness of today's high temperature titanium metal alloys. *(Pete West/AIR International).*

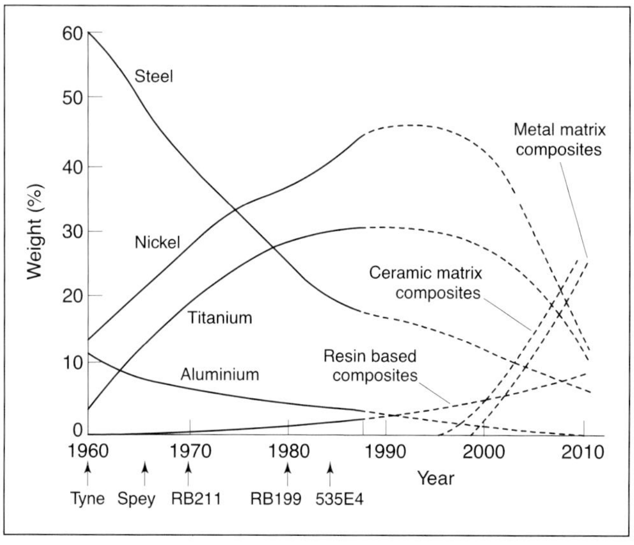

Predicted trends of material usage in jet engines. The composition of future engines will depend critically on the composite materials fully meeting their potential and the creativity of the design engineers. *(Pete West/AIR International).*

The Sukhoi Su-35 'Flanker' in 1996 with foreplane controls and its thrust vectoring nozzles attached to its Saturn/Lyulka AL-31 afterburning turbofans with which it was able to regularly perform the somersault manoeuvre at airshows around the world. *(Ryszard Malachowski via author).*

Eurojet EJ200 for Eurofighter's Typhoon. Originally, the UK's MoD was reluctant to use a convergent nozzle which represents a heavy item far aft of the aircraft's CG. However, the increase in specific excess power at supersonic speed and reduction in afterburner sfc outweighed the cost and weight penalties of the nozzle. The EF2000's requirement is dominated by the interception and air superiority fighter role, this drags down the bypass-ratio from that of the RB199 (BPR=1) in Tornado, to 0.4 of the EJ200. It won't get you so far but produces high manoeuvrability when you get there. *(Eurojet via author).*

Chapter 5

Structures and Materials

The earliest aircraft were made from materials that were cheap and easily worked. The late 1920s marked the start of a transition from strutted and braced biplanes to the cantilever monoplanes of stressed skin construction of the 1930s. Since then major improvements in materials and manufacturing have occurred. These have helped minimise structure weight which, after ninety years, still remains the primary goal of structural design.

Glossary

Aeroelasticity: The aerodynamic effects arising from structural flexibility.

Flexural axis: For a wing, this is the locus of a series of points on the surface where, if a vertical load is applied, only bending, not twisting, occurs.

Flutter: Violent flapping of a wing or tail in the airstream rather like a flag. It arises due to the surface's flexibility in both torsion and bending.

g: The term g relates to acceleration due to gravity, which for the purposes of atmospheric flight has the value of 9.81 m/sec^2 (32.2 ft/sec^2). When an aircraft is flying with no acceleration in any direction, it is referred to as being in 1g flight. During a turn the aircraft experiences an acceleration towards the centre of the turn. The force that pushes the pilot into his seat is equal to his weight times the number of g that the aircraft is pulling.

Limit load: The maximum load that an aircraft is expected to experience in normal, including combat, operation.

Ultimate load: The limit load $\times$ 1.5. The structure must support the ultimate load before it fails.

Load factor (n): Expressed as a pure number of g: n = lift/weight.

Stress: Load per unit area.

Wing loading (W/S): Aircraft weight/wing area.

Pre-World War One

The main factor that governs the choice of materials and structural form is the ratio of the load on the structure to its dimensions. Early aircraft operated at low speeds and hence the loads were low in relation to aircraft size with wing loadings (W/S) of 25 to 50kg/m^2 (5 to 10lb/ft^2). Under these circumstances it was best to concentrate the compression loads into a few small compact rod-like members and to diffuse the tensions into fabric and wires. Any other arrangement would have been heavier, more expensive and inconvenient. Aircraft dimensions were, in many cases, not remarkably smaller than those of equivalent modern aircraft, but their weight was much less than a tenth of modern metal-skinned aircraft. The construction of fabric over a wire-braced framework of wood was logical and efficient, and sometimes nothing else would have got off the ground with the power available.

The biplanes of the period were made from materials that were the cheapest and most easily worked: wood, piano wire and fabric. The wing spar of a biplane was usually a continuous beam on three or more supports to give a very efficient torsion box. The spars, of I or box section, and tapered interplane struts of streamline section were of wood, and fuselages were wire-braced lattice-work, the compression members again being of wood. The solid members needed only to take compression, and since the principal danger in this condition was buckling, such components were required to be as thick as possible for their weight. For this, spruce and birch were especially suitable. To this skeleton was bonded a suitable fabric covering, which was then treated with dope to shrink the fabric giving a tight, rigid structure. The skin covering for wings, control surfaces and fuselages was for aerodynamic reasons only.

This philosophy produced light and robust aircraft so long as their designers were sure which members were in tension and which in compression. In some of the more elaborate biplanes it wasn't always easy to tell which way the loads were going. The feeling of security about flying in strutted and braced aircraft was comforting but unfortunately did not always extend to the engines. Cantilever monoplanes felt much less safe.

There was official resistance in Britain to monoplanes, indeed they were banned because of misunderstood causes of several monoplane crashes. These were probably due to the aeroelasticity of the externally braced thin wing, reverse lateral control in a turn or the effect of download on the front spar and leading edge. This undoubtedly aided the eclipse of the monoplane and retarded research on the aerodynamic and structural problems that would have promoted the type. Consequently, the monoplane was comparatively rare in the UK from 1912 to 1930.

World War One

Little attention was given to early aerofoils with regard to the structural desirability of thickness. Up to around 1912, wing sections had been laid out by the individual aircraft designer, and *if* wind-tunnel tests were carried out it was seldom to discover the aerofoil characteristics. The RAF6 section of 1912 marked a transition in the UK from the old to the new in aerofoil design. This was the beginning of the tendency to flatten the lower surface for increased spar depth. It was a great improvement over anything previously used and was in common use up to 1918. The RAF15 (1915) section had thickening carried further and appears to demonstrate an appreciation of the aerodynamic benefit of rounding the leading edge. Its lower surface was thickened in the region of the probable spar location and left concave between the spars. On biplane aircraft, which had enough

Compared to their biplane contemporaries, early monoplanes felt much less safe. Indeed, there was official resistance in Britain to monoplanes and they were banned because of misunderstood causes of several monoplane crashes. The Morane-Saulnier Type N (of which this is a replica) was known by the French as the Monocoque, a misnomer as the fuselage was constructed from stringers and fabric. *(Philip Jarrett).*

external wing bracing that thick sections were unnecessary, use of the later RAF15 section with its more rounded leading edge was not significantly superseded in the UK until the 1930s. The next step was to greater camber, more for structural reasons in the UK rather than aerodynamic ones, as in Germany. The greater spar depth enabled the great maze of external bracing, characteristic of 1916 designs, to be reduced.

Before the war it was recommended that an acceleration of 3g should be assumed with a safety factor of 2 giving an ultimate load factor of 6g. Early structural failures in the air led to a study of the loads that could arise on aeroplanes during manoeuvres. Calculations made of the loads to be expected in pulling out of a vertical dive in a BE2 showed that after a drop of 1,000ft an acceleration of 9.5g was possible, and at the terminal velocity 11.5g. During mock dogfights between an RAF SE5A and FE8, manoeuvre load factors up to 4g were recorded and 3g was quite common. Flying such lightly loaded aircraft (W/S=10lb/ft^2) in ordinary bumpy weather could give accelerations of +/− 0.5g.

At this time, structural strength testing involved turning the aircraft upside down on the ground and loading the wings with bags of sand or lead shot, to simulate the manoeuvre loads, and noting when the wings failed. As speeds increased so did the loads, which meant heavier internal structures were put near aerodynamic surfaces where the stresses were the highest. With the wings' upper surfaces in compression and lower surfaces in tension, near the neutral axis stresses were low, and this is where lightening holes could be used.

Wood largely persisted through World War One because demand was immediate. There was no time for the theoretical and experimental work that would have entailed changing over to metal, even if it was thought desirable, which few designers did. However, the large amount of wood required for military aircraft construction revealed one of its major limitations. The most suitable timbers were imported from overseas, requiring a large volume of shipping which was otherwise needed for the transport of troops and food. As high-grade spruce became scarce and poorer quality timber was substituted, thoughts turned in the direction of a change from wood to metal. Steel tubes began to appear for interplane struts, while high-tensile steel wires (streamlined RAF wires, if external) were used for bracing. It was during this period that the fail-safe philosophy was first introduced. Lift wires were duplicated and there was always an alternative, though tenuous, path for the load through the drag and interplane strut bracing. It used to be a joke that the best method of checking the adequacy of the bracing was to release a bird in the gap between the upper and lower wings. If it escaped, additional bracing was needed.

Between the Wars

Metal aircraft and stressed-skin construction

The first really effective stirrings toward metal construction had begun in Germany before 1914. Although Duralumin (accidentally discovered in 1908) was available with a yield strength of 210N/mm^2 (13.5 tons/in^2), there was no guidance as to its use.

The UK remained the last citadel for biplane construction. Requirements for new British military aircraft

Result of testing to destruction by compression failure of the upper wing of the Military Wing's Sopwith D.I No. 243. *(Philip Jarrett).*

up to the early 1930s always started with the words 'A new biplane is required. . . .' Despite the fact that Duralumin was being developed to greater strength, workability and corrosion resistance, it was argued that fabric covering, especially when it could be re-doped, gave long service, and recovering encouraged overhaul. Nevertheless, in 1924 the Air Ministry said that all future RAF aircraft had to use metal for primary structure. The traditional fabric-covered biplane was retained and metal members were merely substituted for wooden ones. This was a handicap as it was not the most efficient way of using the new material. Fresh thinking was vital for so radical a change.

The aerodynamic potential of monoplanes had not been realised due to the inadequacy of the early versions. However, improvements in cantilever technology in Germany and the USA led to the inevitable eclipse of the biplane. The lead came from firms like Junkers, Rohrbach, Messerschmitt and Dornier. Rohrbach used the centre box spar, which was the main stress-bearing member of the wing, from which the leading and trailing edges were cantilevered. It was copied by many other designers. Monoplane wings tended to be thick at the root and highly tapered in planform. However, if the wing thickness exceeded 15% of the chord, the profile drag began to increase rapidly, so a compromise between wing weight and drag had to be struck. Most British designers and Air Ministry officials did not heed the pioneering efforts and it was a decade before the UK accepted monoplanes.

During the late 1920s, speeds and hence loads were increasing. Furthermore, a smooth surface was essential. The obvious way to meet both these demands was to use a sheet-metal skin. However, this was heavy, since there was a limit to the thinness of the sheet that could be held to shape with a reasonable number of ribs. Moreover, the metal skin needed to take its share of the load if it was to be viable. There was no difficulty about taking tension in a thin membrane, the problem was how to take compression without the skin buckling. In practice this was done by a semi-monocoque shell, where the thin skin was stiffened by attaching to it spars and stringers, with which it shared the load. The whole elaborate surface formed a hard-skinned shell, stiff in bending.

The ability of sheet metal to carry an increasing load *after* it had begun to buckle, which for conventional structures was regarded as failure, was crucial to the development of metal aeroplanes. It had first been discovered in 1925 by Dr H. Wagner, who was then working for Rohrbach. The Wagner theory of the diagonal-tension field beam, as it is called, formed the basis for further development of aircraft structures. Northrop in the US had devised a similar type of construction in 1928 for a flying wing he was then building. Aluminium sheet was formed into a channel section and the channels were then riveted together to make up the wing. The horizontal members formed the skin while the vertical members formed the wing spars, and additional sheets were riveted in to form the ribs. It was a relatively cheap form of construction, which was why Northrop had chosen it. Even after it developed permanent buckling, and so had failed by normal standards, it was still able to carry around 90% of its 'failing' load.

Aircraft designers of the 1960s, let alone the 1990s, would hardly comprehend with what paucity of technical information and insufficiency of official guidance their colleagues in the early 1930s had to perform the switchover from familiar diagonally braced structures to shell construction employing stressed skins. The emphasis shifted from determining stresses by classical engineering methods to securing stability against buckling with the minimum expenditure of material and weight.

Wing design

Because of the high speeds reached by the new monoplane fighters of the mid-1930s, the stiffness of the structure started to become as important as the strength. The ideal wing structure would be one in which a suitably reinforced skin took all the bending and torsion loads, and light web members took the shear. But gun doors, wheel wells, holes for fuel tanks and attachments for the concentrated loads from the landing gear made this impossible. For the sake of ease of maintenance, and to restrict transport sizes, it was required to make the wings detachable from the fuselage. The joints used to take the concentrated loads from the wing into the fuselage were very heavy.

There were two ways of tackling the wing structure problem. The first was to build the basic monocoque described above and cut holes in it where necessary, then reinforce the edges of the holes. But stiffness drops dramatically when holes are cut in a shell structure. The holes, therefore, had to be small, since large ones weaken the structure too much, but then guns and ammunition boxes became inaccessible. The other method was to build a structure around each of the openings, the covers of which could be made rapidly detachable as they were no longer part of the load-bearing structure. This type of construction was usually the two-spar type. It gave less stiffness and was certainly heavier than the monocoque construction, but it had the advantage of accessibility. Fuselage attachments were easier and gave strong points for jigging, making interchangeability easier. Another form of construction was to make use of the wing leading edge as a D-spar to form a torsion box (with the vertical part of the 'D' located at maximum wing depth) and put the equipment behind the spar.

This had been developed by Messerschmitt and was used on the Spitfire, but little use was seen of it elsewhere, which was surprising since it permitted a favourable disposition of the wing axes of importance in flutter and divergence prevention. The fighter wing ended up as a very mixed type of structure and a combination of all three types was common. Outboard of the equipment the wing could be of monocoque construction, while inboard a combination of

Begun as a private venture in 1931, the Boeing Model 248 was of all-metal construction with a semi-monocoque fuselage and a two-spar wing with light alloy skins. The first P-26A flew in 1934 powered by a 600hp Pratt & Whitney R-1340 Wasp and became the USAAC's first monoplane fighter. The rod projecting forward from the open cockpit of this P-26B is a tubular gunsight and the two tubes projecting forward between the engine cylinders are the gun barrel blast tube extensions. *(Boeing via author).*

two-spar and D-spar construction could be used. The remainder of the wing's aerodynamic shape was maintained initially by a fabric covering (where high stiffness was not important) over conventional ribs as before. Later wings were metal-covered to meet the needs of higher speeds, both in respect of the maintenance of aerodynamic form locally and to provide torsional stiffness, which in turn ensured freedom from flutter.

Production methods

Serious attempts at aircraft mass production were not visible in the UK before 1915 when it was realised that other countries were ahead in both quantity and quality. However, little emerged during World War One that can be regarded as an advance in production technology, and post-war aircraft were still being built by carpenters, fitters and riggers from spruce, metal tubes and fabric. This remained the case up to the end of the 1920s.

From both technical and manufacturing viewpoints, the period 1930 to 1936 saw the beginnings of aircraft production. Metal structures, retractable undercarriages and low-wing monoplanes (with varying chord and thickness) began to pose problems which could not be solved by carpenters. For a while, even with fairly advanced designs, skilled designers converted wooden structures into steel and alloy assemblies without losing structural weight advantages to the new monocoque designs. Other designers managed to achieve similar success with welded steel tubes in heavy production jigs.

With the advent of all-metal aircraft the need was intensified for comprehensive jigging and tooling. The so-called split construction technique was introduced, which has been used since the 1930s. The splitting up of aircraft into components of convenient size using parts manufactured as sub-assemblies outside jigs, and of producing units that could be fully equipped before assembly on aircraft, was efficient. Manual labour had to be replaced by machines and panels of complex curvature shaped with the help of rubber and soft-die presses. The use of castings and forging was encouraged to replace milling and turning, though under strict control and inspection. Drawing office work was fortified by full-scale layout methods ('lofting'). This made for greater accuracy and helped in tool manufacture. Gradually fuselage and wing internal construction was modified by design and production engineers to take advantage of the stressed skin. As aircraft speed increased, dimple riveting gave way to flush riveting and was used early on for the Boeing P-26A of 1932. Wind-tunnel tests carried out by NACA at Langley Field T confirmed the aerodynamic penalty incurred by protruding rivets. For a hypothetical single-engined aircraft, for example, eliminating the rivet head drag was estimated to increase the top speed by 5kts.

Structural trends

The above changes were introduced while maintaining structure weight percentage. The 1930s metal-covered wings were much heavier than the older wood and fabric type, but they were not necessarily heavier in relation to the weight of the aircraft. The older type of wing weighed about 5kg/m^2 (1 lb/ft^2) and was used on aircraft with a wing loading of 30 to 50 kg/m^2 (6 to 10 lb/ft^2). The metal wing weighed 10 to 15 kg/m^2 (2 to 3 lb/ft^2) but the wing loading of the late 1930s fighters was higher so that the ratio of structure weight to total weight was unaltered, even though the ultimate load factor was much higher (Table 7).

Table 7 Comparison of structure weight

Aircraft	*W/S (lb/ft^2)*	*Structure (% TOW)*	*Ultimate load factor**
SE5 (1917)	7.6	29.7	6
Spitfire IA (1938)	24	33	10
Spitfire IX (1942)	30	29	11
P-51A Mustang (1942)	46	31	12
Beaufighter 6 (1942)	42.3	28.3	8.3
Mosquito 2F (1941)	43.7	29.6	8
Hunter F.6 (1956)	70.6	32	11.25

* In 1945 the outdated concept of a safety factor of 2.0 was replaced by a proof factor of 1.125 and an ultimate factor of 1.5 based on more realistic values of normal acceleration.

The trend for the wing loading and ultimate load to increase while the structure weight percentage remained unaltered is significant. Some designers had too readily been impressed by the fact that, for geometrically similar structures, the load carried, assumed to be proportional to the vehicle weight, varies as the cube of the linear dimensions, whereas the cross-sectional area of the structure, which resists the load, varies as the square of the linear dimensions. This is the square-cube law.

That the square-cube law suffered a defeat was not due to the fact that designers fought it. Instead they built different structures by using the advances in the strength of materials, with working stresses much higher than they were in the 1920s. Advances in aerodynamics and improvements in powerplants have led to higher speeds and so to progressively higher wing loadings. From about 1930 there was a steady increase in the general level of about 25lb/ft^2 per decade.

World War Two

Structural design

By the outbreak of World War Two the all-metal stressed-skin monoplane fighter had arrived only just in time for the RAF. There were fabric-covered monoplanes like the Hawker Hurricane and Vickers Wellington, but apart from these, most other UK monoplanes were of stressed-skin construction. Apart from the wooden de Havilland Mosquito, no new structural design concepts or materials were introduced during the war. Any developments were rather to speed up and simplify production.

As speeds increased, consideration had to be given to compressibility and aerofoils of low camber and thickness, with the maximum thickness at about 40% chord. This created wing strength and volume problems for which the best solution was plenty of thickness well forward. Volume in the wings had to be found for fuel tanks, guns and undercarriage. Aircraft had to operate from grass strips and tyre pressure could not be too high, the standard being 2.4 bar (35 psi). Thinner wings would have forced up tyre

The Hawker Hurricane, first of the RAF's monoplanes (1937) had a wing root thickness ratio of 19 per cent. Its structural design and fabric-covered, metal tube construction made it cheaper than the Spitfire and it could take a lot of punishment and was more easily repaired. The construction methods used in the Hurricane retained the Warren girder square-tube construction and the wing had conventional Hawker dumbell spar with girder rib Warren bracing in plan to give the necessary wing stiffness. Fabric covering was used over metal ribs in the wing and wooden frames in fuselage. While it was recognised that stressed-skin construction was theoretically better (later series Hurricanes had metal wings), the urgent need for the aircraft was uppermost in Sydney Camm's mind and the design was determined by what could be made in the Kingston factory without undue changes to their equipment. The result was the availability of the Hurricane in sufficient numbers by the time of the Battle of Britain to determine its outcome. The maximum weight of the Hurricane nearly doubled over the years of its production with a structure weight increase of only 5%. This is a Hurricane Mk 2B of late 1940. *(BAe via author).*

pressure and restricted operations to airfields with prepared runways. As an example, for a wing of aspect-ratio 6 and a taper ratio of 0.4, a W/S of 40lb/ft^2 and tyre pressure of 35lb/in^2, the minimum t/c ratio was 17%. To reduce this to 15% would have meant raising the tyre pressure to 52lb/in^2. As wing loading continued on an upward trend, and fuel consumption for the higher powered engines went up, there was a need for wing thickness to grow. But if the wing root chord was kept large, the thickness ratio could be kept reasonably low. This encouraged the tendency towards more highly tapered wings. It was reckoned that if thickness ratio did not exceed 15% at the root and 12% at the tip, no serious loss in speed would be encountered up to 350kts (640km/h).

It was important to give a fighter as great a range as possible and the logical way of doing this was to carry as much fuel as possible, and this was achieved by putting fuel in the wings. However, there was then a contest for space between fuel, undercarriage and armament. It thus seemed preferable to carry the main proportion of the fuel in the fuselage. Not only is the vulnerable area reduced but the fuel system could be simplified. In the Focke-Wulf Fw 190 all fuel was normally carried in the fuselage and the wheels were very small with high-pressure tyres so that they could retract forward of the front spar. The wing was built in one piece that gave a large weight saving by keeping a light and efficient monocoque structure right across the fuselage, which eliminated the need for heavy wing/fuselage joints.

Aeroelasticity

Advances in aerodynamics and improvements in powerplants gave higher maximum lift, higher speeds and progressively higher wing loadings. The typical wing loading of World War Two fighters was about five times that of the earlier conflict. A cantilever wing, because it can have only limited spar depth, tends to be inherently flexible. Early cantilever wings, therefore, were made as thick as possible, particularly at the root, and had cantilever ratios (i.e. semispan to root thickness) of the order of 10. Wings of this low cantilever ratio were relatively stiff in bending but their torsional (i.e. twisting) stiffness was another matter. By World War Two cantilever ratios had increased to over 15 and serious consideration had to be given to aeroelastic effects in the design of fighter aircraft. Wing twisting in response to aileron deflection decreased the available rolling moment as a function of dynamic pressure (i.e. speed^2), because as speed doubled the associated loads quadrupled. In extreme cases, aileron reversal occurs, where the control surface loads deflect the wing rather than the air. Under these conditions the rolling moment obtained is in the opposite direction to that commanded. It was vital to establish, for each aircraft, that its aileron reversal speed was beyond the maximum speed attainable.

The problem of aileron reversal became a critical design problem for World War Two fighters. Rapid rolling at high speeds became a common combat manoeuvre and any loss in roll response put a fighter at a serious disadvantage. The advantage does not necessarily lie with the fighter that can

The Grumman F8F Bearcat was designed specifically with lightly constructed wings as a means of minimising the aircraft's weight. A novel feature of the F8F-1 was that the outer 1.5ft (0.45m) of the wing was designed with weak points that would break if they became overstressed, to prevent the whole wing failing. To ensure post-incident symmetry, early Bearcats were fitted with explosive bolts to automatically discard one wing tip if the other had already departed. Its pilots were equipped with g-suits, operated by compressed air from a tube on the right hand side of the cockpit. First flying in August 1944, the F8F was too late for World War Two so was the US Navy's ultimate propeller-driven fighter. *(US Navy via author).*

Flexural aileron flutter. Consider a wing having an aileron which, in the absence of constraint imposed by the control wires, would tend to descend by reason of its own weight. Imagine that as the aircraft travels forwards, its wing tips meet an upgust. They will rise suddenly, lifting the aileron hinge but the aileron surface itself will lag behind because its CG is aft the hinge line. This implies that the wing has had its camber increased, so that the lifting power of the wing tip is even greater than before, in other words it is bent more than before. The increased aerodynamic forces tend to lift the aileron too. But before the aileron is well underway, the wing tip will have reached its highest point and will be springing downwards, so that the aileron, relative to the wing tip, is now effectively deflected upwards. This produces a download on the descending wing, still in advance of the aileron. Here then is a boosting mechanism to produce a periodic bending of the wing. It was remedied by arranging the CG of the aileron to lie on the hinge line or slightly ahead of it. Both divergence and flexural aileron flutter can be observed by the leaves on a tree or by the flapping of venetian blinds. There is a critical speed below which flutter will die out and above which it will be maintained. The investigation of flutter in the UK arose in 1925 because of accidents which seemed attributable to its cause. Flight experience had shown that aileron flutter caused the control column to be wrenched from the pilot's grasp. This type of flutter which is only relevant to aircraft without powered (i.e. irreversible) controls was and is cured by mass balancing the control surface.

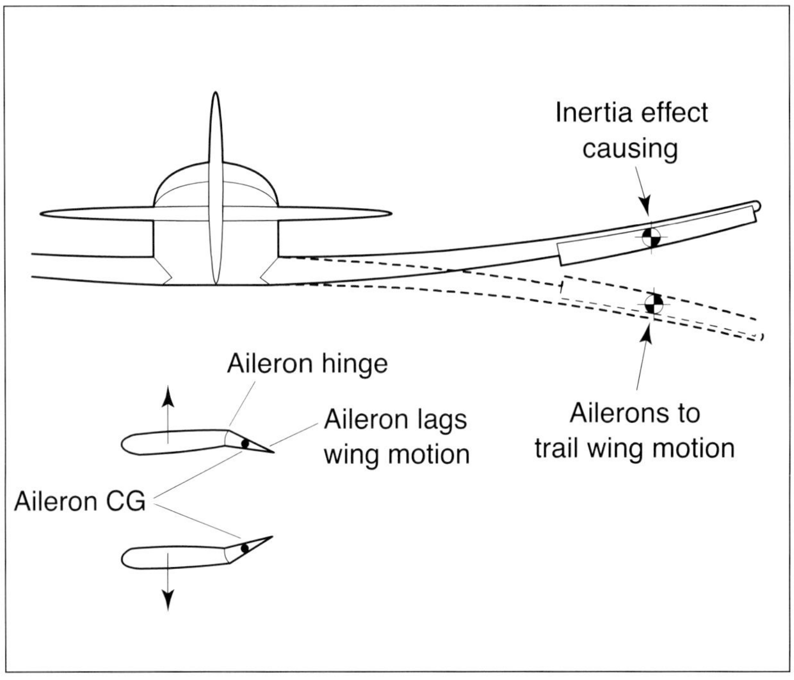

out-turn its opponent speed for speed but the one with the fastest rate of roll or, even better, rate of acceleration into the roll, so that it can get into the turn faster. The designers of the Supermarine Spitfire, Focke-Wulf Fw 190 and the North American P-51 sought higher and higher roll rates. This led them to shorten their wing spans, balance their ailerons for larger deflections at high speed and stiffen their wings. Despite the excellent manoeuvrability of the Spitfire, it was unable to match the superior roll ability of the Fw 190.

The Jet Age

Structural efficiency

Being limited by pilot tolerance, the ultimate load factor of fighters has not varied significantly over the past fifty years: the P-51 was designed to 12g, current fighters are designed to 13.5g. Material properties have improved over the years and have permitted the use of higher local working stresses in compression to allow higher wing skin loadings. However, tensile strength increases have been offset by imposed limitations in stress levels to preclude flutter, provide a longer service life (see later) or sustain increased structural temperatures due to kinetic heating. Wing thickness ratios had to drop drastically between 1945 and 1960 to permit higher speeds, but adequate bending strength had to be maintained. Moreover, the swept wing introduced a greater structural span than a straight wing of the same area and aspect-ratio. Sweepback also introduced additional torsion because the applied loads on the wing act significantly aft on the wing root. All this required more structural material. It might reasonably be expected that the weight of wings relative to the take-off weight would have increased dramatically. This has not been the case, because of the inventiveness of structural designers. Indeed, the reverse has occurred, as shown in Table 8.

When actual wing weights are normalised to the P-51 baseline span (taking into account the effects of planform, section and materials) the weight reductions are even more striking. The normalised or effective wing weights of today's fighters are about 3% of take-off weight compared to 14.5% for the P-51. Structural efficiency has certainly improved. The wings of current fighters could only have been made in almost solid aluminium alloys or steel using the manufacturing, materials and design technology available in the 1940s.

Table 8 Comparison of the wing weights of the North American P-51B and McDonnell Douglas F-15A

	P-51B (1943)	*F-15A (1974)*
Take-off weight (TOW)	8,660lb	41,500lb
Span	37ft	42.8ft
Wing area	236ft^2	608ft^2
Wing loading	36.7lb/ft^2	68.2lb/ft^2
Wing root thickness	15%	5.9%
Quarter chord sweep	0°	38°
Wing weight	1,260lb	2,890lb
Wing weight/TOW	14.5%	7.0%

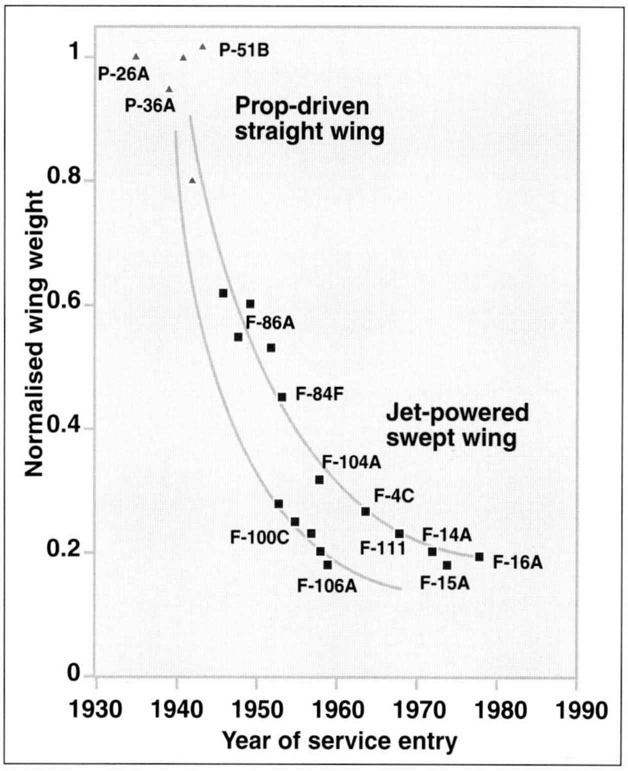

Relative wing weights of current fighters compared to propeller-driven fighters (normalised to the P-51 baseline span and taking into account the effects of planform, section and materials).

Over the past fifty years, fighter aircraft have grown significantly in weight and size. Size increases should allow more opportunity for design improvement and structural tailoring. It might therefore be expected that increased size, in conjunction with technology advances, would reduce the ratio of structure weight to the take-off weight. This has not occurred. The structure weight of World War Two fighters was about a third of take-off weight (Table 7). This remains true of today's fighters because design load factors have increased, and while relative wing weight has reduced dramatically, the weight of other components of the structure (e.g. fuselage) have grown markedly in proportion to the wing, and they have to last much longer.

Integral construction

Up to the end of World War Two structures were mainly fabricated by wrapping comparatively thin skins over an internal framework structure and allowing them to spring to shape. With speeds and wing loads gradually increasing much thicker skins were needed that were difficult to wrap. By the 1950s even thicker skins were needed which had to be pre-formed by rolling, later by stretch forming. For high-speed aircraft very thick skins were needed with internal webs rather than stiffeners, making riveting extremely difficult. Hence machining became the logical production method, especially if tapering in thickness was required.

By the late 1950s integral construction was starting to be used to form the skin and stringers together in one part. This involves machining from a solid slab of metal, with perhaps 95% of original material being removed, which cuts out joints between components. This form of construction involved no new design principles and its advantages were at once apparent. A good surface finish, free from irregularities, was beneficial for drag. It also allowed more efficient use of material since skin thickness is more easily tapered to coincide, for example, with the spanwise decrease in bending stresses. In addition, the structure so formed had far fewer stress-raising holes per surface area and was less prone to fatigue. Wing torsion boxes, with thick skins used as the upper and lower surfaces of integral fuel tanks, were less prone to leaks, easier to seal and eliminated internal bag tanks. This led to a significant increase in fuel volume, something many fighters have been notoriously short of. An additional advantage of integral construction, particularly for the tight constraints of fighters, was the ability to combine many functions into a single complex part to fit within the available space.

Fatigue

Practically all components of an aircraft's structure are subject to fluctuating loads that occur during the life of an aircraft. Fatigue of a structure is really slow death over millions of cycles of applied loading, though it is unlike that of a human being in that rest does not bring recovery. It is unlike human fatigue too in that there are no external symptoms until cracks appear, so knowledge of the loading history is therefore vital. It had been known for many years that materials break at much lower values of stress than their normal static failure stress. (The everyday example is the fatigue failure of a paper clip through repeated, admittedly high stress, bending.) Prior to the late 1940s little attention had been paid to fatigue considerations in the design of aircraft structures. It was felt that sufficient static strength would eliminate the possibility of such failure. The designers of the previous generation of aircraft were oblivious to the effects of time on their structures, because during the 1930s they had been scheming short-lived aircraft for wartime purposes.

When materials with higher allowable working stresses became available, designers seized upon these as a means of meeting the insatiable demand for minimum weight while maintaining the strength levels of their previous designs. Certainly, the new alloys had greater static strength

Evolution of wing construction.
(a) Typical wing of World War One. Load bearing members were positioned near aerodynamic surfaces where the stresses were highest. With the wings' upper surfaces in compression, lower surfaces in tension, near the neutral axis stresses were low and this is where lightening holes could be used.
(b) Fabric-covered Warren truss wing of the mid-1930s with two built-up steel spars similar to those used on early Hawker Hurricanes.
(c) With the advent of metal skins, wings became an all-light alloy fabricated structure. Typical multi-cell metal wing of the late 1940s.
(d) Supersonic aircraft of the 1950s and later, required thin multi-cell wings to withstand high loadings. Skin milling is used on large billets of light alloy to cut away approximately 95% of material to produce wing skins with integral stringers.
(e) The advent of CFC has allowed designers to vastly reduce the metal content of the wings. Shown is the wing of the BAe EAP. Dark areas illustrate the use of CFC and light areas show metal usage, including three of the six titanium wing-attachment joints. For very thin wings the number of spars had to be increased to enable vertical shear loads to be reacted. If the wing is also of low aspect ratio and is required to carry a large number of stores, the structural configuration takes the form of an eggbox rather like that of the EF2000 and, in extremis, the Dassault deltas.

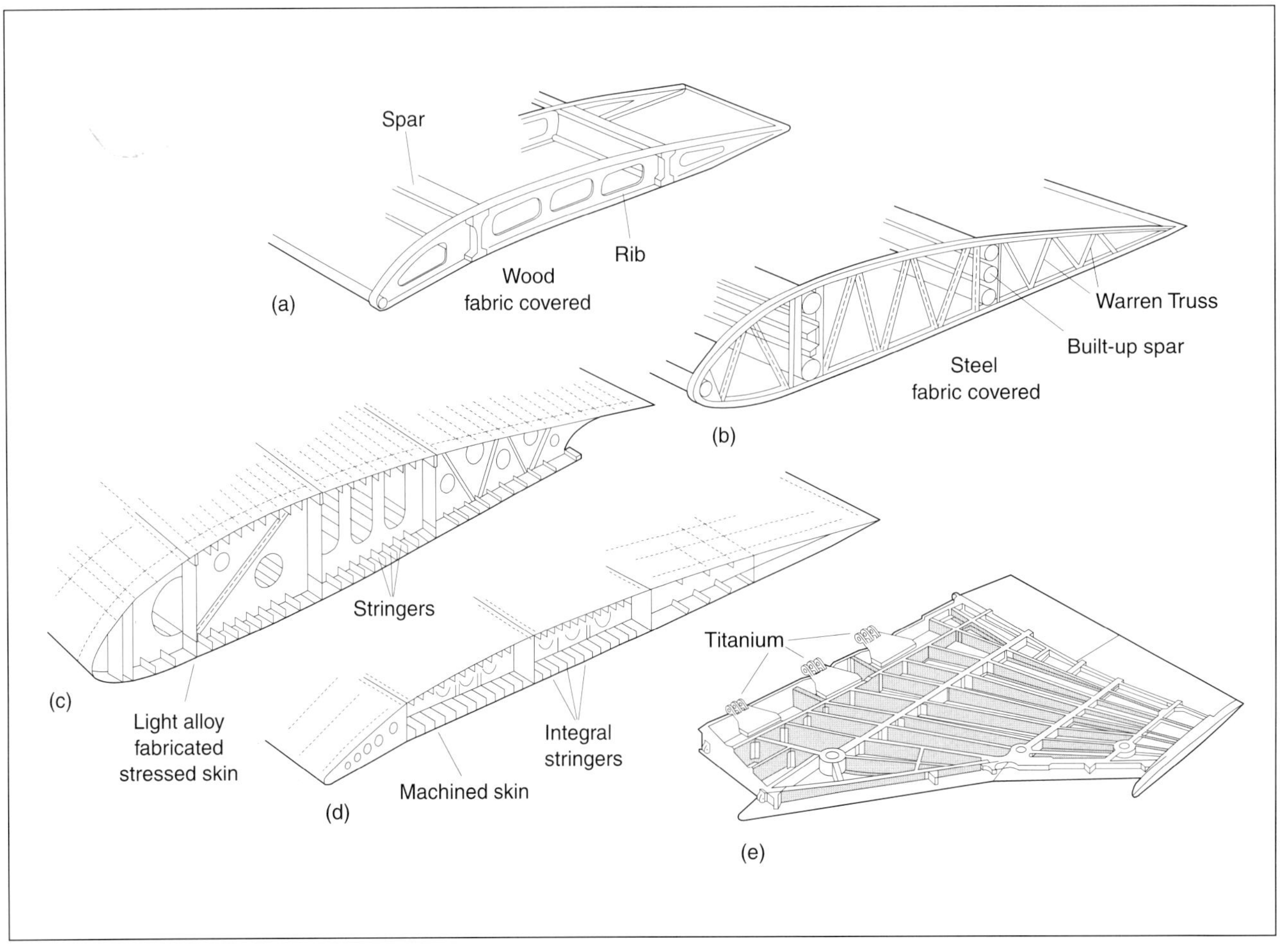

Modern structural testing is in stark contrast to that used 80 years ago when it involved turning the aircraft upside down and loading the wings with bags of sand or lead shot until the structure failed. This photo is of a Saab JAS 39 Gripen structural test specimen in a rig with loads being applied to replicate those that will be experienced in flight. Stresses and deflections are accurately measured and compared with predictions. *(Saab via author).*

but not proportionally (if any) greater fatigue strength. The long-term effects of fatigue, corrosion, crack propagation, stress corrosion and creep were largely overlooked. Since the new-strength aluminium alloys were prone to at least one of these weaknesses, many aircraft suffered frequent minor failures, even if they escaped catastrophic ones. Moreover, evidence began to accumulate that several aircraft crashes had been caused by fatigue failure. The operative causes of this emergence of fatigue as a major factor in design were not far to seek. With the revival of civil aviation after World War Two, airliners had greater utilisation and a longer working life. Higher speeds meant more violent bumps in turbulence. Mean stresses had risen with higher wing and structural loadings.

The seriousness of the situation was dramatically highlighted in the early 1950s by the catastrophic fatigue failures of two de Havilland Comet airliners, the root cause of which was a rivet hole an eighth of an inch in diameter. Small holes and notches are a good way of starting fatigue failures because they induce enormously high stress concentrations and do well for starting ordinary static fracture. When we cut glass we don't attempt to cut right through but instead make a shallow scratch on the surface. The weakening effect of the scratch has little to do with the amount of material removed but has a lot to do with the sharpness of the cut.

These troubles with fatigue further provoked the rival philosophies in structural design. Assuming that failure is in the end inevitable, one school favoured the 'safe life' concept. This requires a study of the load variations and reversals thought to be likely in the operating conditions of a component. One can arrive at the length of time in flying hours (the safe life) after which the component is likely to become unsafe. At this point the item is scrapped whatever its apparent condition. Defining safe life is relatively straightforward; ensuring that it remains valid is not. It depends on the way the aircraft is operated and maintained. The alternative 'fail safe' approach was first used in biplanes when flying wires were duplicated. It provides an alternative path for any load, should there be cracking, corrosion or other trouble.

The concept of damage tolerance originally introduced by a US military requirement showed a shift in structural philosophy from the fail-safe stance of the late 1950s. Since it had become possible to calculate crack growth as an alternative to predicting when cracks will start, each critical location was assumed to have a minute flaw from the day the aircraft was built. In other words, 'Think of a crack and design your aircraft around it'.

The financial consequences of an aircraft suffering fatigue problems can be extremely severe if it causes the grounding of a whole fleet of the type. Current combat aircraft are designed to have service lives of around 6,000 hours, which equates to approximately twenty years (300hr/year), compared with airliners having 60,000 hours plus at 3,500 hours/year. Although adequate precautions are taken to ensure that an aircraft's structure possesses sufficient strength to withstand the most severe expected manoeuvre load, the problem of fatigue still remains.

The quest for stiffness

Higher speeds, thinner wings, stronger materials and more refined design caused a change in the concepts that once prevailed on the subject of stiffness. Previously, when an aircraft structure was designed to meet a specific strength requirement, the resultant stiffness was usually adequate. Sometimes it was necessary to add some material for more stiffness, to correct flutter or resonant vibration problems. Until the modern aircraft arrived, strength not stiffness was the major concern. Flexibility, though, has adverse consequences and the designers' task became that of providing both the necessary strength *and* stiffness at the least weight. Material stiffness depends solely on the chemical nature of the solid and cannot, as a rule, be changed by tinkering with it. In order to use a stronger metal effectively it has to be stretched more. This means that the deflections of the structure will be higher; if the stress has to be raised a lot, so as to save a useful amount of

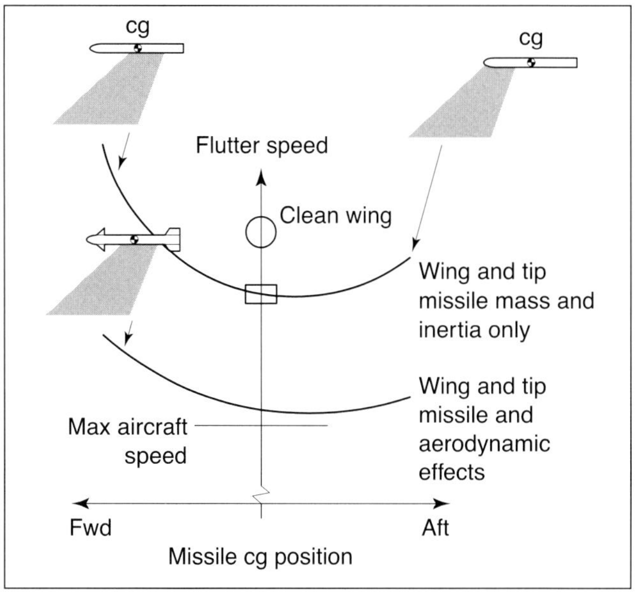

Effect of a wing tip missile on flutter speed. Note that the 'clean' wing has been given a flutter speed higher than the flight speed of the aircraft. Adding the missile mass reduces the combination's flutter speed, whilst allowing for the missile aerodynamic surfaces reduces the flutter speed further. *(Pete West).*

The McDonnell F3H Demon that first flew in 1951 was a single-seat single-engine fighter with 45° leading edge sweep. Handling was poor with pronounced Dutch roll, low rate of roll, low endurance and its Westinghouse J40 turbojet was a source of concern. Even when given a new engine, the Allison J71 with 40% more afterburning thrust, the Demon's problems were not over, due to wing distortion during rolling manoeuvres at high dynamic pressure. This was cured by adding spoilers. Shown is F3H-2 of VF-121 from NAS Miramar. *(US Navy via author).*

A desirable feature is a wing very stiff in torsion. This gives a great advantage but often at the cost of more weight than necessary from the pure strength point of view. There is another consideration which demands great torsional stiffness and that is the efficiency of the aileron control. If a wing is too flexible in torsion, the application of the aileron twists the wing to such an extent that the rolling moment due to the aileron, is largely offset by the moment of opposite sign due to the twist of the wing. What was needed, as also for the prevention of flutter, was a type of control which applies a rolling moment without twisting the wing, or in other words a control that applies an increment of lift at the centre of pressure of the wing instead of at the trailing edge. In the design of the F-100 Super Sabre, North American realised that spoilers rather than ailerons would be an acceptable solution to the lack of torsional stiffness of the aircraft's thin, 45° swept wing. These are F-100s of the USAF Thunderbirds of the late-1950s. *(via author).*

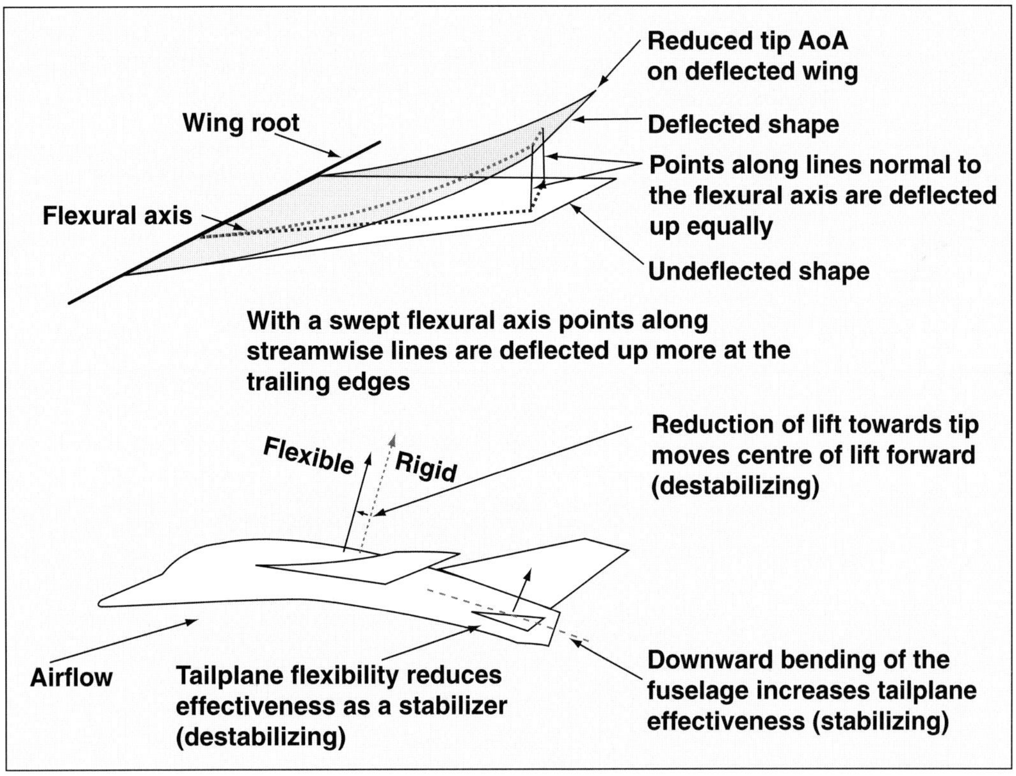

Aircraft are expected to bend but very large deflections are unacceptable. This figure shows the effects which airframe distortions can have on pitch stability. *(Pete West).*

weight, then the deflections will be *much* higher. Aircraft are expected to bend, but very large deflections are unacceptable. Because much of the structure of an aircraft is in compression, it fails by buckling, and the cause of such failures is not lack of strength but lack of stiffness. By simply increasing the strength of a material it will soon run out of stiffness and it is therefore as important to increase stiffness as to increase strength.

Rules of thumb for ensuring a design is relatively free from flutter:

- make sure everything is as stiff as possible
- aim at high torsional stiffness
- although sweep, taper and aspect-ratio are usually dictated by other considerations, given the choice, high values of the first two and a low value of the third are beneficial
- place all equipment close to the leading edge rather than the trailing edge
- ensure all underwing and wing tip mounted store CGs are kept as far forward as possible

Materials

Before any comparison of the mechanical properties of the various candidate materials can be made it is necessary to define the operating envelope of the aircraft. The most significant aspect that distinguishes the design of the structure and choice of materials of supersonic aircraft from those of lower-performance types is the high-temperature environment caused by kinetic heating in sustained supersonic flight. Leading edges of all components experience the stagnation temperature, which is a function of the square of the Mach number. For example, flying at M = 2.2 at 35,000ft (10,600m) where the ambient temperature is nominally −57°C will cause a leading-edge temperature of around 135°C. Existing fighter aircraft do not cruise for long periods at such speeds and they have metal leading edges (to deal with rain erosion and birdstrike) so that such temperatures do not present a problem.

The major direct influence that a new material can have on aircraft design is that of saving weight. This can then be reflected as improved performance, increased range, improved airfield performance, greater warload or a smaller aircraft to perform the mission. The most efficient means of reducing structural mass is to reduce the density of the material, rather than increasing strength, stiffness or toughness, though these other methods would allow further weight reduction. A material density reduction of 10% gives a structure weight saving of 10% that far outweighs the saving gained by similar proportional increases in mechanical properties. A 10% increase in a material's strength gives only a 3% weight saving. Several factors influence the choice of the structural material for an aircraft, but amongst these, strength allied to lightness is probably the most important. Other properties having varying, though sometimes critical, significance are stiffness, toughness, resistance to corrosion, fatigue and the effects of environmental heating, ease of fabrication, availability and cost.

STEEL Steel has high stiffness and high tensile strength but also high density, three times that of the aluminium alloys. It has been used as a structural material in the form of stainless steel: the swing-wing F-111 has a steel carry-through wing box, and a large amount was used in the MiG-25 'Foxbat'. Steel in its widely varying forms is currently used wherever very high tensile strength and stiffness/volume are required, e.g. undercarriage, wing sweep actuators, wing root attachments, engine supports, armour plating, arrester hooks and flap tracks. However, because of steel's high density, titanium has taken over many of its former applications.

ALUMINIUM ALLOYS The majority of the world's current fighter aircraft are made primarily from two groups of aluminium alloys. The so-called Al/Cu/Mg alloys (Duralumin, US 2000 series) are damage-tolerant, have good fatigue performance with a low rate of crack propagation. They are used in fatigue-sensitive areas, especially lower wing skins. The high strength Al/Zn/Mg alloys (US 7000 series) are used in compression-dominated areas (i.e. upper wing skins).

ALUMINIUM/LITHIUM Although Aluminium/Lithium (Al/Li) alloys are not entirely new and have been around since the 1920s, were it not for significant technical and production difficulties, many of today's airframes would already be manufactured in the material. The great attraction of Al/Li is its reduced density, about 10% less than that of typical 2000 series alloys, whereas it has about the same strength as conventional 7000 series alloys. Another benefit is that it is about 10% stiffer than conventional alloys. Its first major use was to be in forgings and thick plate (e.g. the major wing carry-through frames) on the Eurofighter Typhoon. Continued uncertainty with Al/Li, though, forced a limit on the hoped-for 20% of structure weight to around 5% currently. Nevertheless, combined use of Al/Li and carbon-fibre composites (see below) saves around 10 to 12% of structure weight, or 3 to 4% of take-off weight compared with conventional aluminium alloy usage. The use of Al/Li and CFC on the Typhoon development aircraft is shown in an adjacent figure.

TITANIUM ALLOYS When compared to aluminium, alloys of titanium, though 1.7 times heavier, have much higher strength/weight especially at high temperatures, better fatigue life, and do not corrode under typical aircraft usage. They are, unfortunately, much more expensive (ten to twenty times) than aluminium alloys, and were originally much more difficult to fabricate. This explains why the F-16, conceived as a relatively low cost lightweight fighter, has virtually no titanium. Nevertheless, titanium is used extensively in other fighters. The relatively new process of super plastic forming and diffusion bonding (SPF/DB), has allowed an increased use of titanium. This takes advantage of the inherent superplasticity of titanium at around 900°C (when it can be stretched by a factor of around 8) to allow the fabrication of one-piece complex structures from nothing but flat sheets. These are not only cost-competitive but often more structurally efficient. Cost savings of around 50% have been claimed, though there is a price to be paid for in capital equipment expenditure. Weight savings are not as great, though not insignificant.

CARBON-FIBRE COMPOSITES (CFC) Probably the most important material to be introduced into aircraft structures in the last twenty years. CFCs consist of immensely strong fibres set in a matrix of epoxy resin which is mechanically

The first prototype of the MiG-25 'Foxbat' flew in 1964 and could not use conventional aluminium alloys due to kinetic heating at its design Mach number of 2.8. Much of the structure was built from stainless steel rather than the titanium alloy favoured by American designers. This made it heavy, and along with its apparent lack of advanced avionics, removed much of the myth of this Soviet 'wonderplane', though as Ye-266 it captured a number of speed and altitude records. *(Alexei Mikheyev via author).*

Early applications of integral machining were for aluminium alloy parts. However, in the 1950s many companies including McDonnell Aircraft, anticipated that future aircraft would need large titanium and steel parts. Accordingly they bought numerically controlled machines with adequate strength, rigidity, horsepower and cutting speed controls to handle 'hard metals' as well as aluminium. Over 50% of the F-15E (this one from AFFTC at Edwards AFB) structure weight is from integrally machined parts, 25% of which is titanium. *(Boeing via author).*

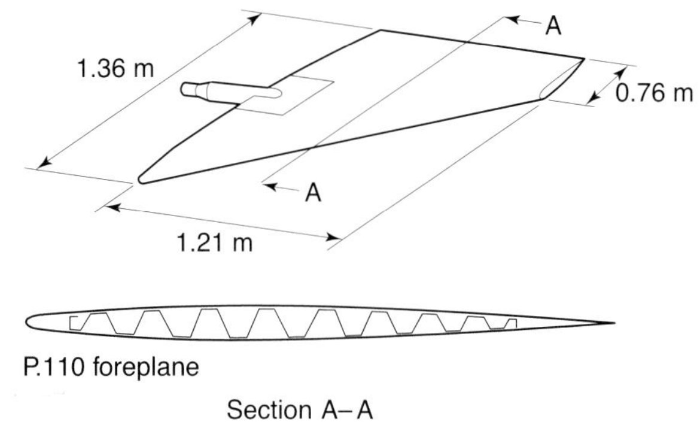

Material	Relative cost*	Mass of foreplane (including spigot)
SPF/DB titanium alloy	0.70	72 kg
CFC	1.00	68.5 kg

* Relative cost based on production run of 300 aircraft

As part of a technology demonstration programme partially funded by the UK's Ministry of Defence during the 1980s, British Aerospace investigated the use of the SPF/DB process which included a full scale foreplane initially designed for its P.110 project. The figure shows the control surface, along with a table giving the cost and weight comparison with a CFC version with its titanium spigot. Despite the slight weight penalty the titanium SPF/DB foreplane, made from four flat sheets of titanium, offered a competitive solution for a future canard configuration, because of the large saving in cost.

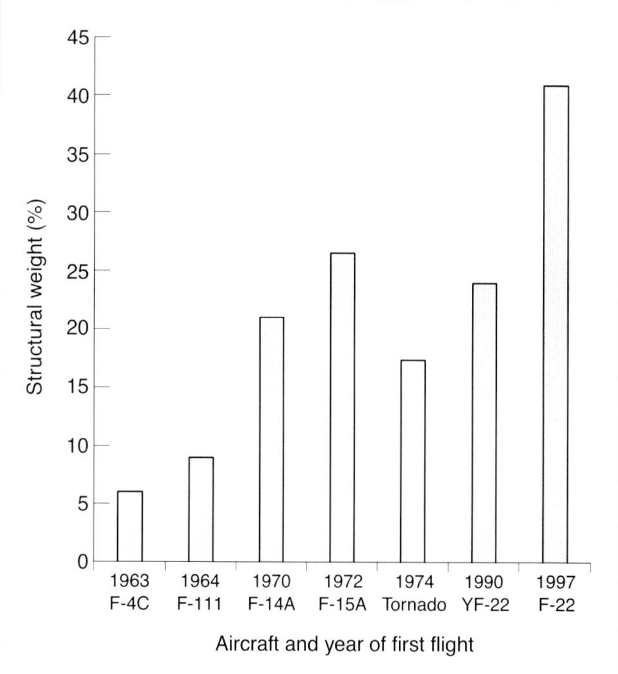

Use of titanium in fighters. Despite the extensive use of composite materials in the Lockheed Martin F-22, it will still employ an increased percentage of titanium.

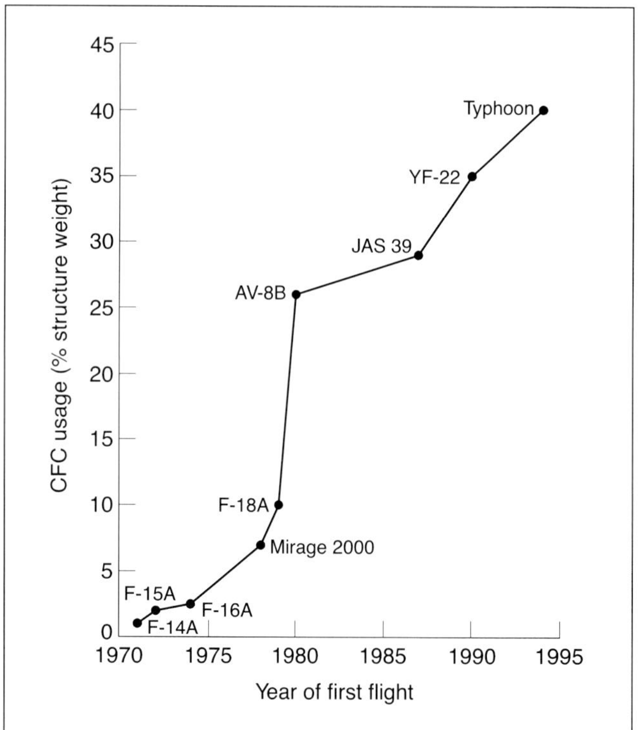

An illustration of the increasing use of carbon-fibre composites in the structures of fighters. Less than 15 years after the announcement of the material by RAE Farnborough, over 25% of the structural weight of the MDC/BAe AV-8B came from CFC, which saved 220kg. *(Pete West).*

and chemically protective. For some decades it had been recognised that higher strength than is exhibited by the bulk material can be obtained when the material is in the form of fine wires or filaments, and is a consequence of reducing the incidence of surface damage or defects from which failure originates. An early example of the application was steel 'piano' wire.

They were developed at RAE Farnborough and announced in 1966. Not only do CFCs possess excellent strength/weight and stiffness but they offer the unusual opportunity to design the structure and the material simultaneously. The directional (i.e. aligned in a given direction) properties of composite materials can be used to aeroelastically tailor structures to obtain, under load, specified wing twist and camber. This has beneficial effects on aerodynamic drag, control effectiveness and air load distribution, leading to increases in warload or range. Such tailoring can be used to obtain a lower weight design that satisfies all of the applicable design constraints such as strength, flutter and divergence. Compared to 2000 and 7000 series aluminium alloys, CFCs offered weight savings of 20%. A further advantage is the ability to mould complex shapes. Still, the material remains expensive and labour intensive in fabricating structures. Other drawbacks include: significantly reduced strength due to undetected damage, reparability problems and environmental difficulties during fabrication and at crash sites. It took about ten years to get CFC into the production cycle. The development of the Harrier into the AV-8B, by British Aerospace and McDonnell Douglas, provided the first major opportunity to use CFC. Despite the increasing use of CFC, especially for thin fighter wings where stiffness is the prime consideration, there will be many parts of future fighters that will be made of metallic materials, in particular areas where there is intense local loading, substructures, areas experiencing high temperatures, and areas where the inherent toughness and ability to withstand damage of metals is needed.

This 1995 photo of an unpainted McDonnell Douglas/BAe AV-8B Harrier II Plus gives an idea of the various material used on the aircraft. The Harrier II uses CFC for 25% of the airframe structure weight and 50% of its wetted area, and was about 225kg (500lb) lighter as a result. CFC is used very extensively in the forward fuselage, the whole of the one-piece wing and the horizontal tail. It allowed a significant reduction in part count: the cockpit of the AV-8B has 88 separate parts and 2,450 fasteners, compared with 237 parts and 6,440 fasteners in the AV-8A. *(Boeing via author).*

Whereas the F/A-18C/D had 50% of its structural weight from aluminium alloys, the F/A-18E/F uses only 29% with the amount of CFC more than doubled (up from 10 to 22%), mostly in the wing and stabiliser skins, centre and aft fuselage. Improvements in manufacturing methods have reduced the number of parts in the airframe by one third. The F/A-18E/F also uses over 150lb of radar absorbing material (RAM) designed to be corrosion-resistant in a salt water environment. The computer-controlled laser-target system shown here is used to precisely align the splice of forward and centre/aft fuselages of Boeing's F/A-18E/F. *(Boeing via author).*

Conclusion

The development of aircraft has witnessed some fundamental changes in structural design requirements which have altered not only the shape of aircraft but also the materials from which they are made. It's noteworthy that apart from one or two exceptions the top speed of fighters has been limited by neither aerodynamics nor propulsion but by the materials designers thought worthwhile to use. Without the improvements in structural efficiency the advances made in these other areas would not have given the performance increases that have been witnessed over the last fifty years.

The Mitsubishi F-2 (first prototype shown in 1996) has a cocured carbon-fibre composite wing box in which spars, major ribs and lower wing skins are bonded together without conventional metal fasteners, yielding weight savings and simpler final assembly. The aircraft has a 25% larger wing than the F-16 from which it is derived. Because it was designed to be used at low level with the higher attendant risk of birdstrikes the F-2 has a separate windscreen in place of the F-16's one-piece canopy.
(Mitsubishi via author).

An advantage of composite materials is the ability to mould complex shapes. This reduces machining costs and very expensive waste material, to get more weight of finished product than with metals, and reduces the number of parts in complex structures by a factor of about 3, so reducing joining costs. However the material is very expensive and fabricating structures was very labour intensive. The Boeing JSF/X-32 inlet duct shown is being wound as a one-piece structure using an automated fibre placement process. This allows large components of complex shape to be produced more accurately, faster and cheaper than traditional methods.
(Boeing via author).

With leading and trailing edge high lift devices deployed and tailplane set for maximum nose-up trim this Dassault Super Etendard of 1977 is about to lose the steam catapult strop that accelerated it to take-off speed. Despite its apparent slenderness (though the weight is off the wheels so the oleos are extended) the undercarriage is designed to absorb four times the kinetic energy of land-based types. The steels in the undercarriage components have become the most refined and strongest in airframes.
(F. Robineau via author).

Chapter 6

Avionics

As the chapters so far have charted, the fighter is a vehicle whose aerodynamics, stability and control, propulsion and structural compromises govern its design. The influence of avionics (aviation electronics) is rarely evident, but the fact it is often obscure does not mean that it is not important. Fighter avionics as defined here includes communication and navigation systems, radar and electro-optics sensors, defensive electronic aids and the ways the information from the various sensors feed electronic cockpit displays. This has encompassed inventions like the transistor and microchip which have had a huge effect on sensor capability, allowing them to play an increasingly vital role in fighter operations. There is every reason to believe that this will continue.

Glossary

Antenna gain: The power radiated in the desired direction divided by the power that the same radar would radiate with an 'all directions' antenna.
Duty factor: The ratio of pulse width to interpulse period. Most (typically 99.5%) of the time, a fighter's radar is listening rather than transmitting, so that duty factor would be only 0.5%.
GHz (GigaHertz): Thousand million cycles per second.
kHz: Thousand cycles/sec.
MHz: Million cycles/sec.
Pulse length: The length (in distance) of a pulse of radio frequency energy as it travels at the speed of light (300,000km/sec (186,000 miles/sec)). Pulse length = pulse width × speed of light. It determines a radar's ability to discriminate between two closely-spaced targets.
Pulse repetition frequency (PRF): The number of radar pulses sent per second (pps), the reciprocal of pulse repetition interval (i.e. 1/PRI).
Pulse repetition interval (PRI): Time between pulses (milliseconds).
Pulse width: The length of a pulse in terms of time (microseconds).
Radio frequency: Speed of light divided by wavelength.

Pre-World War Two

Communications

Aircraft radios were investigated by the UK military in 1910, and the Royal Flying Corps (RFC) tried radio/telephony (R/T) from 1916. Previously communication between aircraft depended on hand signals. The experiments emphasised the need for tight-fitting helmets because of the low power of the received signals. Their comparatively low radio frequency (500 to 1,500 kHz) required quarter-wavelength or half-wavelength (35 to 100m (100 to 300ft)) long trailing aerials, and until pilots remembered to 'wind in' before landing, many wires were distributed over the environs of an airfield. However, radio communication in both the British and French air services was limited to reconnaissance and artillery fall-of-shot spotting aircraft. It was not fully developed for fighters until 1918. Ground-to-air and inter-aircraft communications became established during the 1920s, but reliability and speech quality were poor. To overcome ambiguity words such as 'Bandit' for a hostile and 'Bogey' for an unidentified were introduced.

During this period a direction finding (DF) system applicable to aircraft had been developed. Originally the DF antennas were formed by running wires in tapes around the aerofoil surfaces and up and down the interplane struts, thus giving a loop of large effective area. To make these DF antennas effective with the power of ground stations then available, receivers of high sensitivity were required, and it was necessary to completely screen the aircraft ignition system to curb interference. This was far from acceptable to the engine and aircraft manufacturers. It placed an extra load on the ignition system and added to the difficulties of maintenance and checking of that system. Eventually, sound and totally screened ignition systems became available and one of the obstacles to going to a shorter wavelength was removed. From about 1930 onwards, the 6MHz band became a practical proposition and radio sets working on it remained in service well after the start of World War Two.

Radio direction-finding was virtually all-weather and one of its advantages was that a pilot could home in on the ground station and fly directly towards it. He could then tune into another beacon and proceed towards that. However, these systems were applicable only to bombers or airliners. The day-fighter pilot had to rely on dead reckoning with frequent checking for recognisable ground features. 'Dead' (deduced) reckoning is, in simplified terms, determining where the aircraft is on the basis of where it has been. For the relatively short-range capability of World War Two fighters this was not an enormous problem, though in the heat of combat it was easy to lose one's bearings.

Radar

The inter-war period was one of stagnation so far as night-fighting techniques and tactics were concerned. The drastic rundown of the RAF after the 1918 Armistice and the chronic shortage of funding throughout most of the 1920s and 1930s meant that few resources could be devoted to such a specialised role. Instead, development of these became the responsibility of day-fighter squadrons. The lessons so painfully gained during World War One were largely forgotten. Blind flying training was introduced by the Central Flying School (CFS) in 1930, so that at least some of the hazards of night flying, often more formidable than enemy action, were removed.

However, if the RAF was badly equipped for night defence at the beginning of World War Two, it was no worse off than any other air force. The RAF had at least given thought to the problem of dealing with night bombers. Moreover, important scientific and technology developments were in hand that were to revolutionise the business of night flying. However, the paramount needs of the daylight air defences and the extreme complexity of the new systems meant that for the first year of the war any night-fighters would be groping in the dark.

The impending breakthrough was dependent on the newly devised RADAR (a palindrome standing for RAdio Direction And Ranging), then known under the cover name of radio direction-finding (RDF). In 1936, Watson-Watt put forward the idea of airborne radar to the Tizard Committee for the Scientific Survey of Air Defence. The difficulties of adapting a ground-based system which, in its initial Chain Home version, weighed several hundred tons for aircraft installation were tremendous. By spring 1939, starting in Aberdeen, Scotland and running down the east and south coast of England to Portland, there were twenty radar transmitters, the world's first large-scale radar air defence installation. Whereas Chain Home, initially operating at a wavelength of 26m (11.5MHz), provided range, altitude and a guess at raid size, the requirements for night interception were more stringent. Visual acquisition distance even on a clear night was rarely more than 300m (1,000ft) and often less. To be of any use, a pilot had to be vectored by ground control so that the target was within a 20° cone ahead at a range of between 1.7km and 300m (5,500 and 1,000ft). The feasibility of airborne interception (AI) radar, using much shorter wavelengths than that used by the ground system due to the limit on aerial size, and with much less available power, seemed impossible.

World War Two

Communications

By the outbreak of war all but light aircraft had electrical systems. As operations became more complex, electrics had to supply more lighting, power more instruments, and power radars. The development of radio communications made further demands for electrical generation. Just prior to World War Two, VHF radio (100 to 150MHz) appeared, bringing a greater clarity which played an important part in the plan for controlling fighter aircraft for the defence of the UK. The higher frequency band meant that the antennas could be smaller, initially a springy 'whip aerial', and there were more available frequencies.

Radar

During trials in 1939, a Fairey Battle equipped with AI Mk I radar was able to detect a Handley Page Harrow at ranges of between 5km (3nm) and 300m (900ft). It displayed the relative position of the target in elevation and azimuth on two separate cathode-ray tubes (CRT). Despite this remarkable progress it was still a long way from full-scale operational service, since it was operated by the scientists who had developed it and who could get the best from it. Moisture played havoc with its delicate electronics and it was swamped by ground returns (clutter), making target detection impossible below about 8,000ft (2,400m).

The AI Mk III was the first production model. The Bristol Blenheim, a light bomber, had caused a mild sensation in being faster than the RAF's new fighter the Gladiator. When it came to selecting an aircraft to carry the new AI radar the Blenheim was the obvious choice. The transmitter dipole aerial, operating on a wavelength of 1.5m (200MHz), was fitted on the nose of several Bristol Blenheim Mk IFs, with azimuth receiver aerials on the engine nacelles and elevation receiver aerials above and below the port wing. It was very limited, with a maximum range not exceeding 3.7km (2nm) and minimum range of more than 300m (1,000ft), which was far too long on a dark night. However, a pursuit starting over mid-Kent might end over Northern France. In the rare event of an interception being made, the armament of the Blenheim was inadequate, and often the bombers, although damaged, escaped.

The Fighter Interception Unit was set up in May 1940 to iron the bugs out of the new equipment and develop tactics during operational flying. For the first nine months of the war AI radar offered little but hope, though the first kill was

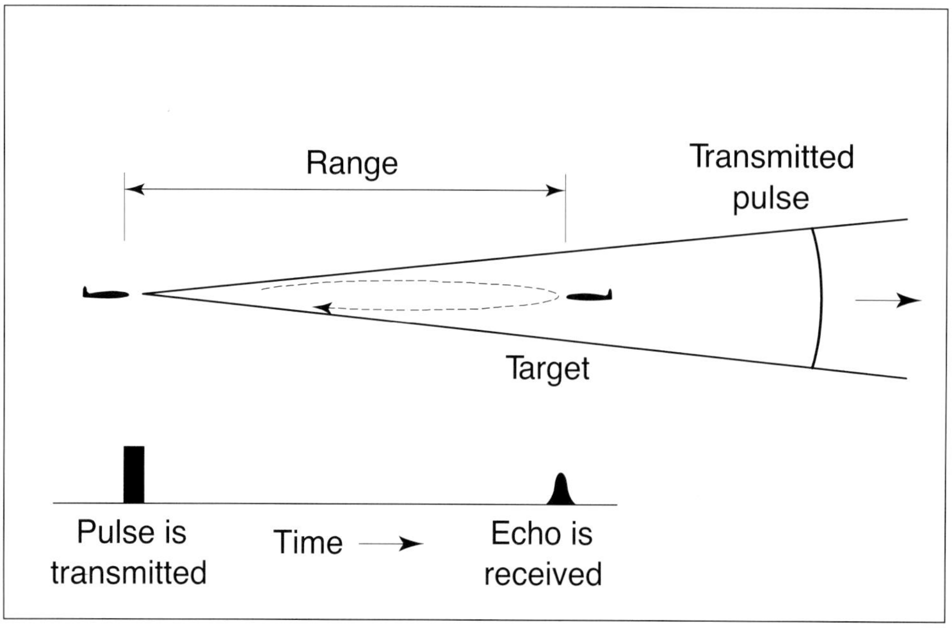

The simplest way to measure range is by measuring the time between transmitting a pulse and receiving its echo. Range = (time ÷ 2) x speed of light. Nearly all of the energy in the pulse carries on past the target.

of a Dornier Do 17Z on a night in July 1940. By May 1941 fifteen specialised night-fighter squadrons were operational, seven flying AI-equipped aircraft.

Transmitter/receiver

A radar operates by sending out some kind of signal using its transmitter and the receiver tries to detect any echoes. If there are any, they give rise to a bright blip on the display, which is arranged so that the position of the blip says something about the target's position. The power of radio waves decays very rapidly; the strength of an echo is inversely proportional to the target's range to the fourth power (i.e $1/R^{1/4}$) and is dependent on its radar cross-section (RCS). Thus, the radar receiver had to be very sensitive in order to detect the extremely weak return from a target, which was a fraction of a millionth of the original output. For example, the radar might transmit 50kW, but receive echoes in microwatts. Because in an aircraft the receiver had to be in close proximity to the transmitter, there was a very real danger of it being damaged by the strong radiated power. The need for high transmitted power, therefore, conflicted with the need for extreme sensitivity at the receiver, and the latter must be shut off during transmission lest it be damaged. This was overcome by using short pulses, with intervals between them long enough for the echo to return to the receiver (i.e. low PRF).

Minimum range

Inevitably, pulses were also reflected back from the ground. Such echoes swamped returns from anything else around that was at a greater distance from the fighter than the altitude at which it was flying. Effective range was therefore limited by altitude, so that no targets could be detected at low level. Whereas altitude determined the effective maximum range of the early radars, they also had a minimum range, within which detection was not possible. This, which has been an enduring problem with all pulsed radars, depends on the duration of each pulse. The result is that echoes from a very close contact would return while the pulse was being transmitted, when the receiver was closed down. The time involved fixed the minimum range.

GCI and IFF

An inland UK form of ground-based radar tracking network was required since Chain Home was early warning only. This led to the ground control of interception (GCI) system, where a ground controller worked at a radar screen tracking both target and night-fighter over a range of up to 90km (50nm). For any reasonable chance of a successful intercept it was necessary for ground control to put the fighter within 5.5km (3nm) of the target, below and astern. GCI started to come into service in early 1941, but one controller could handle only one night-fighter at a time.

Identification Friend or Foe (IFF) systems are an exact electronic equivalent of the sentry's challenge 'Who goes there?' and originated in 1939. The airborne portion of IFF is a transponder (transmitter-responder) which, upon interrogation by the ground radar, instantly replies in code (modifying the echo to give a distinctive blip). This initially crude datalink was used as an effective way of getting aircraft to identify themselves to a radar without their crews having to broadcast for more than a short interval. But like everything electronic at the time, they were not wholly reliable. Moreover, IFF could be triggered by the enemy and systems like the RAF's 'Perfectos', carried by its night-fighters, could provide an accurate range and bearing onto a target's IFF with positive proof that it was hostile.

Radar development

AI Mk IV radar was a considerable improvement whose service introduction coincided with that of the Bristol Beaufighter Mk IF, which outperformed the Blenheim. It was the first really useful AI radar and the only one in large-scale service anywhere prior to 1943. While it was still a 1.5m wavelength set, it incorporated a modulator which, by cutting the time duration of each pulse, overcame the problem of the transmitted pulses being fed directly to the receiver. When night-fighter and target were both at 15,000ft (4,500m), the maximum range of AI Mk IV was 5 to 7km (3 to 4nm) and minimum range 120m (400ft). Below about 5,000ft its range dropped to 1km (0.5 mile), with the ground return almost filling the entire screen, making target detection impossible. Thus any bomber flying low was immune to radar interception so the standard method was for fighters to approach from below and pick out their targets against the sky. Even then, on a very dark night, it was possible to get within AI Mk IV's minimum range yet still be unable to see the target unless its exhaust flames were visible. For German aircraft, with fuel-injected engines, this was unlikely. Because of the risk of a collision, the technique was to synchronise speeds and stay at the minimum limit of the radar while searching visually. Even when visual contact was made, the 'bogey' had to be positively identified.

Centimetric radar

The quest for improved radar performance hinged on three factors: shorter wavelengths, greater average transmitted power and greater signal-processing capability. Despite the rapid progress made, it was apparent that AI radar ought to operate at much shorter (centimetric) wavelengths. Such microwave radar could be miniaturised and designed to send out a beam that could be directed by a reflector like a searchlight. It would reduce the problems of minimum range, clutter and many other difficulties, as well as give a sharper blip. It would also be difficult to jam.

The breakthrough came in 1940 at Birmingham University with the building of a cavity magnetron to give high power, something previously unattainable. This was an oscillator providing enormous amplification to generate power at short wavelengths. Quite early on it was developing 50kW at a wavelength of 9cm. Use of a high-frequency switch enabled the smaller antenna to serve alternately as transmitter and receiver. The single antenna was small enough to be housed within a streamlined nose made of dielectric (non-conductive and therefore radar-transparent) material, reducing drag and improving aircraft performance.

By locating the antenna at the centre of a metal parabolic reflector dish, like that of a car headlight, the emissions could be focused into a narrow 12° beam, thereby directing the energy where it was needed. However, in practice no antenna does this; rather it sends out a pencil beam, flanked by a succession of progressively weaker concentric beams called sidelobes, which could even extend around to the rear. These have been a perennial problem with all radars, which designers try to eliminate, since they not only waste power but are also susceptible to clutter and enemy

Mosquito II in an overall black finish for its night-fighter role, fitted with AI Mk V radar with nose-mounted arrow-head antenna. Compared to the AI Mk IV, this differed in having a pilot's radar 'scope' in addition to those of the operator. *(Philip Jarrett).*

jamming. So that the main beam could look in directions other than straight ahead, the reflector was pivoted and mechanically driven in a pre-set pattern, scanning over a cone-shaped area ahead of the fighter. One of the most difficult mechanical engineering tasks was to achieve sufficient precision, speed and angular movement to point the beam.

Not only did the new AI Mk VII (10cm wavelength) radar provide better definition and angular accuracy, but by training the beam ahead, typically 45° around the centreline, it ensured that ground returns would only be picked up from a relatively long distance ahead. Targets lower than the fighter were still lost in the clutter making their interception no easier, but by eliminating its worst effects, reasonable detection distances could be achieved. Very low-level targets were still undetectable unless the fighter could descend to a lower altitude so that its radar was looking upwards. However, flying low at night or in bad weather was a hazardous business. The greatest disadvantage of centimetric radar was that the volume of sky scanned was far less than that of earlier radars, though the effective range at normal operating altitude was greater. Beaufighter Mk IFs with the GEC AI Mk VII installed in 'thimble noses' were operational before the end of 1941.

Production model AI Mk VIIIs gave appreciable improvements: 15km (8nm) maximum detection range, 100m (300ft) minimum range and reduced ground clutter. Whereas with AI Mk IV, GCI had to direct a fighter in from astern to make contact, the new longer-ranged radar allowed it bring the fighter in from ahead, handing over as soon as contact was gained. GCI involvement thereby became much shorter, enabling individual controllers to handle more interceptions. It started to reach squadrons in summer 1942 but for reasons of secrecy was released for operations over enemy territory only in May 1944 as part of a bid to attain air supremacy over the Normandy D-Day invasion area.

German radar

Night-fighting had been neglected during Germany's rise to power in the 1930s. By the outbreak of World War Two only a small number of units were assigned to this mission that relied on elite corps of anti-aircraft artillery and searchlight troops, although experience soon showed that their effectiveness had been overestimated. Consequently, although the RAF started night bombing in May 1940 and within five months Germany was able to introduce a limited radar-directed ground control for its night-fighters, it was not until the end of 1942 that Germany possessed a force able to seriously threaten the RAF.

Virtually all German radar effort was applied to large sets for ground and ship applications, until late 1940, when Telefunken began work on a radar suitable for aircraft. German research into centimetric microwaves in the 1930s had yielded disappointing results and work had been abandoned, but for this new application Telefunken used a wavelength of 62cm (490MHz). By 1942, the production version FuG 202 Lichtenstein B/C with a minimum range

Mosquito NF XIII HM428/'RO-K' of 29 Squadron, with thimble-nose housing a 10cm wavelength AI Mk VIII radar. The shorter wavelength radar allowed much smaller transmitter and receiver aerials which permitted a reflector dish to be fitted behind the aerial. This focused the beam into a relatively narrow 12° beam. *(Philip Jarrett).*

Arguably the Heinkel He 219A *Uhu* (Owl) was the best German night-fighter of World War Two and certainly the only one capable of meeting the Mosquito on equal terms. Typically, it had two radars: SN-2 aerials surrounding the FuG 212 antennae. Flying on its operational debut on the night of 11–12 June 1943, one shot down five Avro Lancasters within 30 minutes. Despite this, the aircraft never entered quantity production with fewer than 300 being built. *(Philip Jarrett).*

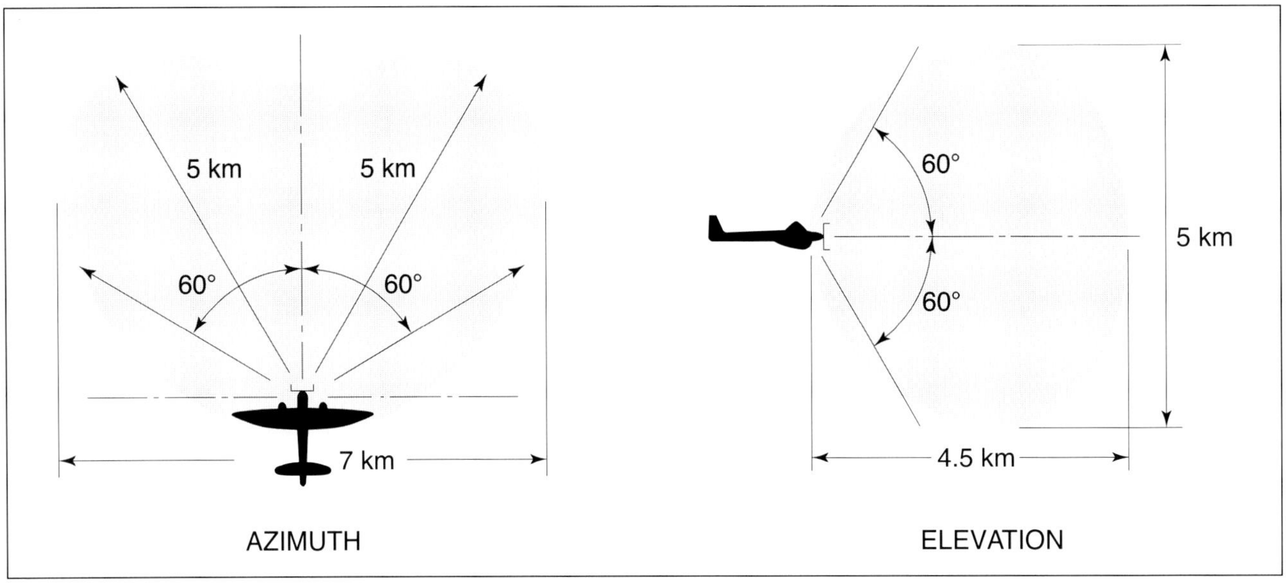

Angular coverage of the Telefunken FuG 220 Lichtenstein SN-2 radar. Weight of the internal units was about 70kg, and being of 3m wavelength it needed an extensive array of dipole aerials. The adverse effects on drag and handling, especially of the Bf 110G-4 series, was quite considerable. Power was 2.5kW, angular coverage was good, but its major drawback was the minimum range of 400m, which most crews considered too high for visual sighting of targets.

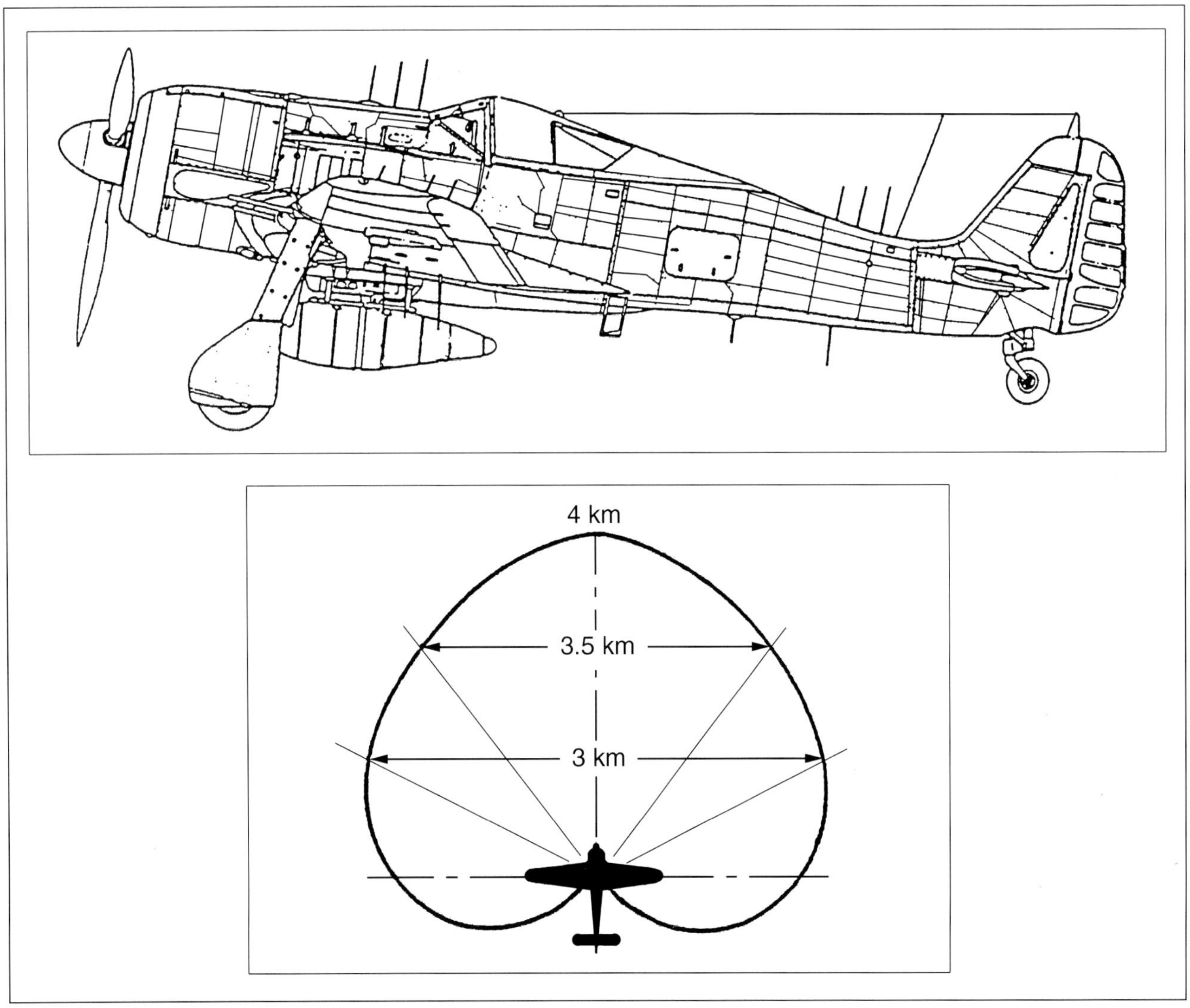

The FuG 217 'Neptun' was fitted to some Fw 190 night-fighters from early 1943, making it one of the first single-seat aircraft to become operational with an AI radar. It had been developed from the FuG 216 tail-warning radar, and used an array of dipole aerials on upper and lower wings and fuselage. A single radar 'scope displayed target range, azimuth and elevation. The FuG 217's radiation pattern and range are shown above.
(Reference: SAE 965631, Los Angeles, 1997).

The US had a limited requirement for night-fighters. The Northrop P-61B Black Widow with two Pratt & Whitney R-2800 Double Wasp radials with a maximum speed of 360mph was armed with four 20mm fixed cannon in a ventral bulge and four 0.5in machine-guns in a remotely trainable dorsal barbette. This was often deleted due to severe buffeting and even when retained it was locked in the forward firing position. Despite its size (that of a B-25) and weight (heavier than a fully loaded He 111), it had good performance above 25,000ft, better than the Douglas P-70 that it replaced. Initial pilot reaction was unfavourable however, because of its lack of manoeuvrability. It was first delivered in May 1944 and carried the SCR-720 centimetric radar. *(P-61B-15 via author).*

of 182m (600ft) enabled night-fighters to operate effectively in conditions of poor visibility. It was fitted to Messerschmitt Bf 110Fs (though its high-drag aerial array, arranged in four groups each consisting of four vertical dipoles backed by four identical dipole reflectors, knocked 25mph off the Bf 110's speed) and the Junkers Ju 88-6bs, and became widespread in 1943. A good radar for its day, the FuG 202 was marred by unreliability and many other shortcomings that caused it to be widely scorned (it required a third crewman to be squeezed into a Bf 110). Nevertheless, RAF bomber losses climbed steadily, with the attrition rate reaching 4% by the end of the 1942, although 10% was needed by the *Luftwaffe* to force a change in the RAF's tactics.

In late 1943, Telefunken began delivering a new radar altogether, FuG 220 Lichtenstein SN-2, which was unaffected by the version of the 'Window' electronic countermeasure (ECM) then in use by RAF bombers. ('Window' was metal foil cut to appropriate radio-frequency length; it was known as *Duppel* by the *Luftwaffe* and is now known universally as chaff.) Rather than using a shorter wavelength and a smaller antenna, SN-2 used a longer wavelength, greater than 3m. This meant that it needed an antenna array even more cumbersome than before, with attendant adverse effects on performance and handling, especially of the Bf 110G-4 series. A further major drawback was its minimum range of 350m (1,150ft), but angular coverage was good and maximum range was about 7km (4nm).

American radar

America was given the magnetron valve in December 1940 and the US Office of Scientific Research and Development considered the gift so important that in its official history it described the device as 'the most valuable cargo ever brought to our shores'. Within two months the Radiation Laboratory at the Massachusetts Institute of Technology (MIT) had a prototype 10cm radar working, with the British magnetron uprated to 100kW. Before the end of 1941 pre-production sets called SCR-520 were flying in Douglas A-20 Havocs, and in early 1942 production switched to the SCR-720. It was a marriage of the AI Mk VIII and the SCR-520, but with the important addition of anti-jam circuits, two display modes and the ability to close the scan (reduce the field of view) once the target had been acquired. This ability to search a small sector of sky and concentrate on one target allowed it to be 'painted' far more often, making tracking easier even in the presence of jamming.

Whereas AI Mk VIII had a fixed aerial and only the dish moved, in SCR-720 the aerial was fixed inside the reflector dish. The scanner rotated rapidly through 360° about the longitudinal axis, and the whole assembly was in turn gimbal-mounted, which allowed it to nod up and down as it rotated. Nod angle was adjustable in flight, and against low-flying targets the down angle could be reduced to minimise ground returns. After tests on a de Havilland Mosquito fitted with SCR-720 against another fitted with AI Mk VIII, the US radar was adopted by the RAF as AI.10. It was the most advanced AI radar of the war. SCR-720 was also fitted in the nose of Northrop P-61 Black Widows that became operational in 1944. As AI.10 it remained the RAF's 1950s night-fighter radar on Gloster Meteor NF.11s and de Havilland Venom NF.2s.

The US Navy had championed the concept of the single-seat night-fighter (the only sort suitable for carrier-based operations), and in early 1941 issued a far-seeing requirement. This was for a radar operating on the even shorter wavelength of 3cm, and by late 1942 the resulting APS-4 radar was in production for Grumman F6F Hellcats. It weighed only 135kg (300lb) and the antenna, transmitter and receiver were compactly housed in a pod that could be fitted to the fighter's wing leading edge. Both elevation and azimuth information were displayed on the same CRT. Maximum acquisition range was about 9km (5nm) but early experience showed that contacts were usually made at about 4km (2nm). The radar featured an ability to reduce scan width from 100° in search to 15° during attack. The high speed of the Corsair, to which it was also fitted, often resulted in overshooting as it was difficult to slow the fighter quickly. Tactics were changed to bring the Corsair in behind and below the target and then slow down in the pull-up to engage.

Electronic warfare

To counter radar, the whole new science of ECM was created. It was known that radar could be jammed by metallic strips, cut to a length matching the wavelength of a hostile radar. When dropped, the strips (chaff) produced returns similar to those of real aircraft, thus cluttering up radar screens with multiple false contacts. It was regarded as so powerful a ruse that both sides refrained from using it for a considerable period, in order to avoid retaliation in kind. Nevertheless, the RAF was determined to introduce a radar that was less susceptible to jamming of this nature, and this duly arrived from the US under Lend-Lease as the AI.10.

Having established that German FuG 202 could not operate successfully in a chaff environment, the RAF at last started using its 'Window'. On 24 July 1943 the RAF launched a large raid over Germany whose early-warning and tracking radars filled up with contacts as over 700 bombers suddenly appeared to blossom into many thousands. Just twelve RAF aircraft went down that night from all causes, an attrition rate of 1.5%, and subsequent raids produced similar results. While 'Window' did not prevent the German radar from following the bomber streams it did prevent tracking individual targets. Freelance day-fighters could, however, infiltrate the stream on its way to and from the target. It seemed that in seeking to defeat the night-fighters the RAF had forced them to adopt a more efficient *modus operandi*. Spoofing of raids and R/T and jamming was used to overcome this. The *Luftwaffe*'s '*Duppel*' was introduced in October 1943, but the German raids of early 1944 still suffered 8% attrition.

An obvious choice for the night-fighter role, the P-82B Twin Mustang entered production in 1945, but was too late for the war against Japan. Whereas in the escort role it carried two pilots, the starboard cockpit was fitted out for the radar operator. A radar pod was suspended beneath the central wing, protruding far enough ahead to bring the antenna ahead of the propellers. In all, three night-fighter variants entered service. The P-82F carried AN/APG-28 tracking radar and the P-82G (seen here in Japan during 1949–50 with the 347th FG) and H both had SCR-720C search radar. *(via Philip Jarrett).*

The conversion of the Gloster Meteor T7 into a night-fighter was undertaken by Armstrong Whitworth and entailed quite radical changes. A lengthened nose housed the radar, and to offset the directional instability this caused, Meteor F.8 tail surfaces were adopted. However, the revised nose displaced the four 20mm Hispano cannon which were relocated in wider span wings outboard of the engines. When fired the recoil forced the wings to flex with a loss of accuracy. The visibility from the cockpit was poor due to the surplus of heavy framing. The Meteor went through four night-fighter variants ending with the NF.14 (as shown belonging to 85 Sqdn, West Malling, 1954–57), which had a clear canopy, improved environmental control, more powerful engine and the Westinghouse APS-57 radar, known as AI.21 in British service. *(via author).*

Soviet radar

During World War Two very little progress was made by the Soviet Union in radar, and apparently none in AI. The first recorded use of radar by the Soviet Air Force is 1944, in the form of primitive mobile ground units. However, with the cessation of hostilities a great deal of German data and equipment fell into Soviet hands, with two workable types appearing by 1948/49. One was the forerunner of the *Izumrud* (Emerald) series which became very widely used to give day-fighters (e.g. MiG-17PF and MiG-19) a marginal amount of adverse weather capability. The Soviet Union's first real night-fighter radar was a larger one (a 0.75m diameter dish operating at about 3cm wavelength) with more than one mode offering high transmitter power and reasonable range. In 1953 the Soviet Union had flown the prototype of its first adverse-weather jet fighter, the Yak-25, but by the end of the decade it still had no effective adverse-weather fighter force, nor had it a comprehensive command and control system.

The Jet Age

Radar lock-on

After the ability to close down the radar scan width came lock-on, which did not enter service until after World War Two. In its earliest form, the antenna was controlled by the operator who pointed it directly at the target. This was done by projecting a small circle onto the display, then driving the scanner until the target was enclosed in it. The lock-on mode was then activated, commanding the set to track the target automatically, by keeping the signal strength of the echo, which of course varied during the radar's conical scan pattern, at a maximum. However, it was the development of lock-on that led to deception jamming. If lock could be broken at a critical moment in the interception, the operator was forced to start again.

Fire control systems

Historically, up to and including the Korean War, just 5% of fighter pilots accounted for 40% of kills, demonstrating that pilot quality is a dominant factor in air combat. Of the rest 50% achieved no kills. The problem was how to make the remaining 95% more effective, for which the only answer had been better training. Electronics now promised an additional solution. A newcomer to the electronics field, Hughes Aircraft Company, had designed its E-1 fire control system in which a pioneer computer was coupled to the APG-32/33 radar. It went operational in the Lockheed F-94 Starfire in 1950. By telling the pilot when he had a firing solution, it enabled interception and attack to be made blind and marked an important milestone in the development of the adverse-weather fighter. The E-1 was later modified to allow attacks from any angle including collision course (see Chapter 9). If attacks could be made from ahead or abeam where the target RCS was much bigger, the jockeying for an astern position and the ensuing long tailchase could be eliminated.

A refinement of the Hughes E-5 fire control system as

The two-seat Yak-25 'Flashlight', which first flew in 1953, had two slim Mikulin AM-5 axial-flow turbojets set in nacelles beneath a 40° swept untapered wing. The Soviet Union was a late starter in the field of AI radar and the Yak-25 had to use the pursuit course, visual contact method of interception. The large blunt radome housed 'Spin Scan' radar but the Yak-25 would have been hard pressed to intercept a Canberra let alone a V-bomber or B-52. One thing in its favour was the claimed long range of 1,620nm by virtue of its 12,000lb fuel load giving it a fuel fraction of 0.33. Like most Soviet designs it went through a series of modifications including more powerful AM-9 turbojets, a larger wing and more streamlined nose. It remained in production until 1958 after 1,000 had been built and was not phased out until the mid-1960s. This one was engaged in ejection seat trials. *(Alexei Mikheyev via author).*

In 1955 a limited adverse-weather capability was provided to the Soviet Union by the MiG-17PF, fitted with RP-5 *Izumrud* (Emerald) AI short-range radar. This operated in S-band (2-7GHz), initially in just one mode and for this reason it was given the name 'Scan Fix' by NATO. The scanner dish was very small (housed in the air intake splitter) and the receiver was separated from it (in the intake lip extension). This is one of the 129 Lim-5PM (MiG-17FPM) built in Poland during 1959–60. *(Ryszard Malachowski via author)*.

The Gloster Javelin, which first entered service in 1956 with the Westinghouse AI.22 parabolic dish radar antenna, was armed in its later versions by de Havilland Firestreak IR missiles. It was a heavy two-seat aircraft whose performance did not approach contemporary day-fighters. In the mid-1950s, the important innovation in Britain's GCI was the introduction of master radar stations, which combined the functions of the early warning and GCI stations with sector operations. This improved the reaction time of the air defences since the same unit that received warning of attack became responsible for scrambling fighters to meet it and directing the subsequent interception. Javelin FAW.9 (XH887) of 23 Squadron carrying four Firestreaks with two 30mm cannon in the wings, each just outboard of the inner pylon. *(via author).*

fitted to the Lockheed F-94C was that the radar was linked with the autopilot so that attacks could be made semi-automatically. This was a revolutionary step. Once the radar had acquired the target and started to track it, the computer would continuously calculate speed, distances and angles, and give automatic steering onto the target, so that the rockets could be fired (it was not accurate enough for guns). However, it gave a rough ride with high g and roll rates, and most pilots preferred to fly manually. For all its advanced features and perhaps because of insufficient development it was unreliable, and the F-94C was not a success. Yet such were the problems of finding a satisfactory successor that it remained in frontline service with USAF's Air Defense Command (ADC) until early 1959.

Datalinks

Notwithstanding its problems of acquiring satisfactory adverse-weather interceptors, ADC had grown in strength to an impressive degree by the mid-1950s. There were fifty-five fighter interceptor squadrons under its control by the end of 1954. An important increase in ADC's strength was made possible by the Hughes E-4 fire control system. Its computer linked the Westinghouse APG-36 radar for use in a single-seater, allowing adaptation of the much higher-performance North American F-86 Sabre day-fighter for the adverse-weather mission. The resultant F-86D entered service in early 1953, and by the end of that year 600 fighters of the type were in service.

By mid-1955 over 70% of the 1,400 interceptors in ADC's inventory were F-86Ds, most of which were converted to F-86L configuration. The principal change in this model was the fitting of a datalink. This is non-speech radio communication, whose genesis can be traced to IFF. The datalink enabled the receipt of information from the semi-automatic ground environment (SAGE) system. This was an overlapping system of radars and GCI to form a computerised command and control system that could monitor an entire air defence battle over the USA. It marked the peak of integrated ground radar and control centre development in the mid-1950s.

Automatic interception

Despite the difficulties of adverse-weather fighter development, the US Air Force kept its sights on its requirement for the 'ultimate' interceptor, first formulated in 1950. With the high-flying supersonic bomber just over the horizon, the next generation of fighters had to fly faster and higher. Interceptions would become even more a matter of split-second timing, calling for more highly automated fire control systems. Moreover, guns and unguided rockets were about to be supplemented by guided missiles. The first fighter designed to meet these requirements was Convair's F-102A Delta Dagger. It entered service in 1956 with the Hughes E-9 fire control system, an improved version of the E-4, and was retrofitted with the MG-10 fire control system in 1958. Digital guidance information was uplinked from the SAGE network into which it was hooked. This was converted into analogue form for display to the pilot and guidance of the interception. Once locked on, the MG-10 was designed to automatically guide the fighter, open the weapon bay doors and fire its missiles.

Potentially much more capable, the Convair F-102B (later designated F-106A Delta Dart) needed to have an electronic navigation and fire control system to give virtually total automation of each intercept mission. Target information was datalinked from SAGE and fed directly into the aircraft's Hughes MA-1 fire control system. Starting by translating the target-vectoring information, the MA-1's computer calculated steering instructions and range closure rates. It guided the aircraft before and after radar lock-on, once the pilot had designated the target on his radar scope, by superimposing a gate symbol over the radar return. A bonus was that the tie-in to SAGE now included the autopilot that could thus be controlled from the ground. SAGE therefore managed the attack and recovered the aircraft back to base, navigating by Tacan and homing in on pre-selected beacons. The pilot thus became more or less a systems manager. However, the F-106A's development was fraught with delays. Even after it entered service in 1959, a decade-long programme of modification and modernisation was necessary before it matured into a reliable interceptor.

Pulse repetition frequency

When radar is referred to as operating in a certain band, this refers to the transmitter's oscillator frequency. The one selected for a radar tends to be related to the antenna's physical size. Thus airborne radars, which have to be compact, tend to operate at high frequency with most AI types operating in bands between 8GHz to 20GHz, far above normal radio-listening bands. This frequency range offers the best compromise. It is not too high that the pulses suffer from excessive signal absorption by the atmosphere, and yet it is sufficiently high to generate adequate gain to produce a finely controlled radar beam without unwanted big sidelobes. This frequency has no connection to pulse repetition frequency (PRF). PRF is a totally different, and much lower, frequency. As its name implies, it is the number of radar pulses sent per second. Nowadays, PRFs vary from 2kHz to 30kHz, and since the factors affecting their choice are complex they are only briefly addressed here. Another important measure is pulse repetition interval (PRI), which may be anything from 100 to 10,000 microseconds. Being able to vary PRI is an extremely powerful way of exposing all targets.

The first of three Lockheed YF-12As flew at Groom Lake in August 1963. Its fire control system was the Hughes ASG-18, originally developed during 1956–59 for the North American F-108 Rapier. It was probably the first AI radar to have a slotted-waveguide planar antenna and the first fully developed coherent PD radar to fly in a combat aircraft. The technology had been made possible by replacing the magnetron with the newly developed TWT, allied to advances in digital computers. The ASG-18 provided not only the knowledge for the AWG-9 produced by Hughes for the F-14A, but also the starting point for the APG-63 for the F-15A and (with a programmable processors and other updates) F-15C, the APG-65 for the F-18, the APG-70 for the F-15E and the APG-71 for the F-15D. Though these are totally digital, have vastly greater computing power and many other improvements, they all work in basically the same way as the pioneer PD sets of the late 1950s. *(Lockheed Martin via author).*

Pulse-Doppler radar

Until the late 1950s, the magnetron had been the heart of radars in sending out powerful short pulses at low PRF. It had the advantage of providing unambiguous range data, because all the echoes from each pulse were received before the next was transmitted. However, magnetrons do not maintain coherence from pulse to pulse (i.e. each pulse is not exactly in phase with its associates), so that velocity data could not be extracted. Radars using them detected high-flying targets well because range data rather than velocity data are the criteria. Moreover, the need for only modest signal processing resulted in little signal loss and good detection efficiency. However, the low PRF magnetron-based radars had no way of extracting velocity data, so when they looked down for low-flying targets they were completely overwhelmed by clutter and could not distinguish the target from the ground. The need to detect targets at low level assumed very great importance from about 1960 onwards. This was where the crews of the new 'under-the-radar' strike aircraft such as the Blackburn Buccaneer knew the defensive systems had their greatest difficulty of detecting them, so that's where they were going to be.

The search for a solution to look-down capability had already begun in 1949 for the USAF's IM-99 Bomarc long-range interceptor missile programme. With its own active radar it was to intercept enemy aircraft far beyond the horizon and GCI, using a seeker that could extract target echoes from ground clutter. The solution was found in 1951, leading to the first pulse-Doppler (PD) radar, virtually universal in today's fighters.

Doppler shift is the apparent change in frequency of a signal caused by the relative motion between the target and the radar. As a target approaches more reflected signals per second reach the radar's antenna, and as it recedes the rate decreases. Thus, relative target velocity creates a tiny shift in frequency that the radar had to detect and discriminate from ground returns. What was needed was to measure very accurately the Doppler shift from the ground passing under the target, then filter out all echoes moving in conformity with it, within pre-set limits. Anything left over would almost certainly be a target. The problem was one of precision. While the incoherence of the magnetron was tiny, it was too large to provide the level of accuracy required to make use of the Doppler effect.

The solution was the travelling wave tube (TWT), not a new invention since it dated from 1942, but its potential for AI radar could only be exploited when higher-speed computers became available. By passing the emissions from a continuously running ultra-stable oscillator through the TWT (which, like the magnetron, vastly increases the power of any signal fed into it), then chopping it into very precisely timed pulses (at rates up to 10,000 pps), each pulse would be exactly in phase with the others. This gave the needed coherence to enable Doppler shift to be extracted.

The problem of ranging

The other advantage of the TWT was that while it could not produce the high peak power of the magnetron, it could be run at a much higher PRF. By emitting more pulses per second it could radiate higher average power than the magnetron, thereby giving longer detection range. However, to create the necessarily clutter-free space around the target, the PRF had to be about 100 times greater than that of contemporary fighter radars. Such a high PRF would make range finding almost impossible and be virtually useless in beam attacks where there is little Doppler shift.

The MiG-25 'Foxbat' was designed as a counter to the CIA's Mach 3 Lockheed A-12, and possibly the SR-71 with its much smaller RCS. The Soviet answer was high transmitter power as in 'Smertch-A' which in the absence of transistors used thermionic valves, put out 600kW that could reputedly kill rabbits at 200m. Even so, it was as good a way of detecting low RCS targets and could burn though jamming. Radar range was about 50nm, half that of the YF-12A's ASG-18. Smertch-A, although far from being a coherent PD type had a certain amount of look-down capability. Intended to fly a fully automatic interception, 'Foxbat' was tied to ground control by digital datalink, which took it from shortly after take-off to just prior to landing, with the pilot acting as systems manager. *(Alexei Mikheyev via author).*

This is because the traditional ranging method, based on measuring the time between pulse and echo, meant that with so many pulses being emitted, the radar would not be able to distinguish which pulse was responsible for which echo. So while there was now a radar that could detect targets in look-down, it could not tell their range nor help beam attacks. The solution was to impress a lower frequency modulation (FM) on each pulse making each echo instantly recognisable from another. This, however, is at the expense of reduced detection range because of the complex processing required of the received signals. High PRF pulse Doppler modes were first introduced by the AN/AWG-10 radar of the F-4J Phantom.

Medium PRF

By the early 1970s AI radars combined the accurate velocity discrimination of high PRF (PD) with the accurate ranging of low PRF. The next requirement was for a waveform that combined both features. Use of medium PRF, between 6 and 16kHz, might have looked rational, but filtering the greatly increased number of signals was made possible only by the digital signal processor. These can be automatically adjusted to match the changeable waveform being transmitted, in terms of pulse rate and actual frequency. The higher the PRF the greater the detection range, while the lower the PRF the more accurate the velocity data. A succession of medium PRF bands is used to provide high-quality tracking data, while high PRF with FM can be interleaved with medium PRF for long-range detection. It is thus more flexible than high PRF, able to look down, reject clutter and match rapidly changing target speeds. That said, signal-processing losses are high,

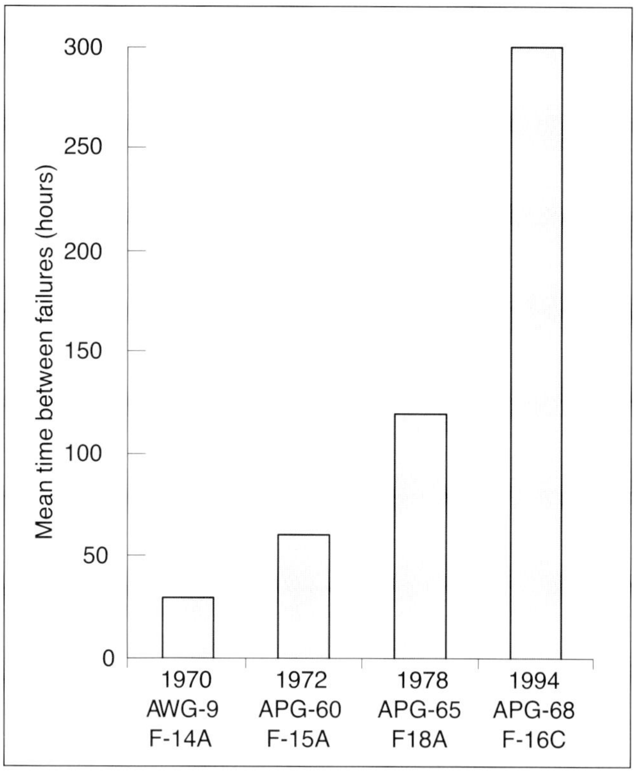

Reliability improvements over time of modern radars. Whereas the figures shown for the first three radars refer to the date of aircraft first flight, the APG-68 is the version with an advanced programmable signal processor as fitted to the Block 50 F-16C/Ds.

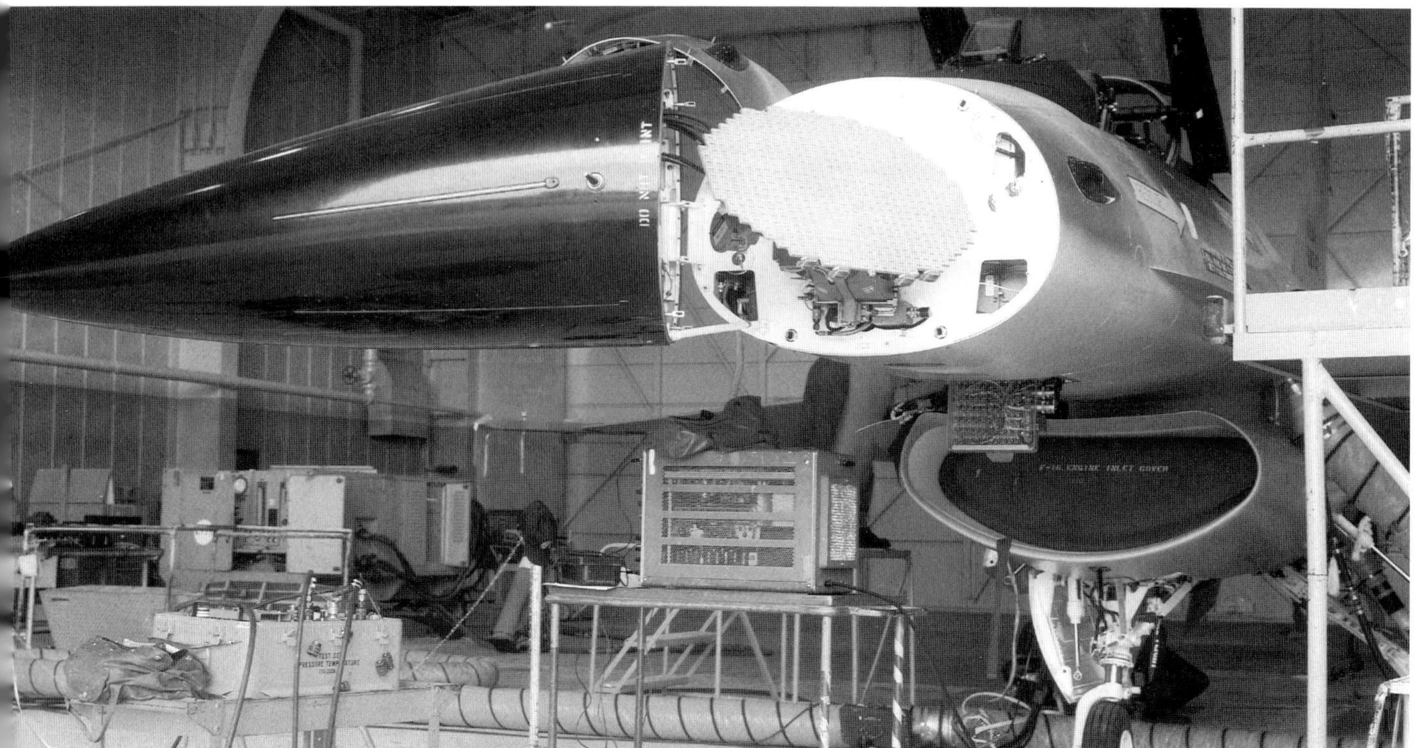

Northrop Grumman's Electronic Sensors and Systems Div. (ESSD), formerly Westinghouse Electronic Systems, has produced about 4,000 AN/APG-66 and 2,500 AN/APG-68 radars since 1976, primarily to equip the F-16A/B and F-16C/D, respectively. The APG-66 equips a total of 16 platform types, however, and is the baseline equipment for the BAe Hawk 200. The radar will remain in production until at least the year 2000 and ESSD has committed itself to providing support a further 20 years beyond that date.

Table 9 Simplified characteristics of the various (chronological) PRF modes

PRF (Type)	*Advantages*	*Disadvantages*	*Early example*
Low 2kHz (Monopulse)	Precise ranging, minimum signal loss due to simple data processing, long look-up detection range, cheap	Poor in look-down due to massive ground clutter in which targets can hide	IA.23 (Lightning), Cyrano II (Mirage IIIC)
High 30kHz (PD)	Good in look-down against high closure rate targets, long-range detection	Poor at ranging, poor for low closure rates (e.g. beam attacks), needs low duty factor since receiver is blind while transmitter emits	AWG-9 (F-14A), AI. 42 Foxhunter (Tornado F.3)
High (PD with FM)	Better in ranging and beam attacks	Signal loss due to signal processing reduces detection range	
Medium (6–16kHz)	OK in look-down, best compromise for medium-range combat	Needs high data-processing capability resulting in high signal loss which reduces detection range	APG-63 medium & high PRF (F-15A)

Three PRF ranges are used by the best air-to-air radars. In the past, fighter pilots knew which mode worked best under given conditions of altitude, speed and attack scenario and selected them accordingly. Today, the pilot flies the fighter and leaves the radar to decide how it scans. In the search mode the processor may choose to use medium PRF looking down, a high or medium PRF on the level and a low PRF looking up. When the processor has decided a target has been detected it puts a clear bright symbol on the head-up display (HUD) or dedicated radar display.

reducing maximum detection range, but medium PRF offers the best compromise for medium-range combat.

Medium PRF is the cornerstone of fighters like the F-15, the first to add this capability with its Hughes AN/APG-63 medium PRF-focused PD radar in 1974. Microprocessing technology is largely responsible for changing this, and providing truly capable multimode radars.

Major features of radar antennas

The parabolic dish remained in widespread use well into the 1960s, and was hydraulically driven. The next step was a variation on a theme, the inverse-Cassegrain antenna. This was again of parabolic type, but with a sub-reflector placed in front. The most common antenna in current fighters is the planar array which is a flat plate constructed so that it emits through slotted apertures across its face. Almost all the radiation is in the main beam, with very small sidelobes. The latest type of antenna is the phased array in which the beam is pointed electronically and does not have to rely on mechanical movement of the scanner to point at the target. This is called electronic scanning, which has many advantages over conventional antennas. Smaller

The Su-27 'Flanker's' large radome houses a multimode PD radar with an antenna only slightly smaller than that of the MiG-31 'Foxhound'. Estimated search range was 130nm with tracking at 100nm for a large target. Troubles with the radar delayed service entry until 1986, and it was reported to have a high alarm rate and a short time between failures. Only one target at a time could be engaged and discrimination between targets in formation is reputedly poor due to its lack of onboard signal processing capacity. As for the MiG-29, an IRST is an integral part of the fire control system, with the seeker mounted to starboard, ahead of the windscreen, where it hardly obstructs forward field of view. *(Su-30 of the Russian Air Force via author at IAT'97).*

LEFT and BELOW: The radar for the Dassault Rafale is the RBE2 developed by Thomson-CSF and Dassault Electronique and was the first phased-array type in Western Europe. Electronically-scanned antennas comprise a large number of transmitter/receiver elements across the face of the antenna, each of which can transmit or receive a portion of the total power. They do not need to move since beam pointing and shaping is done nearly instantaneously by electronic phase shifters. These control the amount of delay from the radiating elements. Of look-down/shoot-down multimode type, RBE2 is stated to have a detection range against fighter-size targets of 54nm (100km) in the look-down mode.
(Dassault Electronique via author).

Table 10 Antenna types

Antenna	*Advantages*	*Disadvantages*	*Early example*
Dipole (TV aerial-type rod)	Limited by RF wavelength and all that was appropriate at the time	High aerodynamic drag, separate transmitter and receiver antennas, unsteerable	AI Mk III (Blenheim IF) 1940
Parabolic dish reflector (car headlight type)	Smaller and could be internally housed in radome, steerable, higher gain, smaller sidelobes concentrating energy into main beam	Complex gimbal system needing mechanical drive	AI Mk VII (Beaufighter IF) 1942
Planar array (flat plate with slotted emitter apertures)	Higher gain for given size, much smaller sidelobes, lighter so lower inertia enables faster pointing	Narrow bandwidth, higher cost, problems with polarisation	AWG-9 (F-14A) 1972
Phased array (mechanically fixed, electronically steered by many individually controlled transmit/receive modules, likely to have limited mechanical movement in future)	Eliminates mechanical drive giving instantaneous pointing, multiple modules so less prone to total failure, multiple frequencies to allow several scanning modes simultaneously, reduces time for emitting and increases time for listening, reducing the chance of being detected	Performance degrades at extreme-look angles but will be married to a mechanism to give physical movement	SBI-16 'Zaslon' (MiG-31) 1983

integrated circuits allow more smaller transmitter/receiver modules to be fitted with more per unit area (perhaps 1,000 on a fighter-sized unit). Each module can be individually controlled, with transmission and reception split 50/50, allowing multiple frequencies to be used simultaneously. This gives little time for transmitting and a long time for listening, which together with controlled beam shaping to reduce unwanted sidelobes yields a low probability of intercept, thus reducing the chance of being detected or jammed. Active array radar has comparatively little thermal noise, which increases detection sensitivity and reliability. Resolution is fine enough to give non-cooperative target recognition and is the type to be used on the Dassault Rafale and Lockheed Martin F-22.

Growth of avionics

For many years, making radars more capable made them bulkier and heavier. They needed more volume, more power and vastly more maintenance manhours. With over a thousand thermionic valves the cooling requirements also grew. The situation was aggravated when the US began to couple radar and autopilot to produce fire control systems linked to SAGE. Invented in 1948, the transistor took several years to become available for operational hardware, but when it did the results were remarkable. The ensuing miniaturisation allowed capability and reliability (even under adverse environmental conditions) to be vastly improved, and bulk, weight and power requirements to be reduced. As can be seen in Table 11, though, even with the mighty microchip, the demand for enhanced potential has increased the weight fraction allocated to avionics.

Electro-optic sensors

In close combat, the human eye has always been the favoured method of instant identification, but it is limited by distance and ambient light. Since 1940, avionics has increasingly made use of ever-shorter wavelengths, moving into the field of electro-optics (EO), which is electronics at or near optical wavelength. EO devices have been used to expand both range and vision at low ambient light levels.

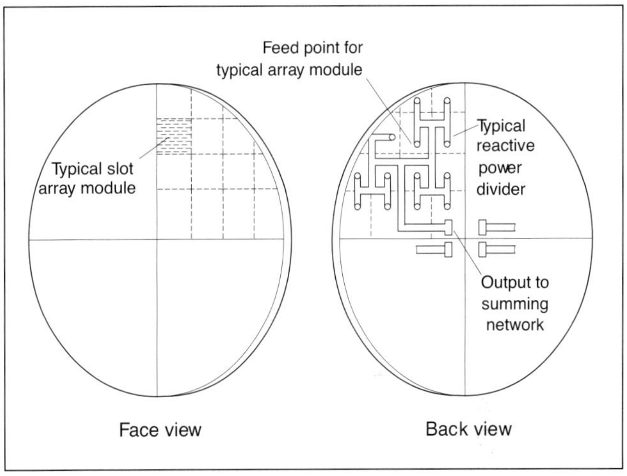

Instead of a dish reflector, most modern radars use a planar-array or slotted waveguide antenna. This is flat and radiates signals of equal phases from accurately positioned slots in its face. Typical examples are Westinghouse APG-66, GEC/Marconi Blue Vixen and Phazotron Kopyo.

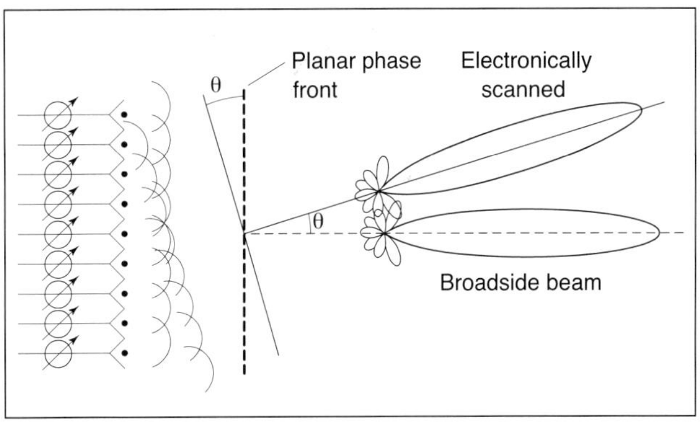

Electronically-scanned antennas comprise a large number of transmitter/receiver elements across the face of the antenna, each of which can transmit or receive a portion of the total power. Also called phased-array antennas, they do not need to move since beam pointing and shaping is done nearly instantaneously by electronic phase shifters. These control the amount of delay from the radiating elements. Used on Thomson-CSF/Electronique Serge Dassault RBE 2 PD for Rafale and Grumman APG-77 for F-22.

Table 11 The growth of avionics

Aircraft (year)	*Avionics weight kg (lb)*	*% Aircraft empty weight*
F3H-2 (1956)	410 (900)	4.2
F-101A (1956)	450 (1,000)	3.8
F-102A (1958)	770 (1,700)	7.1
F-101B (1959)	650 (1,425)	5.0
F-106A (1960)	820 (1,800)	7.4
F-4A (1962)	620 (1,360)	4.9
F-15A (1974)	750 (1,650)	5.7
Rafale D (2000)	800 (1,750)	8.6

Electro-optics devices have been used to expand both range and vision at low-light levels and some early and purely optical telescopes were used. For example the target identification system/electro-optic (TISEO) was mounted on the port wing on over a hundred F-4Es in the early 1970s, and the TV-linked system on Grumman F-14s which was useful for giving target identification beyond visual range (up to twenty miles).

Infra-red (IR) detection of aircraft by searching for hot targets dates back to 1940. After World War Two, various design teams researched IR homing for air-to-air missiles. However, it was not until 1960, a gap of twenty years since their limited wartime use by the *Luftwaffe*, that IR sensors reappeared on fighters, the Convair F-102A Delta Dagger, F-106A Delta Dart and F-4B. Both the Convair aircraft were equipped with infra-red search and track (IRST) sensors with the same seeker head as their Hughes AIM-4D Falcon missiles. IR sensors are compact and well suited to military use because they are passive and simply look for contrasts in normal radiation which they can do either at day or night, though they have serious limitations in bad weather. Although easily overlooked, they are now increasingly recognised as an invaluable device on a fighter. Because they work at the much shorter IR wavelength they have better angular discrimination than radar and can pick out individual targets in a tightly packed formation. A further possibility is to use the IR image to identify targets. To be fully effective the IRST must be integrated with the aircraft, so that track data is available to the missiles for accurate targeting. They lack any ranging capability in their own right, but this can be overcome by using a laser ranger (which is more precise than radar) as, for example, on the MiG-29 and Su-27.

Though flying since 1977, when the MiG-29 'Fulcrum' was first revealed in 1985, what really surprised Western observers was its fire control system. This consisted of the Phasotron N-O19 'Zhuk' PD multimode radar, with maximum search range of 54nm and capability close to that of the Westinghouse APG-68 of the F-16C. The radar is interlinked through the fire control computer with an IRST and laser ranger system (just ahead of the windscreen), and a helmet-mounted sight. IRSTs had been in service in the West for many years, but were widely regarded as supplementary equipment. In the MiG-29 it is an integral component allowing medium-range interceptions to be made in radar silence. This is a MiG-29SMT, the first variant designed from the start to incorporate a Mil Std 1553B equivalent databus. The rectangular dorsal fin root extension houses an upward-firing chaff/flare dispenser.
(Alexei Mikheyev via author).

Sensor and countermeasures

All attack sensors are susceptible to countermeasures. The deployment of decoys can saturate an air defence system thereby minimising the loss of aircraft, by forcing the wasteful expenditure of threat munitions. While an aircraft can eject both flares and chaff (now in fibre element form) simultaneously, later separation of the two types of decoys (IR and radar) occurs because of their quite different aerodynamic behaviour. PD radars can easily distinguish the target from a free-floating decoy. The more recent development of the aerially towed decoy overcomes this problem with a corner reflector, to enhance its radar cross-section (RCS) to mimic that of the real aircraft.

So how does a crew know that it's under threat? Radar-warning receivers (RWR) have been in widespread use from the mid-1940s. Current aircraft have several RWRs covering a wide frequency range (typically 0.5 to 20GHz), disposed to give all-round coverage and built into the basic airframe. The same is true of laser warning receivers (LWR) which are the latest type of defensive sensor. The

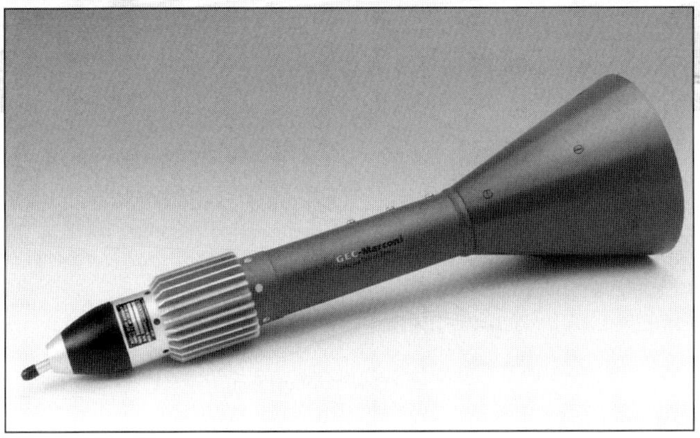

MAIN: Where stealth is absent: Tornado F.3 of the RAF's Operational Evaluation Unit, carrying a Sky Shadow ECM pod with GEC-Marconi Ariel towed radar decoy aboard (port outboard pylon). In addition, two Phillips/Matra Phimat chaff/flare dispenser pods are carried on the wings, with two Vicon Series 210 flare fittings under the rear fuselage. *(GEC-Marconi via author).*

INSET: GEC-Marconi Ariel towed radar decoy which is effective against modern monopulse, CLOS, active and semi-active missiles and home-on-jam systems. Working in the H/J bands, it was flight tested from 150kts to Mach 1.2. In a high-speed dispenser it has optional recovery by parachute after jettison prior to landing. *(GEC-Marconi via author).*

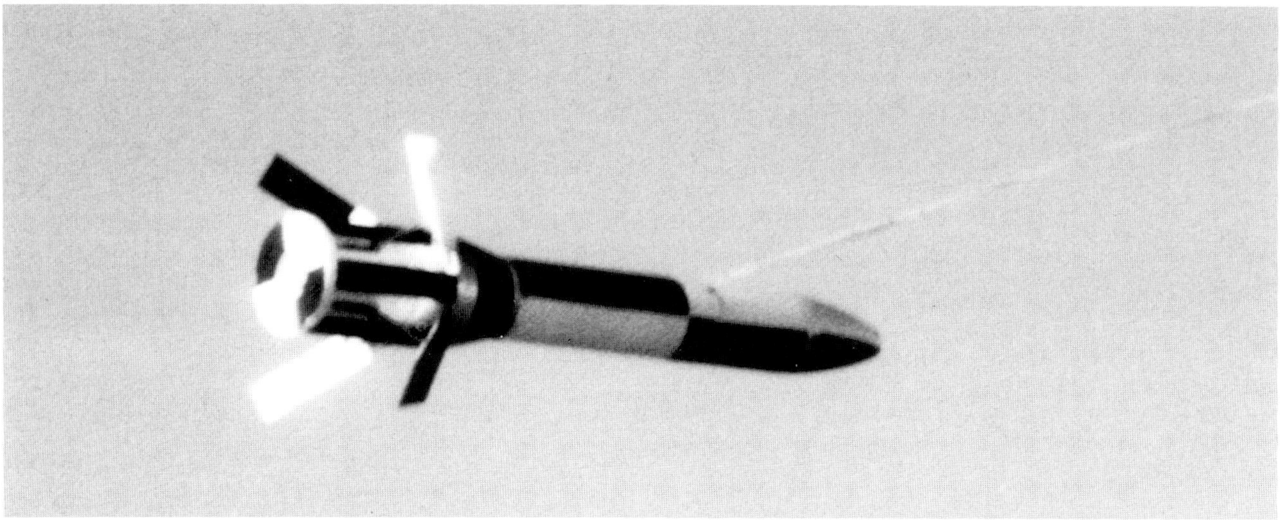

When a pilot decides to launch a towed decoy, based on an alert from a radar or missile approach warning system, the device must deploy rapidly and safely. This requires the device having aerodynamic stability throughout the aircraft's flight envelope. The Sanders Integrated Defensive Electronic Countermeasures RF subsystem of which the 'smart' fibre-optic flight vehicle shown being towed is part, is intended for initial use on the Boeing F/A-18E/F Super Hornet. *(Sanders via author).*

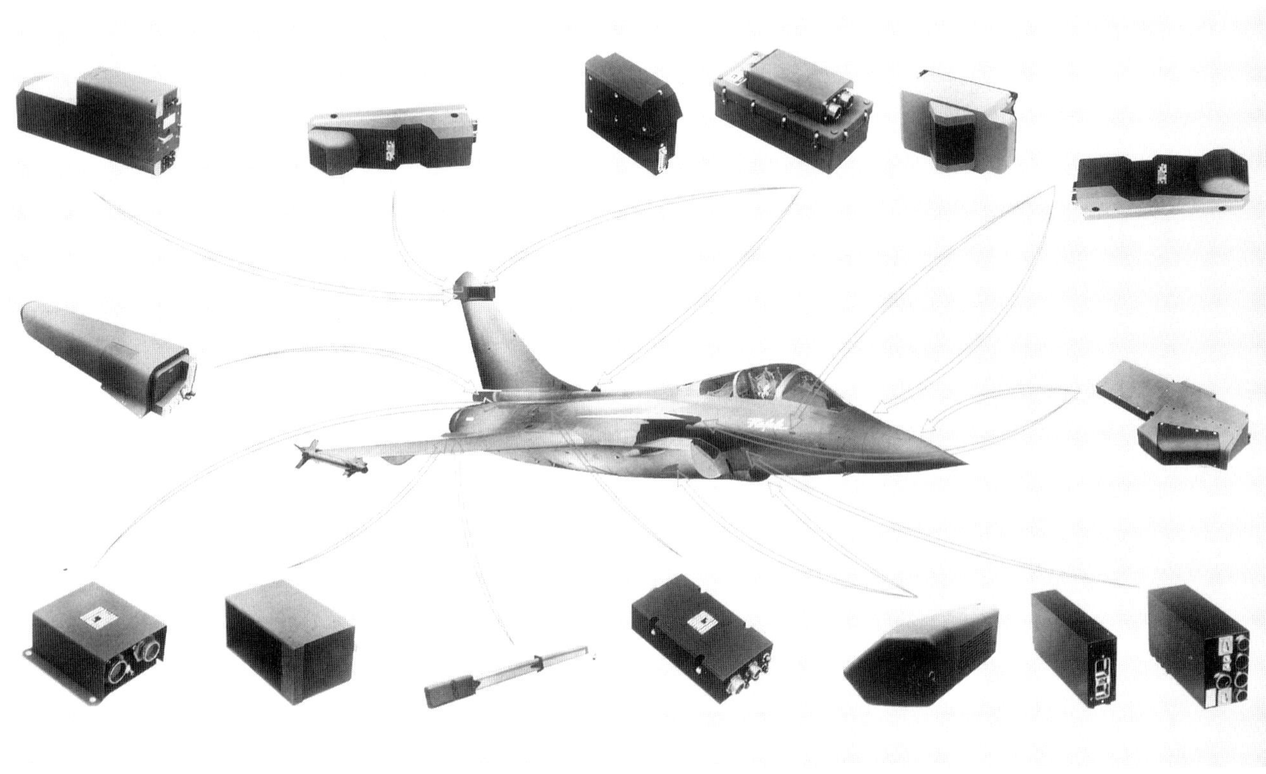

To maximise survival in combat, new fighters are defended by a comprehensive threat warning and self-protection suite. The Dassault Rafale, for example, has Spectra (system for protection against all adverse radiation) which comprises 15 major elements without using any of the aircraft's store stations, as shown. Having a total weight of 550lb, all elements (including IR detection, decoys, laser warning, EM detection and jamming and missile approach warning) are integrated by a dedicated EW databus and central processor. *(Thomson-CSF via author).*

data from these passive sensors are automatically routed to the electronic-warfare (EW) processor, where all detected emissions are analysed (a formidable task) and the intentions of the adversary become known. These can be presented to the crew of the aircraft under threat and, along with missile approach warning systems, used to activate the release of decoys automatically.

Increasingly, these data are used also to activate jammers, which emit electromagnetic radiation that can disrupt an active enemy sensor. These active EW systems fall into two distinct categories: those which emit broadband radiation (virtually 'blinding' the enemy) and those which emit 'deception', radiation, which is the more subtle EW approach. It attempts to hide the user for long enough to gain a tactical advantage, and it is more efficient regarding power. However, it needs very high-speed data processing plus up-to-date and accurate libraries of EW information, neither of which is easily obtained. Only the best-equipped military forces have the appropriate electronic intelligence-gathering capability to deploy such advanced, and expensive, systems.

The prototype Grumman F6F Hellcat powered by a Wright 1,700hp XR-2600 Cyclone first flew in 1942 but production F6F-3s had the 2,000hp Pratt & Whitney R-2800 Double Wasp. Armament comprised six 0.5in machine-guns. Of the 7,870 F6F-5s produced 930 were supplied to the Royal Navy as Hellcat IIs. These are F6F-3s operating from USS *Yorktown* in August 1943. *(via author)*

As though shading itself with its folded wings this Vought F4U-7 Corsair in the colours of the *Aéronavale Flottile* 15F (1954) was one of 94 designed and built specifically for French use in Indo-China. The stripes on wing and fuselage indicate its use during the Suez Crisis in 1956. A characteristic of the Corsair was its cranked or inverted 'gull' wing designed to provide clearance for its massive propeller while keeping the length and hence weight of the carrier-strength undercarriage to a minimum. The Corsair had light and comfortable stick forces, good performance, adequate stall warning and docile behaviour at the stall and made it a weapon of choice. The final F4U-7 variant remained in production until 1953. *(via author)*

The fast manoeuvrable, rugged and reliable Kawasaki Ki 100 was one of the finest Japanese fighters of World War II, although not introduced until 1945. The aircraft was developed from the unsuccessful Ki 61 that had a problematic liquid-cooled engine. This was replaced by a 1,500hp Mitsubishi air-cooled radial engine. *(via author)*

Designed to meet a strategic fighter requirement (Weapon System 105) of 1951 the McDonnell Douglas F-101 was adopted by the USAF Tactical Air Command as a fighter-bomber. The Voodoo was powered by two Pratt & Whitney afterburning J57s. This is one of the 29 pre-production F-101As (53-2426), 'Fire Wall' of ARDC, fitted in 1957 with an afterburner development later used in the F-101B. *(Boeing via author)*

The second generation of turbojet installations featured nose pitot air intakes that had the benefit of taking in clean undisturbed air well ahead of the airframe and expelling the exhaust gases clear of all structure at the rear. Dassault Super Mystère B2 (first flying in 1955 and the first West European aircraft capable of exceeding Mach 1 in level flight) of EC 1/12 Cambrésis and Mystère IVA (first flight 1952) of EC 2/8. *(Sirpa-Air via author)*

The Saab J35 Draken which appeared in 1955, was remarkable for its time in having a 'double-delta' wing planform, with the inboard section swept at 80° and the outboard at the still high value of 57°, giving it the extremely low aspect ratio of 1.77 combined with a very low wing loading. Despite the use of a simple pitot inlet to its Volvo RM6C turbojet (Avon 300 series), the J35D version's high, clean T/W of 0.86 allowed it to reach over Mach 2 and climb at an initial rate of 49,000ft/min (250m/s). Provision for Rb 27 radar-homing (forward) and IR-homing (rear) Rb 28 Hughes Falcon AAMs with the Ericsson-built IR tracker under the national insignia, distinguished the J35F, delivered from 1965, seen here in the colours of *Flottlj* 13.
(I. Thuresson via author)

Close contemporary of the MiG-25 was the MiG-23 Flogger of 1967. It was powered by a single Krachaturov R-29/R-35 turbojet rated at 26–28,000lb in afterburner, as shown by the exhaust shock diamonds giving high acceleration and rate of climb (50,000ft/min). Its swing-wing was to give good airfield performance but at minimum sweep it was very g-limited so was not agile at low speeds. Shown is a MiG-23UB two-seater combat trainer of the Polish Air Force just after take-off with the gear retracting as the ventral fin folds down. Capable of over Mach 2.3 at altitude with its wing at 72° sweep the MiG-23 required extra fin area in addition to an already large dorsal surface. *(Ryszard Malachowski via author)*

Give the military an aircraft and someone always want to hang on more weapons. The original Hawk had graceful Küchemann wing tips of the classic Hawker style, therefore an obvious place to hang yet another AIM-9 Sidewinder was missing. The resulting modification is shown here, on a Mk 208 built for Malaysia, together with a revised leading edge fence and vortex generator arrangement. Another important addition is the SMURF (side-mounted unit root fin) just above the ventral strake. This was developed for the US Navy's T-45 Goshawk to suppress flow separation on the underside of the all-moving tailplane when using full flap deflection. *(BAe via author)*

The great virtue of the delta wing is that its thickness/chord ratio can be kept low for supersonic speed without recourse to thinner but heavier swept wings. However, with a delta wing can come rapid speed loss in turns because of their high induced drag. For its private venture Mirage F.1, Dassault reverted to the conventional swept wing, generously equipped with high lift devices and an all-moving tail. These are Mirage F.1CTs showing the launch rails normally occupied by Matra 550 Magic AAMs, extended ahead of the wing leading edge to act as flutter suppressers. *(Sirpa-Air via author)*

The conformal fuel tanks (CFTs) on the F-15E add 74% (1,216gal/9,728lb) to the aircraft's 'internal' capacity of 1,637gal/13,094lb. This one, belonging to the 494th Fighter Squadron from RAF Lakenheath is carrying four AIM-7M Sparrows on the CFTs, two AIM-120 AMRAAMs under the wings, plus LANTIRN pods. The under-fuselage noise environment, during carriage trials of AIM-120s by the F-15, was far more severe than Hughes designed for and led to failures of the control surfaces on the aft end of the missile, in the late 1980s. A main area of emphasis with the AIM-120 has been to achieve much higher reliability and maintainability, in addition to smoke reduction, increased speed and range, improved warhead fusing and sensing and better ECCM. *(via author, IAT '97)*

The Saab JA37 is powered by a Volvo Flygmotor RM 8B turbofan that achieves a 74% increase in static thrust using its afterburner. The Viggen is unusual in being a fighter that uses a thrust reverser. These aircraft are from F4 Air Force Wing. *(Peter Modigh via author)*

Despite the efforts of aerodynamicists to produce clean shapes someone always wants to hang stores on the outside. It is somewhat difficult to discern the elegant delta wing of this Dassault Mirage 2000-5, first flown in 1990, festooned as it is with three fuel tanks, four Matra Super 530 medium-range and two 550 Magic short-range AAMs. *(F. Robineau via author)*

The MiG-29's Klimov R-33D turbofans have an sfc of 0.77 dry and 2.1 with afterburner. This vast increase in fuel consumption is because afterburning does not occur at the highest pressure in the engine and fuel is not burned as efficiently as in the combustor. Afterburners are used only for take-off and climb to altitude, combat and a short dash to maximum speed. Prolonged use of afterburning especially at low altitude would result in such a rapid fuel burn as to make endurance and range so short as to be useless. This was the case for the MiG-29 (9-12) Fulcrum A of the Swifts aerobatic team. *(Alexei Mikheyev via author)*

As originally conceived the General Dynamics YF-16 was a longitudinally unstable (subsonically) lightweight fighter demonstrator. In 1978 the F-16A became the first production aircraft to feature a full-time (no mechanical back-up) fly-by-wire flight control system. Requirements for departure prevention led to the AoA and g limiter plus control of sideslip during high AoA rolling manoeuvres. Above 28° AoA an F-16 pilot can command no more than +1.0 g. Above 30° AoA, full nose-down tail deflection is automatically applied by the F-16's FCS. From 1981 Block 15 aircraft had tailplanes of larger area to provide for improved air-to-ground capability and the aircraft became an inherently stable design. Shown are F-16As of the Royal Netherlands Air Force. *(Lockheed Martin via author)*

Eurofighter Typhoon DA.2 powered originally (1994) by two Turbo-Union RB199 Mk 104E engines, shown at low speed with leading edge flaps and inlet lips drooped. In service, the Typhoon will initially carry ASRAAMs in place of the AIM-9 Sidewinders in addition to the four semi-submerged AIM-120s. *(British Aerospace Defence)*

The earliest (1988) of the three modern canard-configured European fighters, the decision over the number of engines for Saab's JAS 39 Gripen was limited by availability (i.e. General Electric F404 or Turbo-Union RB199). The more powerful Pratt & Whitney F100 and PW 1120 were considered but the F404 met the requirements at a significantly lower cost. *(Saab/ British Aerospace)*

Head-up displays

When the first small computers arrived they were allied to a high-mounted display that replaced the gyro-gunsight, and became the head-up display (HUD). The HUD, with its large sloping combining glass, set in the pilot's forward line of sight with the images focused at infinity, has become a dominating feature in the cockpit of modern fighters. It enables the pilot to concentrate on the target without having to glance down at the conventional instruments, in order to check not only information from the weapon aiming computer (WAC) but also that needed for navigation and aircraft trajectory control. The symbols and alphanumerics are projected from a CRT through a system of lenses. Within limits the software/database of a HUD can generate a wide range of information to match different aircraft operating modes. The database and the sensors that provide inputs to the system can be extended to include displays of information concerning engine health, fuel state and radio frequencies.

Table 12 Chronological development of the HUD

HUD	*Characteristics*	*Early applications*
First Generation (1960–66)	Analogue symbol generation, more versatile than lead-computing gyro-gunsight	Buccaneer (20° wide) Viggen (30° wide)
Second Generation (1966–75)	Digital symbol generation, more flexible, accurate and reliable, WAC included, PROM (programmatic read-only memory)	A-7D (20° wide, 15° vertical) A-4M
Third Generation (1975–)	PROM to EPROM (erasable PROM), navigation, air data, self-contained tracer line gunsight, holographic, reduced light loss	F-16 (20° wide, 15° vertical) F-16C (36°wide, 20° vertical) Rafale (30° wide, 22° vertical)

The GEC Avionics HUD Weapon Delivery and Navigation System represents a 'drop-fit' avionics update for the Northrop F-5 to greatly enhance, at relatively low cost, the combat capability of older aircraft.
(GEC Avionics via author).

The spectacular view through the frameless windscreen of a Lockheed Martin F-16C showing the wide-angle holographic HUD. The use of holographic optical techniques gives the pilot a far wider field of view than with conventional HUDs. It also provides a display to allow operations day and night. The first of this type for the LANTIRN-equipped F-16 was delivered by GEC Avionics in 1983. *(GEC Avionics via author).*

The need for such a device was recognised even in World War Two, when pilots engaged in air combat and low-level attack survived mainly through sheer experience. Doing the same things at jet speeds during the Korean War accentuated the problem. In 1955, the US Navy experimented with the first HUD in which basic flight, navigation and sighting data were displayed from a large CRT and a reflector glass in front of the pilot. Remarkably, US industry was slow to get involved, and in 1962 the first properly engineered HUD was produced in the UK, and the Buccaneer was the first production aircraft to be equipped with a HUD. Whereas the first ones had analogue computers, within a few years (1966) these were replaced by digital types with enhanced versatility, reliability and performance. With the advent of helmet-mounted displays it may be that HUDs will be dispensed with in the future.

Multi-function displays

The need for more computers in their aircraft encouraged designers to install more CRTs. These appeared in the 1980s as multi-function displays (MFDs). They show data relating to flight vector and attitude (complementing the

A helmet-mounted display (HMD) displays HUD symbology directly onto the helmet visor so a pilot will no longer need to manoeuvre his aircraft to designate an off-boresight target. In addition the pilot's eyes are guided to the target by the display so increasing visual range. This Sextant Topsight monocular (HMD) for use by Dassault Rafale pilots demonstrates the French company's fully integrated approach: building a new helmet around a display system rather than the converse. The head-supported mass including oxygen mask is 1.45kg. *(Sextant Avionique via author).*

HUD), internal systems, weapon status and navigational data. The screens are surrounded by software-controlled keys whose function changes as the display changes. This kind of multi-function use of displays and controls has led to a lot being packed into a small space, with the optimum number of MFDs currently seeming to be three.

The wide range of instruments, electronic displays, buttons, switches, levers and illuminated panels, as well as warning enunciators, are there to enable the pilot to make the most effective use of a machine whose first cost is measured in tens of millions of pounds. There is also the cost of training the pilot that may be another million or more. The 200 or so hours to train a Hurricane pilot in 1940 cost much less (as did his aircraft), at around £50,000 at today's value.

Communications

The simplest fighters use just one type of radio, though with a wide spectrum of frequencies. Invariably, they have two independent sets, for redundancy. The preferred set is a V/UHF (very/ultra-high frequency) radio. This gives short-range line-of-sight and relatively interference-free voice communications. It does not require high power nor a large antenna, typically being a streamlined flat blade only a few centimetres long. For most current types in service, communications can be encrypted in some way (analogous to a scrambler phone) to avoid eavesdropping. They also frequency hop to avoid jamming. Where a fighter has to perform long-range duties, a beyond-line-of-sight radio system is used. This is usually HF (high frequency) radio and benefits from reflections in charged bands of the ionosphere.

IFF

Vitally, IFF remains in service, albeit in a much more sophisticated form compared to previous systems. With the introduction of missile armament potentially capable of engagements at beyond visual range (BVR) in the early 1960s, the distinction between day and night interception became largely irrelevant. What remains, though, is the problem of positive identification. The reliability of IFF systems, essentially unchanged in concept since the pioneering days of radar, had become increasingly suspect, and it became generally necessary to obtain visual identification before engaging. During the Vietnam War and later conflicts, many medium-range missile shots were prevented from being taken at BVR and few night engagements occurred because of the risk of fratricide.

The seemingly intractable problems of target identification are likely to continue to inhibit engagements in some circumstances. The traditional IFF is recognised as being an unreliable means, at any rate in its present form, for a variety of reasons. It depends on all friendly aircraft being fitted with the same equipment, but this is not always the case. Standardisation was one advantage that Warsaw Pact aircraft always enjoyed, though within NATO in the late 1980s there were three different, albeit not entirely incompatible, IFF systems. A failure to respond does not necessarily indicate that the contact is hostile. It could have suffered IFF failure or battle damage. The IFF might even be switched off or have failed due to the enemy jamming or spoofing.

Once an enemy has the equipment capable of responding to a friendly IFF interrogation it only requires careful monitoring to establish the current codes and then mimic the responses of a friendly aircraft. However, to ensure that a spoofing enemy cannot feign the correct IFF response, a more complex system was needed. By using high-speed data processing, it has become possible via the NATO identification system to make IFF interrogation and response so complex that simulation is virtually impossible. In addition, non-cooperative target recognition via signal analysis by the most up-to-date fighter radars is an adjunct to conventional IFF, so that new fighters should be able to make BVR attacks with confidence.

Navigation

By the late 1940s, Tacan (tactical air navigation) was in production for military aircraft. In principle Tacan is like VOR/DME. VOR (VHF omnidirectional range) is based on a fixed network of a series of continuously broadcasting ground stations. One of its advantages is that it enables a pilot to home in on the ground station, but it is a line-of-sight system. DME (distance measuring equipment) works a little like IFF by interrogating the ground-based unit to find the distance from it. Airborne Tacan transceivers (transmitter/receivers) could be tuned to Tacan ground stations within a range of 370km (200nm) but were originally bulky, often weighing 45kg (100lb). With time they became much smaller and lighter with increased range, so that current Tacans often weigh less than 10lb, and can interrogate out to a range of 300nm.

Inertial navigation systems

These are the only totally self-contained navaids, requiring no emissions or external help. Their operation depends on the fact that any mass in motion possesses inertia, and uses Newton's laws of motion, which relate force to mass and acceleration. Integrating (i.e. summing) with respect to time the change of acceleration experienced by a mass will give velocity (speed and direction), and integrating again will give position. Most advanced types of inertial navigation system (INS) have three accelerometers so that accelerations can be measured in the three orthogonal planes with very high precision. However, at least in the early types of INS, they could not just be bolted onto the aircraft. Any unwanted rotation of the aircraft about each of the three axes would cause two of the accelerometers to give false outputs caused by a component of the earth's gravitational field. Accordingly, most INS are mounted on a stabilised platform which remains level throughout any manoeuvres the aircraft may perform.

Following trials in several aircraft, a prototype INS weighing 1,180kg (2,600lb) steered a Boeing B-29 Superfortress across the USA in 1953. By 1958 Litton had created the LN-3 system compact enough to fit the tightly packaged Lockheed F-104G Starfighter. With a mean time between failures (MTBF) of fifteen minutes in 1960, this quickly matured to a figure of several hours. However, the ten minutes it took to warm up was far too long. Litton devised an alert-align mode which could be performed during rearming and refuelling. This involved entering the latitude and longitude not just of the airfield but the exact spot on the airfield.

Laser and fibre-optic gyros

Over the last twenty years the mechanical gyro has been increasingly usurped on all but the lowest-budget

A step change in radar and cockpit systems management came with the McDonnell Douglas F-18, which first flew in 1978. It was given the most versatile multimode radar of its day: the Hughes APG-65. With it, the Hornet pilot was seated in the middle of an information explosion. The problem became how best to handle it. All the information needed in flight and in combat had to be in a readily digestible form. The solution adopted by McDonnell Douglas was to abandon the plethora of dials and switches and channel everything into three multi-function CRT displays (MFDs), which were interchangeable at need. A press of a button would call up the information required at any moment, while critical information was displayed in the HUD. This was the first of the glass cockpits now found in all new fighters and airliners. Twenty years later and No F/A-18E/F cockpit photo shows it with five CRTs. *(Boeing via author).*

programmes by the solid state laser-ring gyro (LRG) which, despite its name, operates on a principle which has nothing to do with gyroscopic action. It is a precisely machined triangular block of quartz in which a laser beam passes simultaneously clockwise and counter-clockwise in small tunnels around its periphery. From measurement of the movement of fringes in the interference pattern where the beams meet, an LRG detects the rate of rotation of the block. It is so sensitive that placing an LRG on a table gives an output caused by the earth's rotation. The LRG has the advantage of having no moving parts (giving it a much higher MTBF), and with no wheel to spin up is ready for use more quickly. Fibre-optic gyros are being developed and promise even greater accuracy because the path length, which dictates the sensitivity, can be as long as the fibre used.

A laser reference system (LRS) will have, typically, three LRGs within it, and a data processor that converts the rotation rate into attitude (pitch and roll angle, and heading) data. This is a vital source of information for both the flight control and navigation systems. If the processor also has a set of accelerometers, its computations can maintain a log of the direction in which an aircraft has been pointing, and the speed it is travelling. When continuously updated (many tens of times per second) this provides precise data, and it is the principle of a laser inertial navigation system (LINS). These are aimed at achieving accuracies of 1.5km/hr (0.8nm/hr), alignment times of three minutes, and replacing the traditional INS to reap the reliability benefits. The first production fighter to be so equipped was the McDonnell Douglas F-15E with a Honeywell LINS having around 3,000hr MTBF, thirty times higher than the old system on early F-15s.

Far-reaching effects emanate from the latest radio-based navaid, Satnav (satellite navigation). The most widely used system is the US-developed global positioning system (GPS). Satnav is usable worldwide, and its three-dimensional position-fixing accuracy, for an authorised military user, is typically better than 5m (16ft). Combined INS/GPS systems have the virtue of both the long-term accuracy of GPS and the short-term accuracy of the INS. They are likely to be the leading fighter navigation data source of the future. It is probable that ground-based local radio navaids (e.g. Tacan and instrument landing systems) will be withdrawn from service gradually over the next quarter century.

Databuses

The development of the microchip at the beginning of the 1960s yielded miniaturisation of electronic components leading to the microprocessor. Equipment that would previously have occupied a small room could now be fitted within a suitcase. The effect was revolutionary, and with each new generation of microchips appearing every three years (now eighteen months) came a four-fold increase in capacity. Better understanding of the capabilities of digital data processing has allowed a completely integrated approach to all aircraft systems. Together with a wholesale transfer of functions from hardware to software, this revolutionised flexibility, reliability and cost.

The aircraft designer has to have a good idea of what the sensors on his aircraft will be doing, and what information its many computers are expected to digest and deduce. This determines what the software residing in each computer will do, and defines what information will pass between units. With 'black boxes' becoming smaller and cheaper, the last two decades has seen a shift towards many times more powerful (due in large part to parallel processing techniques), distributed and task-dedicated computers (forty on the Saab JAS 39 Gripen) and away from the all-powerful central computer. Some functions, such as navigation and weapon-aiming, have affinities, and this is what led to 'nodes' on the digital databuses that characterise modern fighter avionics.

They arrived in the early 1970s in the form of the Mil-Std-1553B databus and have dominated most recent programmes. Unlike its predecessors, it is a very high-capacity multi-source multi-sink bus, very roughly akin to a multi-track railway carrying passengers (the data) and depositing them at their proper destinations. The cost of developing this intricate infrastructure is beginning to dominate the design process itself, but it has been essential to reduce the time and cost involved.

JTIDS

The most well-publicised datalink system in use is the joint tactical information distribution system (JTIDS). It is a multi-billion-dollar programme enabling users in every kind of air and surface platform to know who is where and what they are doing in real-time, across the entire battlespace. It is a high-data rate system that requires an airborne transmitter and receiver of colossal complexity compared to barely ten years ago. It makes a modern fighter effectively a link within a gigantic telecommunications net. Its crew can be in receipt of information from sources such as an airborne warning and control system (AWACS), other fighters, GCI, ships and ground troops within seconds of intelligence becoming available.

To enable large numbers of transmitters and receivers to link into the one network on a real-time basis, one of the basic communication techniques is used. This is time-division multiple access (TDMA), which means that each transmitter sends out its message in short bursts, called time slots, lasting only milliseconds, while there are 128 slots per second. Computers synchronise hundreds of messages from the different sources so that all can use the same network, perhaps all of them transmitting a different message in a single second. JTIDS achieves its security and resistance to jamming by repeating each message up to fifty or more times, each time, via frequency hopping, on a different frequency over a band of several hundred MHz. Previously, the passing of AWACS data could only be done over the radio, usually by voice.

Terminals carried in each aircraft not only receive and display information, they also receive and transmit, thus forming relay platforms to pass the line-of-sight (UHF) data around the network. Class 1 terminals are installed in AWACS aircraft and major ground sites. Smaller Class 2 terminals go into USAF F-15s and RAF Tornado F.3s. Because of the colossal amount of data flying around, a fighter's terminal filters out only what is relevant to it, which could be navigation or weather data. Most importantly, it can greatly increase a crew's situational awareness. For fighters, a God's eye view (not just in front) of friendly and hostile track and targeting information can be provided by an AWACS, as a visual image directly onto their radar screens. This avoids the need for the crew to use their own radar and risk revealing themselves with tell-tale emissions. Furthermore it can

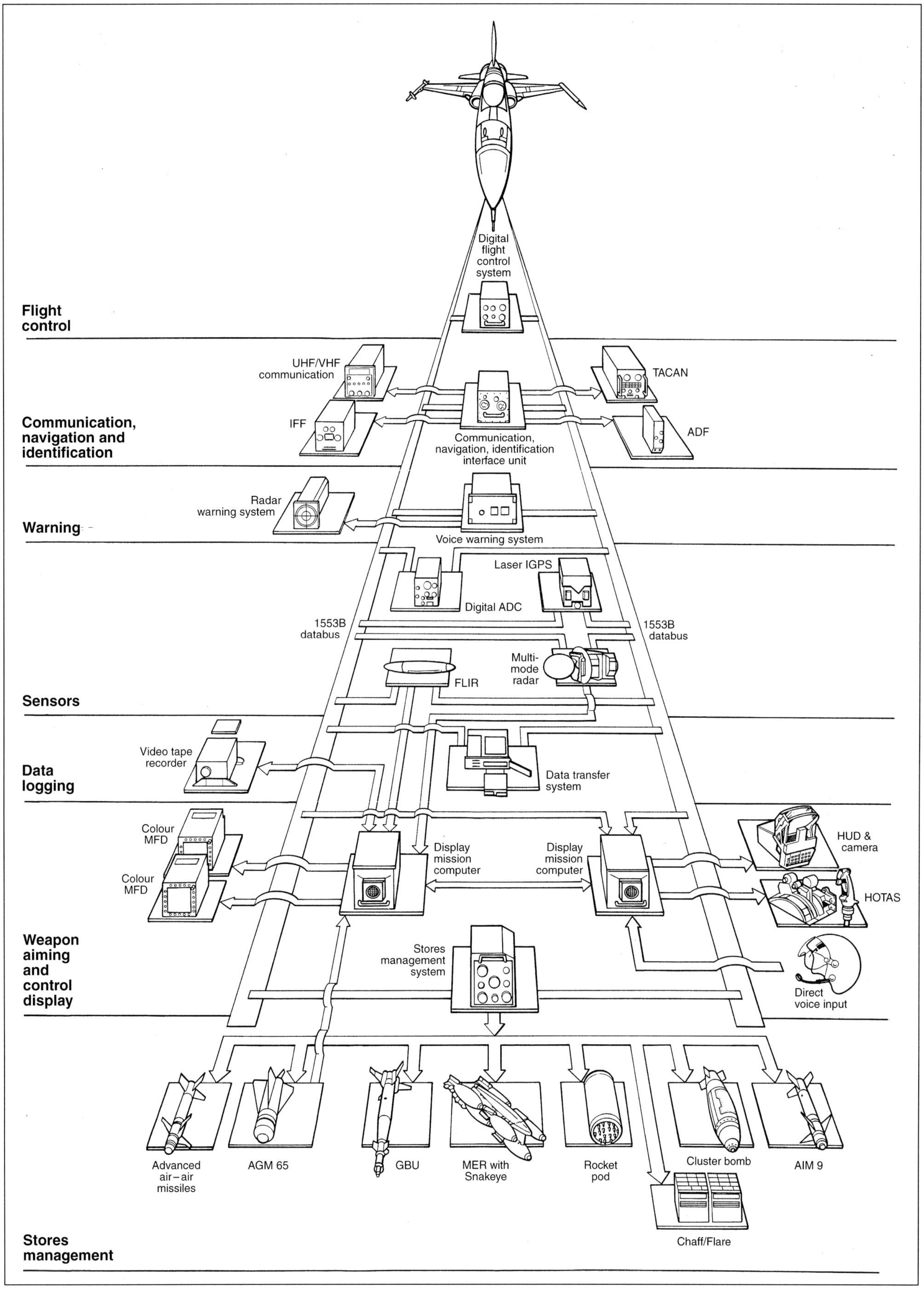

Structural refurbishment and an avionics update can, for comparatively modest cost, extend a fighter's life and provide combat capability approaching that of a new and more expensive aircraft. This drawing shows a simplified arrangement of a dual Mil-Std-1553 databus as part of an avionics retrofit for the Northrop F-5E. *(Smiths Industries).*

allow a BVR attack if the AWACS has already identified the 'hostiles'.

Conclusion

With the increasing number and sensitivity of sensors comes an enormous diversity of warfare activity that a fighter can survey. This is leading to concerns over sensor data fusion. It is not possible, nor desirable, to keep all information on display, and the evolution of the cockpit, hand-in-hand with avionic developments, is a challenging arena.

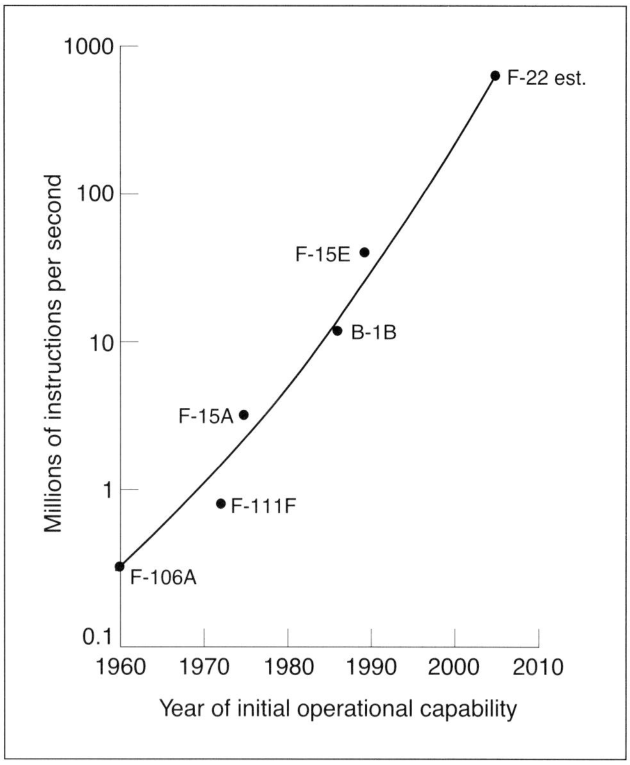

Graph showing the enormous increase in data processing rate built into the avionics systems of leading combat aircraft over 40 years. Note the logarithmic scale on the vertical axis which indicates that the F-22 has 16 times the rate of an F-15E.
(AIR International).

Chapter 7

Stealth

The reduction or other control of an aircraft's radar, infra-red, visual and acoustic signatures can greatly improve its survival, resulting in improved weapons effectiveness. Although radar and infra-red stealth are important, other aspects of low observability (i.e. acoustic and visual) cannot be ignored. The experience for US air arms during the Vietnam War emphasised the need to pay even more attention to aircraft signatures. The topics of benefits of signature control, contributions of an aircraft to its radar cross-section (RCS) and infra-red (IR) signature, methods to reduce RCS and IR levels, penalties/costs of signature reduction in terms of performance, volume, weight and maintenance, use of radar-absorbing materials, and decoys are addressed.

Glossary

Exchange ratio: Enemy to friendly killed.

Micron: A millionth of a metre.

Radar cross-section: Ratio of scattered power density in a given direction (usually backscatter) to the incident power density, normalised to be independent of range. It is here, for simplicity, measured in square metres.

RAM: Radar absorbing material.

It is useful to start with the question: why signature control? The control of signatures lets fighters achieve higher exchange ratios and improve survivability, thus allowing the destruction of high-priority targets without unacceptable loss. By the early 1970s, in the wake of the Vietnam War, it had become obvious that offensive air operations were being severely jeopardised by radar-controlled air defence weapons. The reality was that modern fighters had essentially the same basic susceptibility to being detected, tracked and engaged as did their World War Two predecessors. Moreover, the radar-directed threats had become vastly more lethal and were reaching the stage of being potentially devastating.

This reality was reinforced by the dramatic impact of recently-introduced Soviet surface-to-air missiles (SAMs) on the combat attrition of US-supplied aircraft during the 1973 Arab/Israeli War. Self-protection jamming systems, added during the Vietnam War, could often not deal adequately with such threats. It became clear that major reductions in RCS had to be made if manned strike aircraft were to remain viable weapons platforms. This led to the initiative that funded the experimental survivable testbed (i.e. XST that led to Have Blue), which eventually produced the Lockheed F-117A.

However, while remaining undetected gains time to plan and execute the optimum attack, once combat is engaged it will be difficult to remain unseen. For example, once launched, a missile in flight, unless itself stealthy, will soon be detected by its target. The launch aircraft, manoeuvring for another attack or to escape retaliation, will become distinctly less stealthy.

Radar Signature Control

The threat to aircraft of radar detection and tracking comes from a variety of sources, as Table 13 illustrates.

Since radar range is a function of the fourth root of the radar cross-section (i.e. $RCS^{1/4}$) an order of magnitude reduction in RCS, for example, will give a 44% reduction in detection range:

Table 13 Typical radar threats

Radar system	*Frequency (GHz)*	*Wavelength (cm)*
Early warning	0.15–0.2	150–200
	3–4	7.5–10
Ground control intercept	2–3	7.5–15
Height finders	2–7	4–15
Aircraft	8–20	1.5–4
Air-to-air missiles	10–20	1.5–3
SAM transportable		
Acquisition	0.15–3	10–200
Tracking	5–10	3–6
SAM mobile		
Acquisition	2–6	5–16
Tracking	5–13	2.3–6
Radar-guided AAA	14–16	1.8–2

$$\frac{R_1}{R_2} = \left[\frac{RCS_1}{RCS_2}\right]^{\frac{1}{4}} = \left[\frac{1}{10}\right]^{\frac{1}{4}} = 0.56$$

Similarly, the search area of the radar will be reduced to 32% and search volume to 18%. A *very* large reduction in RCS, not one tenth but one thousandth, is essential to have a tactically significant effect (i.e. an 82% range reduction).

In 1957, well prior to the Soviet Union's shooting down of Gary Power's U-2, the CIA asked Lockheed to conduct a low observable U-2 replacement study. It was rapidly concluded that a combination of supersonic speed and low RCS was required for the strategic overflight reconnaissance mission. Codenamed 'Oxcart' this project was the beginning of Lockheed A-12/SR-71 development effort. However, as can be seen, the resulting single seat A-12, which first flew in April 1962 was predominantly shaped by its Mach 3+ speed, so very great emphasis needed to be placed on active countermeasures. Nevertheless, considerable incorporation in the aircraft's structural design was made of RAM and the 15° canted fins further aid in reducing RCS. *(Lockheed Martin via author).*

The application of a very thin electrically conducting interlayer on the canopy can prevent backscatter from objects within the cockpit. However, such treatment may have only a very marginal effect on a non-stealthy design such as the Lockheed Martin F-16C (of the 363rd Fighter Wing at Shaw AFB, South Carolina) shown here launching a Raytheon AIM-120C AMRAAM. Furthermore, merely launching such a weapon, will destroy whatever stealth has been achieved by the carrier aircraft. *(Raytheon Systems via author).*

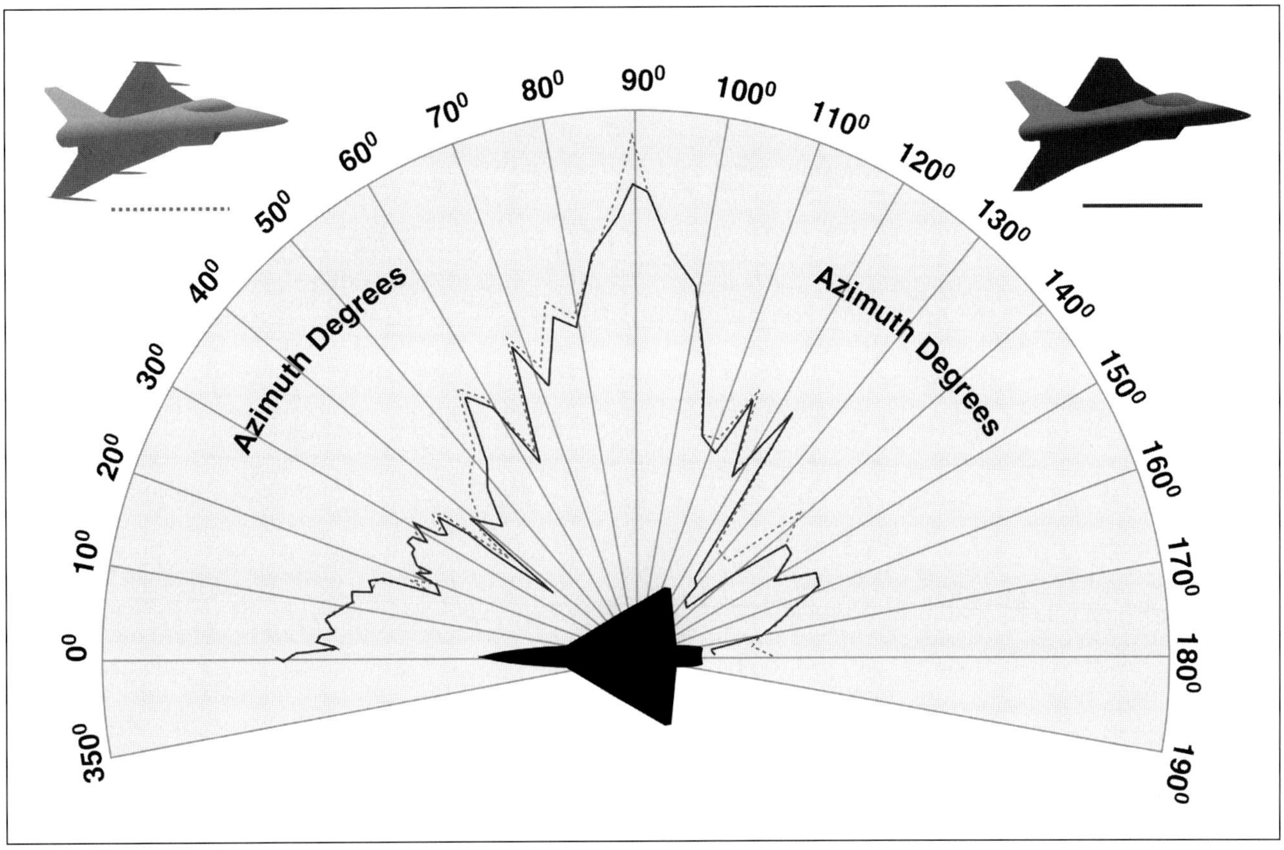

Predicted signature of an armed and unarmed generic fighter illuminated by a 3cm wavelength radar, as it sweeps from nose-on to tail-on. This example includes surface reflections and edge diffractions but excludes corners, tips and double reflections. From the front (and even to 30° azimuth) the box-type air intakes totally dominate the signature since their straight lips cause strong diffractions at their sharp and unswept edges. More than 60% of the RCS at 64° is due to the wing's leading edge. Addition of the weapons and their pylons can be seen to be very large at 39° and 139°. Note: the radial scale is logarithmic not linear. *(Pete West/AIR International).*

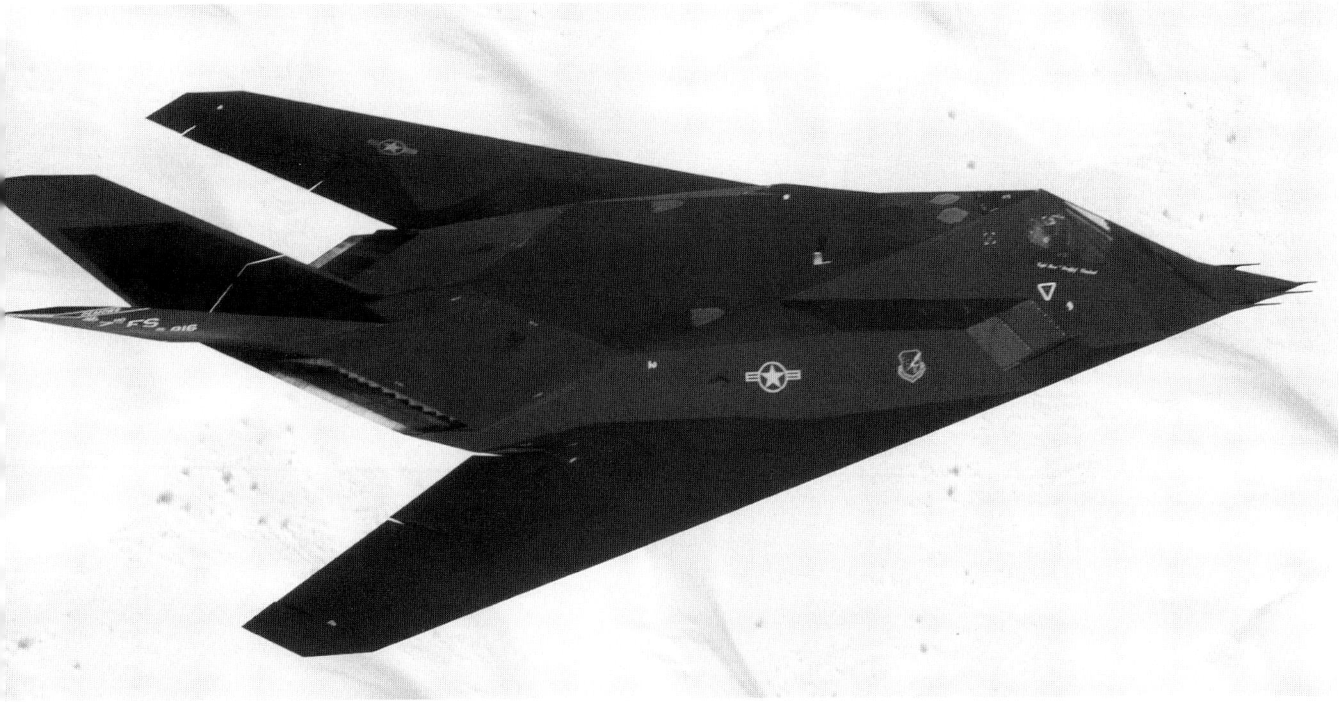

In development the biggest problems with the Lockheed Martin F-117A Nighthawk were how to keep the tailpipe from cracking and the air data system from icing up. The flat tailpipe was not structurally sound under high pressure and easily cracked. With experience of the Gulf War, modified heat shields, seals, airflow paths and thermal protection bricks at the edge of the nozzle tiling were modified to improve its maintainability. The low RCS pitot-static probes, which extended from the nose recorded for the air data computer, pitot and static pressure, AoA and sideslip. If they iced up the aircraft would depart from controlled flight in under two seconds. A non-conductive heating wire, thinner than a human hair, was the solution. To mitigate RAM-related maintenance burdens, the fighters routinely fly training sorties with some RAM removed, and pilots try to avoid weather that could degrade the surface. F-117A over White Sands, near Holloman AFB, NM. *(Lockheed Martin via author).*

Contributions to RCS

RCS depends on: aircraft shape, aspect angle or orientation with respect to radar line-of-sight (LOS), ratio of radar wavelength to target size, polarisation of transmit and receive antennas, surface quality and constitution of the target. Methods to control RCS depend critically on the size of the electrically-conducting component being illuminated compared with the wavelength of the radar signal illuminating it. If the wavelength of the signal is much less than the physical size of the component, and if it is smooth enough, radar waves reflect much as a mirror reflects light.

The RCS of an aircraft is determined by the magnitudes of two distinctly different contributions: (a) Its shape, both overall and in detail (including surface quality). Realistic reductions in manned aircraft size have an insignificant effect on RCS). (b) The electromagnetic properties of the airframe materials.

Aircraft shaping is useful over a wide range of radar frequencies but over a limited range of aspect angles. Typically, for fighter aircraft, the forward cone of angles is of greatest interest and, hence, large returns can be shifted out of this sector into the broadside directions. Aircraft can be shaped to ensure that most radar waves will be scattered and not reflected back to the transmitter. Leading and trailing edges of wings, control surfaces, inlet lips, door gaps, etc., can be aligned to ensure that the energy that is, unavoidably, reflected back to the transmitter is concentrated into a few spikes. This will give the opposing radar one good return when the alignment is ideal, but a much weaker return on subsequent sweeps.

Shaping methods to reduce RCS

There are two different approaches to shaping. Firstly, a faceted configuration can be used with flat surfaces to minimise normal reflections back to the illuminating radar. The key to the revolutionary US work on faceting in the 1970s lay in a paper on diffraction published in the Soviet Union in 1962 by Ufimtsev, a Russian physicist, and translated by the USAF Systems Command Foreign Technology Division in 1971. It allowed Lockheed's Skunk Works to develop an 'industrial strength' method of predicting the RCS of simple two-dimensional shapes built up from a series of triangular surfaces (e.g. Have Blue and F-117A) that was compatible with the available computing power of the time. Secondly, a compact, smoothly blended external geometry can be employed to achieve a continuously varying curvature (e.g. Northrop Grumman B-2A and Lockheed Martin F-22A). This requires far greater predictive ability and enormously increased computational capacity.

By the late 1950s those working in the field realised that the very large RCS reductions necessary to achieve any operational benefit could not be accomplished simply by coating an otherwise conventional aircraft with RAM. Many physically small and apparently insignificant features

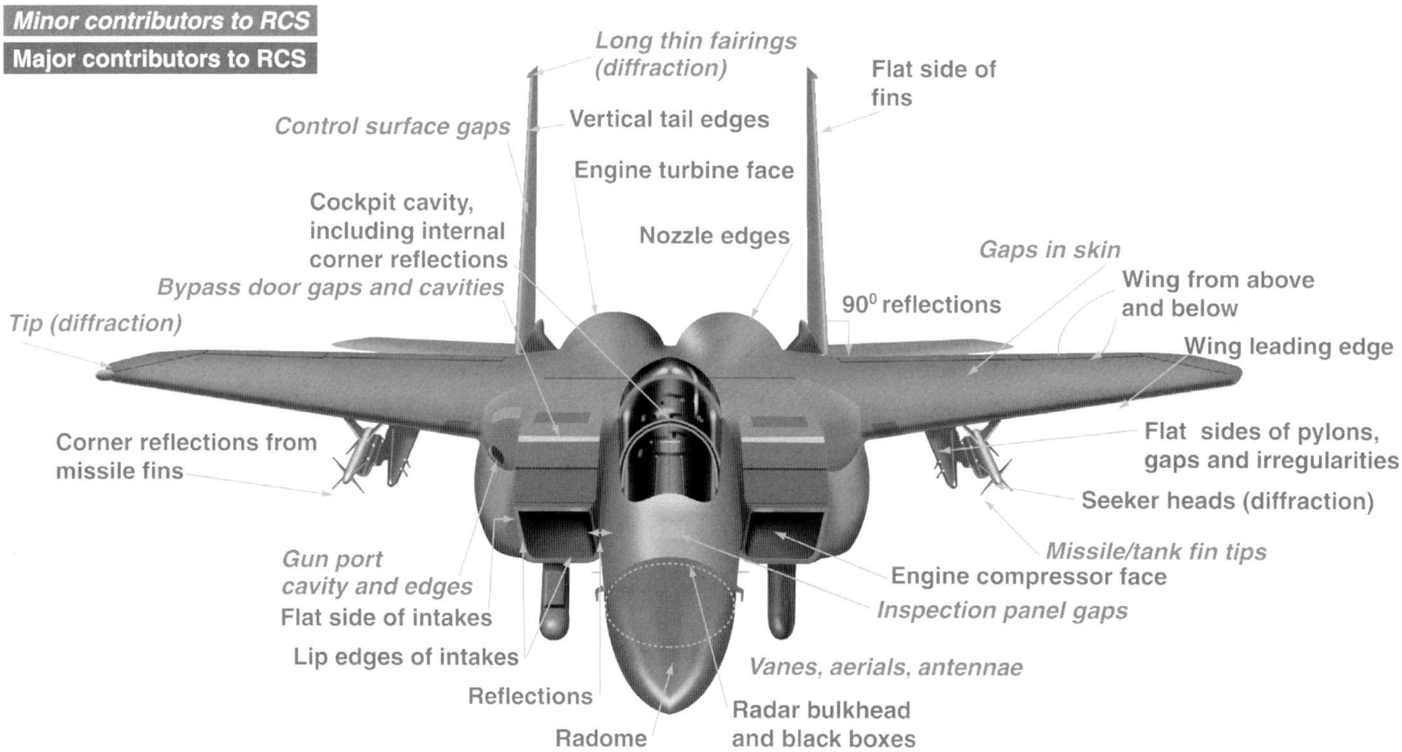

Major contributors to RCS for conventional aircraft: engine compressor faces (forward) and turbines (aft) due to Doppler signature; air intakes for engines; external stores, including missiles' seeker heads; wing leading edge, especially if unswept; corner reflections at intersections of fins and tailplanes; wing from directly above/below; radome and bulkhead, if transparent to illuminating radar; cockpit, including cavity effect due to a very large number of corner reflectors; engine nozzle if viewed from rear; flat, slab-sided fuselage when viewed from side.
Smaller contributors to RCS for conventional aircraft: fuselage in head on view; wing leading edge and control surface gaps which cause scattering; local air inlets e.g. for cooling and air conditioning; local surface protuberances; even the smallest proturberances cannot be ignored and each may become resonant at a different frequency; long thin fairings including missiles; fins and tailplanes.
Note: If the major contributors to RCS are carefully dealt with then the previously minor ones become important.
Stealth is difficult to achieve but easy to lose through lack of attention to detail.

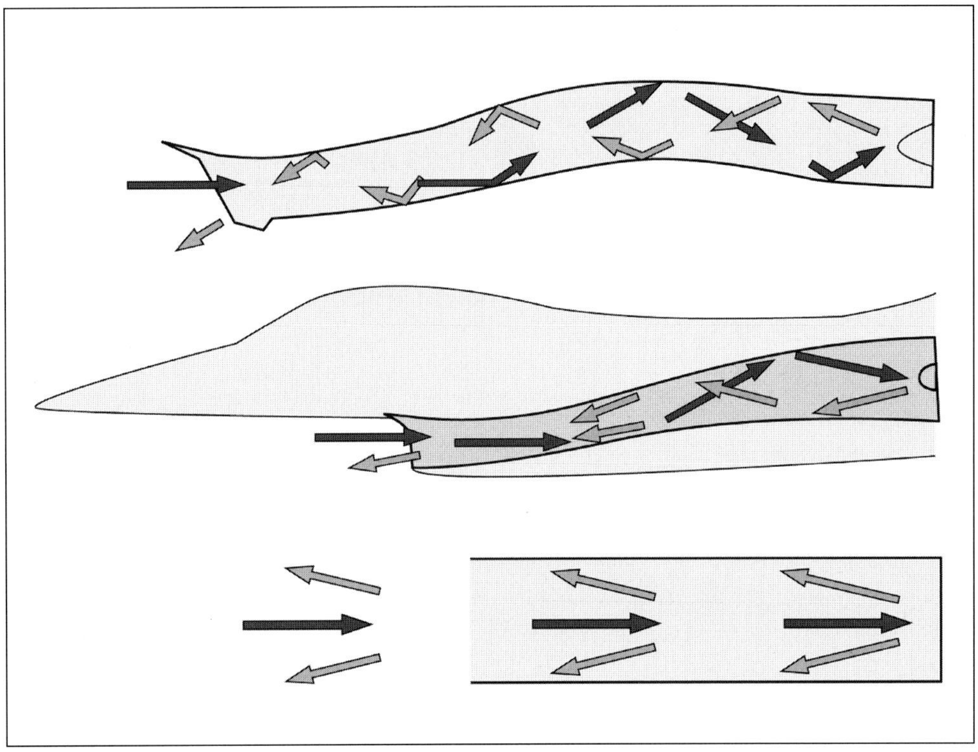

Use of a 'double-S' shaped air intake duct over the wing box on Eurofighter Typhoon (top). The vertical offset from the inlet almost achieves 100% line-of-sight blockage to engine compressor face. With RAM lining of the duct the combination greatly reduces the frontal RCS due to backscattering. The duct is reminiscent of that used on the F-16 (middle) but the latter is a simple single-S shape and exposes about 60% of the compressor face. A straight duct (bottom) has the largest RCS by far.
(Pete West/AIR International).

of an aircraft generate radar returns that are still quite detectable. A very low RCS must be designed into an aircraft from the outset, with rigorous attention paid to all three elements: overall shaping of the airframe and its components, special detail treatments, and RAM.

Summary of RCS reducing techniques:

- reduce the number of radar spikes by concentrating on many similarly aligned reflecting surfaces
- use swept leading edges to scatter reflections away from the source
- avoid vertical surfaces wherever possible, otherwise cant the surfaces to avoid right-angle corner reflectors
- carry stores, including fuel, internally
- minimise or eliminate control surfaces
- eliminate the cockpit transparency cavity by employing an unmanned vehicle, or reduce cavity effects by treating the transparency with a very thin conducting layer to prevent backscatter from objects inside the cockpit
- use a clean external geometry with minimum protuberances such as aerials, inlets, doors, gaps
- avoid flat or re-entrant surfaces likely to be normal to incident radiation
- bury engines, with air intakes and exhausts located over the upper surface of the airframe, to mask the cavity from the major illuminating radar threat
- give the inlet duct an 'S' or 'double-S' shape to hide the compressor face and to force multiple reflections on the RAM-lined duct
- use a screen (with spacing fine in relation to incident radiation) over the air intake, together with gauzes, vanes and deflectors within the diffuser duct, and extend the length of inlet guide vanes and arrange them to cause internal reflections on RAM and mask the cavity
- avoid variable geometry intakes to minimise reflections from the gaps and steps of the compression ramps and eliminate bypass doors by finding other methods to control intake airflow
- carefully shape the inlet lips (including sharpness) and nozzles by sweeping to align with major surfaces
- inlet boundary layer diverter flows should bleed around the engine cavity or into the bypass duct and not be dumped overboard (via doors) separate to nozzle flow
- design and manufacture any internal structure within radar-transparent skins to reduce reflections in given directions as the cumulative effect of the interior reflections could easily exceed the radar return from a metallic skin
- use RAM wherever appropriate (e.g. leading edges, bulkhead and black boxes within radar cavity, on the interior of the inlet and on metallic structure under radar-transparent skins)
- use a very high quality of manufacture to avoid gaps, holes, etc., since attention to detail is vital
- avoid discontinuities in conducting paths since electromagnetic waves induce currents in metal skins
- cover gun port, inlet and exhaust of APU when not in use
- if a radar is to be carried, use a low probability-of-detection/interception type, the features and use of which can be tailored according to the mission
- use an automated mission planner, incorporating terrain features and actual aircraft RCS patterns to optimise mission route profiles, to accommodate rapidly updated radar threat locations and operating parameters

Alignment of surfaces on Lockheed Martin F-22 to reduce number of strong radar spikes. Leading and trailing edges of wings, control surfaces, inlet lips, door gaps, etc., are aligned to ensure that the energy that is reflected back to the receiver is concentrated into a few spikes. This will give the opposing radar one good return when the alignment is ideal, but a much weaker return on subsequent sweeps.
(Pete West/AIR International).

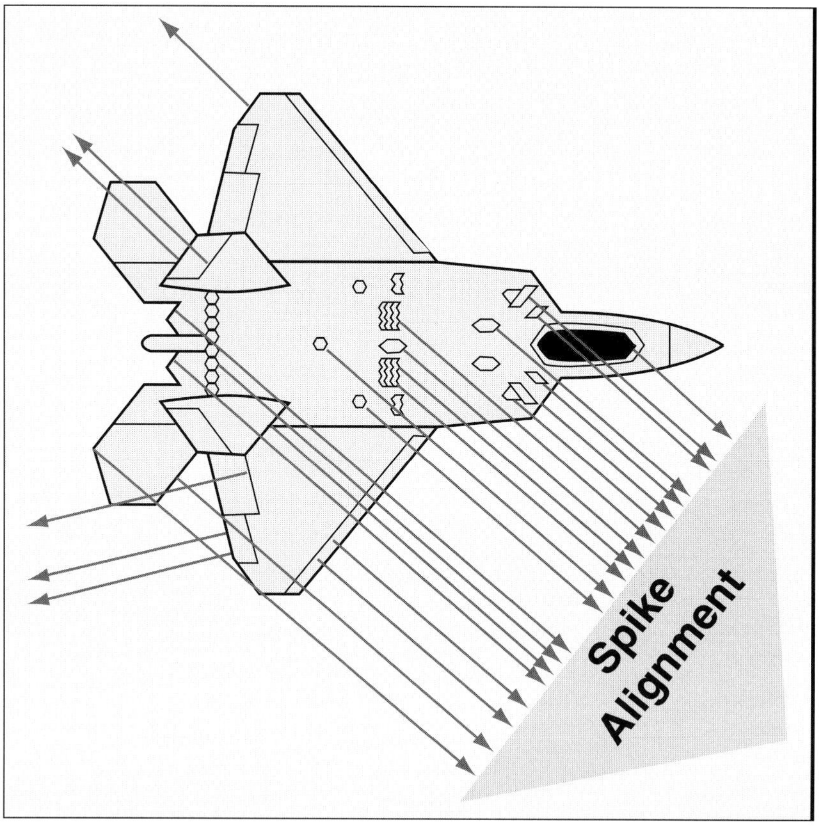

Ground testing of the Pratt & Whitney Low Observable Axisymmetric Nozzle (LOAN). This is claimed to provide a significant reduction in radar and IR signature by the engine through geometrical shaping, a new cooling system and special coatings on internal and external structure.
(Pratt & Whitney via author).

Infra-red Signature Control

Infra-red (IR) signature

Passive (non-emitting) IR detection devices rely on the fact that every atom of matter, including clouds and rain, continuously sends out electromagnetic radiation at an IR wavelength which corresponds to its temperature. It is necessary to think in terms of absolute temperature. Even though a certain object may be regarded as cold – a snowflake, for example, at 0°C – on the absolute temperature scale it is at 273K. For aircraft detection, IR seekers look for *contrasts* between hot parts on the airframe

Although the Eurofighter Typhoon is not a stealth-driven design it is probably the first West European aircraft to have target levels of RCS in its specification. It has a contractual value of RCS for the front hemisphere with a configuration of four AMRAAMs semi-recessed and two pylon-mounted SRAAMs. *(BAe Military Aircraft).*

This photograph emphasises the rear end of the Pratt & Whitney F119 engine (in afterburning setting) for the Lockheed Martin F-22. Not only is the nozzle two-dimensional to give thrust vectoring (±20° up/down) but it is serrated to reduce the installation's RCS. *(Pratt & Whitney via author).*

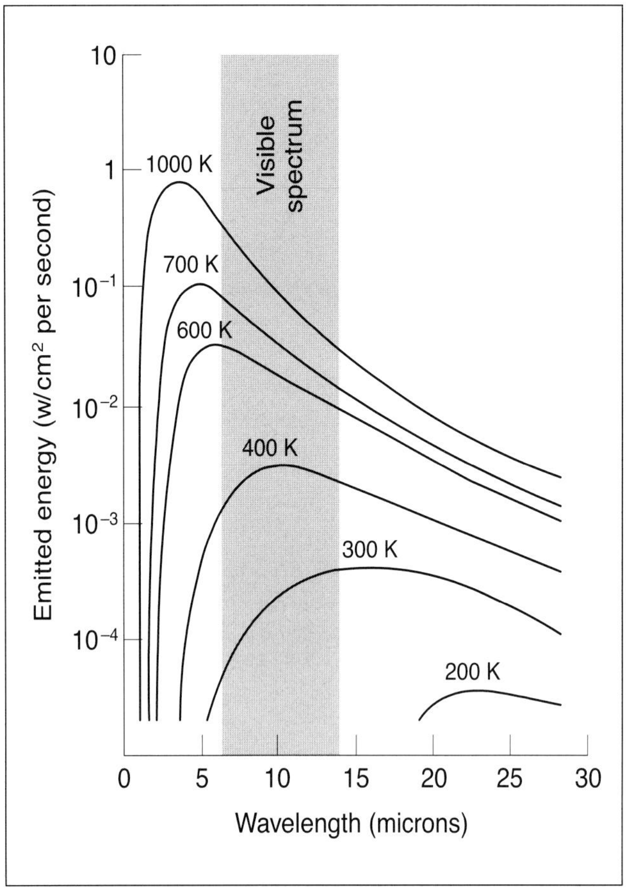

As an object gets hotter not only does the radiation intensity increase but its wavelength gets shorter. The curve showing 300K corresponds to 27°C, while 1000K is typical of a red hot jetpipe of a non-afterburning engine.

such as jetpipes and surfaces subject to kinetic heating, and the background radiation.

In designing IR detectors several things have to be considered: the range of wavelengths emitted by the target, the likely wavelength of the most intense radiation, the ways these wavelengths are affected by the atmosphere, and, because the maximum contrast is desired, the character of the likely background radiation. Many IR devices operate in the 8 to 13 micron band since this is the most IR-transparent band in the atmosphere. In engine exhausts, carbon dioxide produces most of the IR signature at 4.2 microns, so modern IR sensors can 'see' at two different wavelengths (medium: 3 to 5 microns and long: 8 to 14 microns) to provide good target discrimination.

Contributions to IR signature

The radiated IR energy is proportional to the fourth power of absolute temperature (i.e. T^4). With engine turbine entry temperatures currently at around 1,900K and rising, the back end of a fighter is the greatest source of IR radiation. With afterburner on, it becomes more so. Moderate stagnation temperatures (which vary with Mach number squared) are inevitable on leading edges of a fighter's airframe due to kinetic heating at high Mach numbers. As the stealthiness of fighters increases, so their missiles' exhaust plumes play a greater role in early detection. Lower-visibility plumes will minimise detection of both launch platform and missile.

There are several ways of reducing IR signature:

- have the ability to supercruise (cruise at supersonic speeds without afterburning) to restrict the temperature of the nozzle; moreover, supercruising allows the pilot to engage on his terms, increases weapons' envelopes, minimises exposure to SAM threats and not only stretches combat radius but forces an adversary to expend his own fuel in order to get his aircraft to an energy state where he can engage it
- use a high bypass-ratio (BPR) engine to mix in cold air to reduce exhaust temperature, but that said, a BPR greater than about 0.4 conflicts with the requirement of the high dry thrust to achieve supercruise
- use a curved jetpipe to mask the hot turbine stages.
- use two-dimensional nozzles (which increase the surface area of the exhaust plume) or ejector nozzles (which mix in ambient air) to increase the rate of cooling
- increase cooling of the outer skin of the engine bay or insulate to reduce the temperature of the airframe skin
- use a curved air intake to mask, to some extent, forward emissions from the engine
- limit maximum supersonic speed to reduce IR signature due to kinetic heating

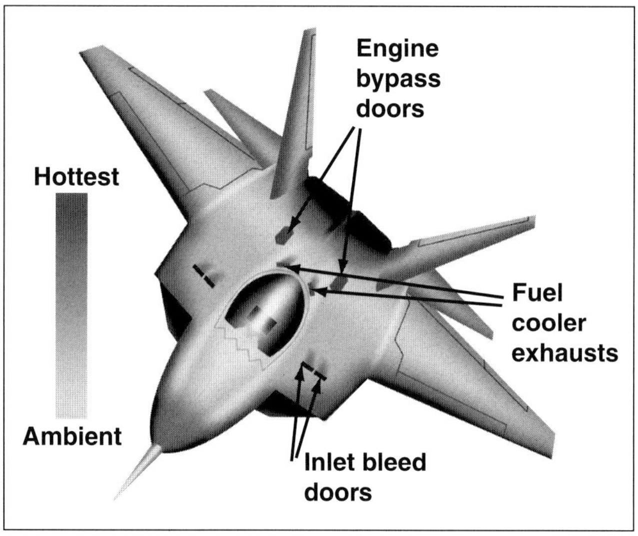

The rear end of a fighter (where the engines are located) is the greatest source of IR radiation. With afterburner on, it becomes more so. In addition, the airframe of a fighter at high speed heats up due to kinetic heating and moderate stagnation temperatures are inevitable on leading edges of a fighter's airframe at high Mach numbers.
(Pete West/AIR International).

The Costs of Signature Control

The requirement for stealth results in unconventional configurations, the performance, controllability, produceability and maintainability of which contain a large number of unknowns. However, the overriding requirement for any future aircraft (e.g. JSF) is affordability. This crucially affects the availability of platforms to perform a mission, as does reliability, maintainability and survivability.

The probability of a kill, P_K, is equal to $P_D \times P_{A/D} \times P_{H/A} \times P_{K/H}$, where
P_D = probability of detection *(signature)*
$P_{A/D}$ = probability of acquisition given detection *(signature)*
$P_{H/A}$ = probability of a hit given acquisition *(signature/ countermeasures)*
$P_{K/H}$ = probability of a kill given a hit *(vulnerability/ weapon effectiveness)*
Thus, signature reduction plays only a part, albeit a major one, in kill probability.

History shows that it is easy to turn a good idea into an unaffordable one. The Lockheed Martin F-22 is a solution to the dilemma, but in its 'gold-plated' form is probably affordable only by the United States. Some of the adverse effects of signature reduction are increased costs, additional maintenance, and added weight and volume leading to performance penalties. In addition, stealth brings with it 'special access' security which is costly in terms of both time and money, and if applied without careful thought it can become a serious impediment to success. For example, the Northrop Grumman A-12 security was so tight that the US Navy and DoD did not subject the programme to normal reviews and were late in learning of weight, schedule and cost problems that led to its cancellation.

Aerodynamic penalties

The aerodynamic problems posed by stealthy aircraft, especially if they are inherently unstable, are:

- significant reductions of cruise lift/drag (L/D) ratio will arise if shaping for low RCS overly compromises aerodynamics. The faceted shape of Have Blue contributed only minimally to its low L/D ratio (7.5). Much more important were the high leading-edge sweep (72.5°) and low aspect-ratio of the wing. This loss of L/D will have an adverse effect on range. Moreover, limitations of the basic wing planform coupled with the lack of high-lift devices or of a separate horizontal tail to help trim the aircraft can also result in very high take-off and landing speeds.
- stability and control due to reduction/removal of control surfaces ensuing from stealth considerations. The need for tailless designs places great reliance on the flight control system (FCS) and may require thrust vector control (TVC) to reduce fin and rudder areas.
- the controllability of novel configurations (e.g. the YF-23, which employed highly tapered and aerodynamically loaded wing tips and tailless aircraft) and the associated effect on the FCS (i.e. increased mass and engine power offtake penalties). Consideration needs to be given to the effect on the FCS of the air data system, the efficacy of which may also be reduced by stealth considerations.
- due to the increased size from the need to carry all stores and fuel internally, basic airframe drag will be increased. Buffet and ride qualities due, for example, to very short-coupled configurations are also issues to be considered.
- the problems of weapon release and weapon bay aerodynamic and acoustic loading, caused by internal carriage of stores. These require extensive engineering effort.

Propulsion penalties

The propulsion penalties due to stealth design include:

- the effect on aircraft and engine performance of inlet duct positioning and shaping, due to stealth requirements. The loss of available installed thrust due to air inlet and nozzle shaping for RCS and IR control can be a major penalty. The insistence of 100% LOS blockage to the engine face to reduce signatures, via a long S-duct to give the required offset, will mean losses in pressure recovery (i.e. intake efficiency) that increase markedly with throat Mach number.
- the air induction system, which can be simplified by using a fixed geometry inlet, although it will lose pressure recovery at high Mach numbers. The wide spectrum of angle of attack (AoA) and sideslip required of the inlet, together with the extreme range of hot- and cold-day operation, leads to special needs for matching inlet and engine air flows. This has been done in the past, quite successfully, by variation of the geometry of the compression surfaces and by spilling air from the subsonic diffuser. If the inlet is fixed the former method is not available, which necessitates either direct spill to the freestream via bleed doors or bypassing air around the engine. The former method is undesirable for RCS reasons, whereas the latter may benefit cooling of the exhaust, as appears to be the case with the B-2A. Inlets that have to operate at much higher AoA than hitherto are likely to suffer severe flow distortion unless, as on the previous generation of fighters, the inlet is shielded by the fuselage/wing strake or has extensive stagger (overhang) or rotating inlet lips (e.g. Eurofighter Typhoon). The F-22's inlet avoids the use of any variable geometry (to minimise RCS) or shielding, apart from about only 30° of stagger.
- RCS-reducing screens, which on supersonic aircraft can only be placed within the subsonic duct, otherwise unacceptable pressure recovery losses occur. The provision of screens, vanes or deflectors within the subsonic portion of the inlet duct will further reduce pressure recovery and add weight. However, there may be a need for vortex generators within the duct to suppress flow separation in a fixed inlet at high AoA, for example, and RAM-coating them may serve a dual purpose.
- pressure recovery, which is vital in the anticipated Mach 1.4 to 1.5 supercruise regime. Lack of it will degrade the installed thrust and will lead to the need for significantly higher aircraft uninstalled thrust/weight ratios. However, elimination of the actuation hardware (e.g. hydraulics and seals) and movable surfaces (and their inevitable gaps) typically found in a variable geometry inlet will

The use of internal weapon bays is not new as shown by this late 1950s sequence showing a Convair F-102A about to launch Hughes AIM-4 Falcons. With the pilot's finger depressing the trigger the fire control system would determine the optimum moment for launch whereupon the missile bay doors flicked open (in less than 0.5sec) and the weapons fired. Trim changes were automatically compensated by the flight control system. Much development was needed to ensure that the fierce blast of the Falcon's motor did not create havoc inside the bay. Many thorny problems, such as the persistent tendency for the Falcons to break range-lock at the moment of firing, were traced to the blast effects of the rocket motor. *(General Dynamics via author).*

increase reliability while decreasing weight and cost, as well as RCS. Moreover, the volumetric and structural efficiency of the inlet will be improved as a result of fixing the inlet and will help pay for the weight of any RAM treatment. It may be that the inlet aperture can have a 3-D geometry. Such shapes offer better aerodynamic performance but have been avoided because their behaviour was difficult to predict accurately. In addition, manufacturing processes have favoured conical or two-dimensional inlets because of their relative simplicity. However, advances in analysis techniques now allow fixed 3-D shapes to be aerodynamically, electromagnetically and structurally integral with the airframe. Normally a high-cost, heavy item, new production methods may make them more affordable, and structural RAM in certain sections of the inlet duct may be a way of reducing weight.

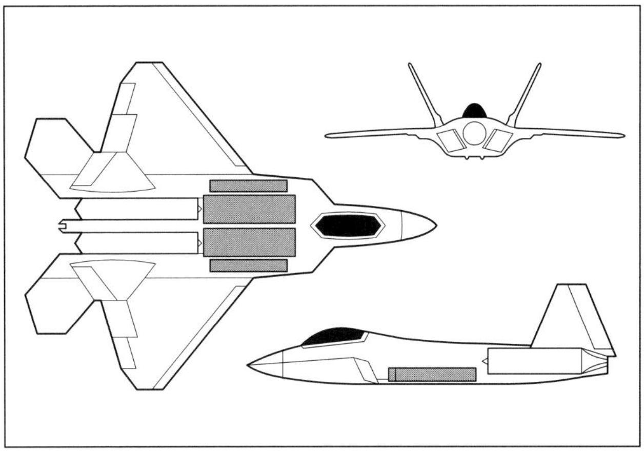

The need for internal carriage of weapons and fuel can mean a high volume penalty. The F-22's missile bays (shown shaded) have a total volume of around 270ft^3. Allowing for airframe structure and systems, this provides enough for about 11,000lb of fuel.

Volume penalties

Fighters are notoriously short of fuel capacity, and volume constraints due to internal weapons carriage do not help in this regard. Large internal weapon bays, their influence on structural performance, their effects on stealth characteristics and their operation in weapon release, all contain some unknowns. However, the USAF has had experience, since the 1950s, with such bays on, for example, the Convair F-102A Delta Dagger and F-106A Delta Dart. Nevertheless, the question arises as to whether very low observable (VLO) strike aircraft will be able to carry enough stand-off weapons in their internal bays to achieve a sufficiently high kill rate per sortie. This may be achieved by the use of a very accurate 450kg (1,000lb) bomb, designed to be as capable as current 900kg (2,000lb) weapons (though this argument is equally applicable to 'conventional' weapons carriage). As an example, the F-22 weapons bay had to be redesigned to contain four of the joint direct attack munitions (JDAM/GBU-29) of 450kg each, as an alternative to six compressed-carriage AIM-120Cs. The F-22's two main weapon bays are each about 0.6m (2ft) deep by 1.2m (4ft) wide by 4m (13ft) long, with the two cheek bays about 0.52m (1.7ft) wide and deep and 3.3m (11ft) long. This gives a total volume of around 7.6m^3 (270ft^3), which, allowing for airframe structure and systems, is sufficient for 5,000kg (11,000lb) of fuel. This is a little under twice that carried internally by an Lockheed Martin F-16 or equal to that carried by an McDonnell Douglas F-15A Eagle. The military always want to carry more ordnance, so the F-22A is to enter service equipped with four wing hardpoints for external carriage of weapons and fuel tanks.

Weight penalties

An aircraft that carries all stores and fuel internally requires a large fuselage which, being a major and therefore heavy component, leads to increased weight. For example, the operational empty weight of the F-15E is 8% greater than that of the F-15D. While a large measure of this weight increase was due to the structural redesign for the strike role of the F-15E, the provision of the extra conformal fuel tankage was a contributor.

In 1994 it was realised, from use with enhanced RCS prediction tools, that the F-22 had problems with its signature. This required modification of some details and increased use of RAM, including the inlet duct. Unspecified signature control changes added 60kg (140lb) plus 45kg (100lb) due to the change to titanium wing spars. By the time of its critical design review in February 1995 the empty weight of the F-22 had increased by 4%, though only partly due to signature control. The chevrons on the undercarriage and weapon bay doors were reduced by 60 to 70% from six or seven teeth per door to two, and made bigger. The bigger teeth required additional structure bracing adding weight, but bigger teeth meant fewer corners to give smaller RCS. Instead of spending huge amounts of money to correct the performance shortfalls due to the weight growth and engine efficiency shortfalls, the USAF accepted them and slightly relaxed the aircraft's performance specification. The F-22's required sustained turn performance at altitude was reduced and load factor marginally lowered.

Manufacturing penalties

Building for stealth requires a level of care and perfection unprecedented in aerospace. As shaping techniques for the major components made it possible to achieve lower RCS than ever before contemplated, smaller details (e.g. the air data system, doors and access panels, refuelling receptacle) became relatively more important. Ensuring that RCS remains low is a process of taking care of many thousands of small details, including changes to seams, corners, holes, steps, gaps and edges. Where designing for affordability is concerned, the foregoing performance requirements, and others associated with the smoothness of the aerodynamic profile required for stealth reasons, are forces driving away rather than towards affordability. There is, therefore, a very real challenge in achieving an affordable airframe against this background of greater accuracy of build, new materials and new structural forms. For example, the wing span of the B-2A of 52.4m (172ft) is held to within 0.015% (i.e. 6mm (1/4in)). With new higher-than-estimated values of its RCS, the F-22 had the number of access panels reduced by one third, and 80% (200 to 44) of the drain holes on its undersurface were eliminated. The inlet's interior geometry

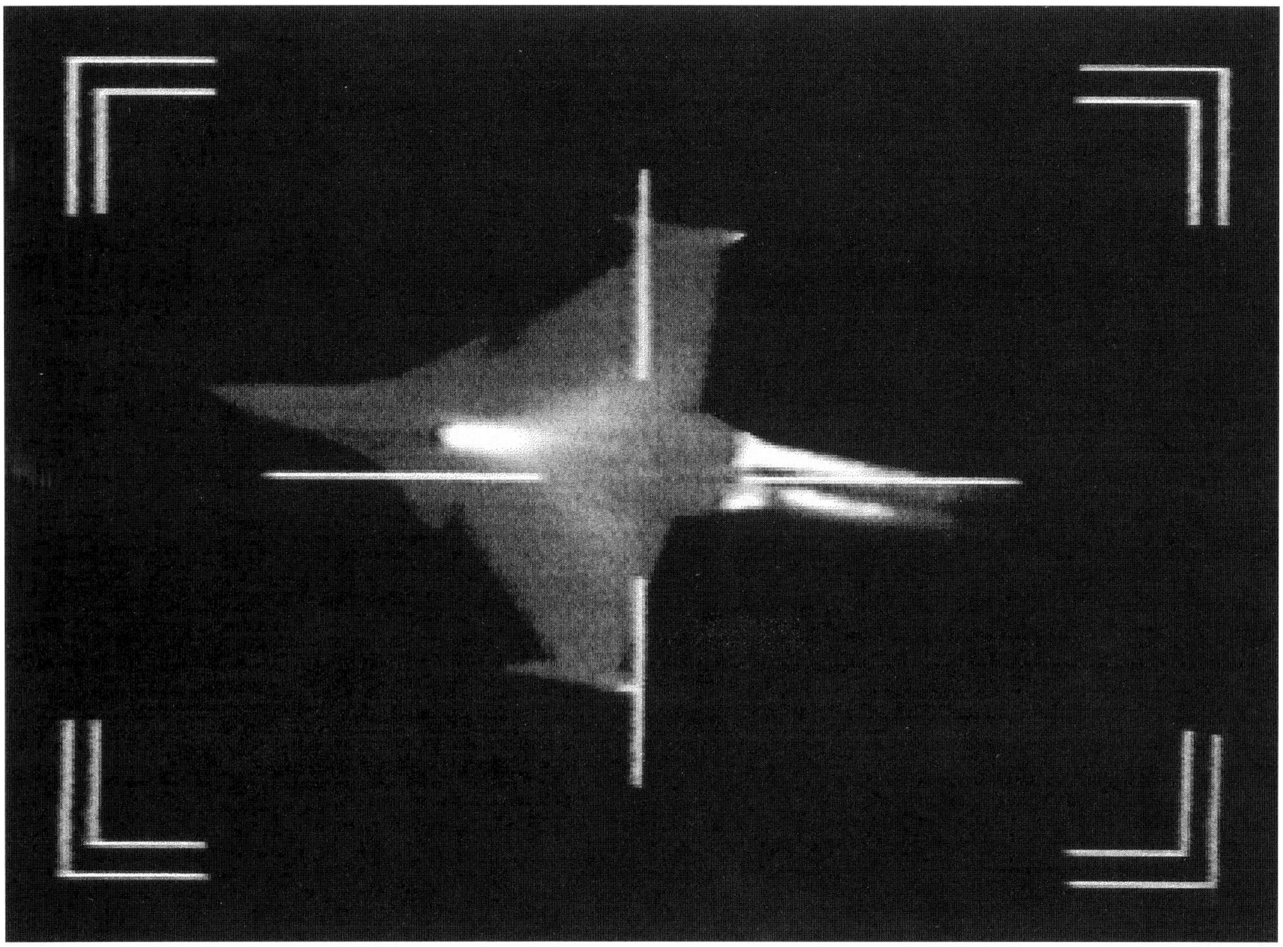

This 8-12 micron wavelength thermal image of a Dassault Rafale shows not only the afterburner plumes and wing tip-mounted smoke generators but also the hot spot marking the environmental control system heat exchanger's exhaust and diffusion outwards over each wing. *(Mark Richardson via author).*

The Lockheed Martin F-22 Raptor on its first flight in 1997 showed some of its stealth characteristics: diamond cross-section forebody, constant-chord control surfaces and shielded engine nozzles. The nose probe contains flight test instrumentation and would not be carried on an operational aircraft. The two diamond-shaped openings ahead of the port canted fin are the exhaust for the APU while just ahead is a similarly-shaped door over its inlet. *(Pratt & Whitney via author).*

needed changing along with some materials. A tightening of tolerances in tooling was also required.

Maintenance penalties

The application of signature control technology will affect an aircraft's support system. The use of RAM, for example, in structures and as coatings, needs extra support, test and evaluation procedures to verify the continued performance of the signature reduction methods, since stealthiness, difficult to achieve, is easy to lose. Just before the contract negotiations leading to Lockheed's F-117A, its Have Blue stealth demonstrator had been picked up at fifty miles flying towards the radar range at White Sands, NM. This was due to three screws securing a panel on the undersurface not being quite tight and standing proud of the surface by less than 3mm (⅛in). Uneven surfaces were coated with a radar-absorbing putty.

Overall design requirements for low observables may overly constrain the layout of an aircraft and its systems. A good arrangement for stealth is not necessarily a good arrangement for maintenance accessibility. In the case of the F-117A, for example, the design philosophy was stealth at all costs. If the aircraft wasn't stealthy, it didn't matter how maintainable it was because it would not be able to perform its mission.

It is claimed that the F-22 will reduce by half the required number of maintenance personnel per aircraft (to 8.7 from 16.6 for the F-15). However, the RCS modifications referred to above included the number of engine access doors being reduced from three to two per engine. This may affect the task of maintenance and make it harder to achieve the reduction to twelve MMH/FH for the F-22, from fifteen MMH/FH for the F-15, and engine change time to 1.5hr from 2.5hr. The equivalent numbers for the Typhoon are nine MMH/FH, and the time for four personnel to change two EJ200 engines is aimed at forty-five minutes.

Radar-absorbing Materials and CFC

A good, low RCS design should exploit shaping to the greatest possible extent. However, there are situations where shaping may be inappropriate, or might fail to meet one's objective in full. In these cases the aircraft designer turns to RAM in either the design phase or as a retrofit. The use of RAM could reduce a Tornado's RCS by a factor of ten. But, as noted previously, because radar signal strength weakens as the fourth power of the distance, this would result in a relatively small reduction of detection range. Nevertheless, RAF fighters employed in the Gulf War had RAM tiles glued inside intakes and radar-absorbing paint (RAP) on the leading edges of fins, wings and tails.

As its name implies, RAM is intended to reduce the scattered signal by absorbing some part of the incident radiation. Microwave energy is converted into heat energy with hardly any noticeable temperature rise because the energies involved are extremely small. Various kinds of material can be made to absorb microwave energy by impregnating them with conducting materials such as carbon and iron.

In the main, there are two kinds of absorbers currently used, dielectric RAM and magnetic RAM. Addition of carbon products in an insulating material introduces electric resistance and changes the electrical properties. Hence carbon-based absorbers are called dielectric RAM. The most familiar examples are pyramidal absorbers found in anechoic chambers. Dielectric RAM is usually too bulky, fragile and unusable where space is limited and severe mechanical vibrations exist. Magnetic RAM uses iron products such as carbonyl iron and iron oxides called ferrites. Iron effectively dissipates radar waves and has been used in paint. It is quite effective against the high-frequency radars used in modern fighters. Unlike dielectric RAM, magnetic RAM is compact, thin and of adequate strength to withstand loads and an abrasive environment. Nevertheless, its thickness does rob volume from volume-limited aircraft; moreover, it is heavy and expensive. The materials are usually embedded in a form of rubber tiles as so-called parasitic RAM, which can then be glued in position, and they are suitable for inlet ducts. For high-temperature applications, such as around the nozzle of the engine, use has to be made of ceramic-based RAM.

From the early 1980s Lockheed experimented with ways to spray on RAM as an alternative to the very conservative method of gluing it on in sheets. By mid-1985, automatic spraying processes were being tested on entire wings. The current emphasis in RAM development is placed on finding lighter, cheaper and wider band absorption stealth coatings to allow thinner layers to be used.

Little information has been released on radar-absorbing paint (RAP), which is a ferrite paint (often referred to as iron ball paint). This is, apparently, a polyurethane-based sprayable coating that has the advantage of being able to be sprayed in varying thickness and can additionally provide an electrical bond between adjacent panels. Since all iron-based materials are liable to oxidation then, certainly for shipboard-based aircraft, effective measures have to be taken to minimise the extra maintenance required. An additional problem is the familiar one of galvanic corrosion due to the dissimilar materials.

Carbon-fibre composites (CFC) have been employed for some time to give increased strength/weight and stiffness/weight in airframes, the benefits of which can counter the weight increase due to signature control. However, one of the consequences of CFC usage is that lightning attachments do not have the current paths available on metallic structures. This gives rise to potential skin damage and sparking. Such effects are unacceptable in integral wing tanks. Lightning protection for the British Aerospace Experimental Aircraft Programme (EAP) was designed into the wing. This involved aluminium strips on the external surface, thin metallic mesh over the thin skin areas, and other insulating and conducting measures. It was the first aircraft with CFC integral fuel tanks to be designed to withstand the 200kA lightning strike threat. However, mixtures of conducting metals and non-conducting materials causes radar scattering at joints, such as the canopy, radome and doors.

Decoys

Although stealth offers significant benefits for beyond-visual-range (BVR) combat, close combat is still inevitable. Any benefit is likely to be limited to the increased effectiveness of countermeasures that will not have to disguise a large radar or infra-red signature but, rather, decoy the threat.

The deployment of decoys can saturate an air defence system, thereby minimising the loss of aircraft by forcing the wasteful expenditure of threat munitions. Decoys are used not for signature reduction, but, rather, signature *control* for deception purposes. Decoy vehicles have to mimic the signature characteristics of the actual aircraft, including not only the absolute levels they have (so the decoys have artificially enhanced signatures) but also signature fluctuation rates. As threat sensor capabilities have improved there has been a requirement to mimic even the signature information that can be extracted via signal-processing techniques, such as imaging IR.

While an aircraft can eject both flares and chaff simultaneously, later separation of the two types of decoys (IR and radar) occurs because of their quite different aerodynamic behaviour. Pulse-Doppler radars can easily distinguish a target from a free-floating decoy. A towed decoy overcomes this problem and can provide a single source using a flare and corner reflector.

The Future

On the basis of a programme of much independent research carried out during the 1980s, the USAF stated in 1989 that no defensive concept would negate the value of stealth technology in the near future. Over forty potential means of defeating stealth were investigated in the programme. It is interesting to note that by 1995 it was recognised that radar systems (especially low-frequency ones) and techniques do exist which can detect at tactically useful ranges even the B-2. Because the required thickness of RAM is so large and shaping has little effect at the lower radar frequencies, current stealth capacity is limited to the shorter wavelengths (frequency range of 3 to 10GHz). Long wavelengths are less affected by small details of shape and absorbent structures. Though current stealth technology may frustrate modern air defence radars, the same is not true of older, long-wavelength (lower-frequency) systems that have been kept operational worldwide. Some countries

When everything else fails! A Tornado F.3 of 111(F) Sqdn ejecting flares. The deployment of decoys can saturate an air defence system thereby minimising the loss of aircraft by forcing the wasteful expenditure of threat munitions. However, they have to closely mimic the signature characteristics of the actual aircraft. As threat sensor capabilities improve there will be a requirement to mimic even the signature information that can be extracted via signal processing techniques, such as imaging IR. *(BAe via author).*

Technicians at Northrop Grumman Corporation's Electronic and Sensors Division prepare an F-22 radar antenna for tests in an anechoic chamber. The AN/APG-77 uses an electronically steered phased array of 2,000 transmit/receive modules to provide the agility, low RCS and wide bandwidth necessary to support the F-22's air superiority mission. *(Northrop Grumman via author).*

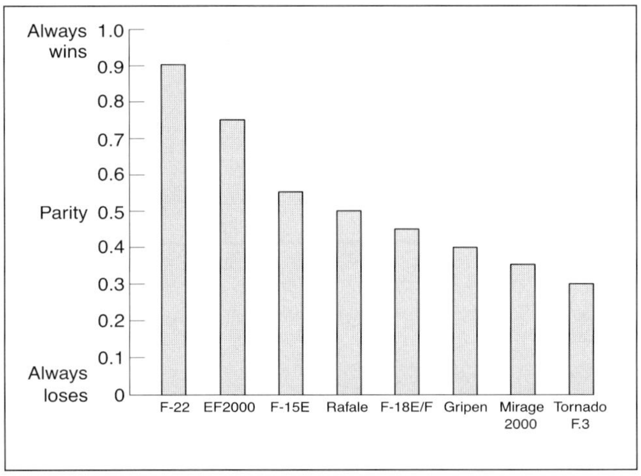

Combat simulations carried out in the UK by the Defence Evaluation Research Agency (DERA) for up to 8 v 8, produced the scores shown when the aircraft were matched against the upgraded Su-27 'Flanker' equipped with AA-10 Alamo missiles. It can be seen that the Typhoon is superior to all fighters except the F-22, which is three times better in terms of exchange ratio. Though the data should be treated with scepticism, it does point out the advantages of low RCS and stealthiness in general.

were prompted to do this not because of low RCS aircraft but to avoid over-reliance on one type of radar and to overlap many different types to make their air defence system more difficult to jam.

However, all airborne targets detected by long-range surveillance radars must eventually be passed over to fighters or SAM sites. These are equipped with high-frequency tracking and targeting radars that can be defeated by shaping and RAM. How effectively surveillance radar systems could hand over to shorter-wavelength sensors is questionable, and is one of the main arguments for investing in stealthy designs.

There are differences among stealth advocates over what degree of invisibility will be needed in the twenty-first century. There are advocates on both sides of the argument. To some, referring to low-observables (LO) means *very* small signatures. They argue that this cannot be achieved by adding new materials and coatings to existing platforms. Other defence officials do not believe that an aircraft has to have very low signatures to be effective. The US Navy, for example, has staked its immediate aviation future on an $80 billion investment

On the McDonnell Douglas F/A-18E/F all major apertures are aligned with the wing leading and trailing edges, and the fuselage sides and air intakes are canted at the same angle as the vertical tails. Antennae and vents are aligned with the main wing planform. Main undercarriage and engine bay doors have serrated edges. Because the air intake ducts do not provide line-of sight blockage to the engines, they house a radar absorbent baffle which reduces intake efficiency but decreases a major source of radar return. *(Boeing via author).*

in the F/A-18E/F Super Hornet. This is not a low-signature design but one that uses the addition of new materials, some reshaping and coatings. Stealth advocates note that much of the effect will be lost because stores hung from wing pylons as they are on the F/A-18E/F will increase the RCS greatly.

It may be that, though a few years ago it was thought a large fleet of VLO aircraft would be needed, a more modest investment would be more sensible. It may be, as the advocates of the Eurofighter Typhoon feel, that the use of VLO fighters and missiles is not the most reasonable nor affordable answer. Vehicles with some frontal stealth, including reduced-signature versions of today's aircraft backed up with a modest stealth fleet of future strike fighters and stand-off jamming and deception, could 'do the job'. A long-range stand-off weapon coupled with high terminal accuracy diminishes the need for VLO, which is only required some of the time anyway, and there are many more targets that do not justify VLO. Investment in stealth is heavily influenced by declining defence budgets; less stealthy assets may be sufficient and much cheaper. Instead of concentrating on airframes, the focus may be on the external carriage of stealthy weapons (conventional shapes with RAM treatment and/or conformally mounted). Beyond that there are advocates of the many unmanned combat aerial vehicles that have appeared in conceptual form over recent years.

While signature control is necessary to achieve acceptable exchange ratios and improve survivability, an enduring tenet of aircraft design is that, as the introduction of the swept wing showed, you do not get something for nothing. The key is how to keep new aircraft affordable. This will only happen if there is no 'requirements creep', though past experience is not encouraging. Stealth is difficult and expensive to achieve, but easy to lose through lack of attention to detail.

Chapter 8

Systems

The extent of the various systems used on aircraft is enormous and represents a large fraction of the value in terms of cost of the modern fighter. It is not possible in one chapter to do full justice to the topic and only the major items of fuel and power systems (hydraulic and electrical), along with undercarriage design, are considered.

Fuel Systems

Fuel supply

Up to the 1930s the supply of fuel was simple: a gravity tank above the engine on the upper wing, either alone or as a gravity header tank supplied from the main tanks in the fuselage by a windmill-driven fuel pump, usually mounted on a centre-section strut. When the low-wing monoplane fighter arrived, there was no alternative but to fit engine-driven pumps. While efficient in themselves, they were handicapped by their location relative to the tanks. When bolted to the engine, as was customary, they became heated, which caused problems of vapour lock due to high temperatures and low atmospheric pressure. In addition, they became increasingly inefficient as the pumping pressure to be developed was raised, and this increased with manoeuvring flight due to the g loads. By the middle of World War Two, larger engines with greater fuel demands led to the use of immersed electric pumps in the tanks to provide an artificial head at the engine pump. With the introduction of the jet engine the use of immersed pumps (fuel-lubricated, with the ability to run dry safely) was retained.

Fuel transfer pumps are needed for moving fuel between the tanks. This requires a fuel supply to a collector tank, holding only enough for a few minutes' flying, for each engine before being fed into the engine feed lines. The pumps also need to move fuel around the aircraft to keep the centre of gravity (CG) within acceptable limits, to maintain pitch and lateral trim balance. With increasing aircraft speed, the centre of lift moves aft, but trimming the aircraft with the tailplane increases drag. To avoid this, the CG can be shifted by pumping fuel from forward to aft tanks.

Fuel tanks

Whereas World War Two aircraft were refuelled at overwing points, the modern fighter has a single underwing/fuselage refuelling receptacle. In most cases an aircraft is not fuelled to capacity, rather the ground crew selects the fuel load and sets the appropriate levels at the refuel/defuel panel. As the fuel enters, large quantities of air have to be displaced very quickly (around 1,000lt/min), since the fuel is pressure-fed (up to 0.34MPa (50psi)) to speed up the process. To avoid the tanks bursting the excess air has to be vented overboard. The vent system also allows air to leave the tank as the aircraft climbs and enter the tank on descent; the differential pressure would otherwise cause the tank to burst or collapse.

Flexible (bladder) fuel cells were used widely in the 1950s, and are still used as fuselage tanks. Their advantages are ease of installation via access doors, ease of repair by turning the tanks inside out, self-sealing for small-calibre weapons, crash and vibration resistance. Their disadvantages include high cost, damage by folding and unfolding and insertion through sharp-edged holes, additional weight and hence cost due to the inverse relationship between aircraft empty weight and warload. Whereas the shapes of wing tanks are fairly simple, fuselage tanks are complex. They and their systems have to compete fiercely with other components for the limited volume within the fuselage profile. The segregation of fuselage tanks into a large number of small ones is due to the need to avoid air intakes, structure and engine.

Integral wing tanks were introduced in the 1950s when aircraft structures had become more stiff due to requirements of higher flight performance. This, coincidentally, fulfilled a much-needed fuel tank function without the extra weight of flexible cells. More recently, the use of carbon-fibre composites (CFC) has provided complex-shaped integral fuselage tanks with the advantage of increasing fuel capacity by 10 to 15%.

While fuel tanks and systems are required to have zero leakage, this is rarely achieved. It is the major disadvantage of integral tanks. Continual problems with leaks have been experienced, leading to high maintenance costs. Whereas wing-tank leaks can be located in many cases by walk-around inspection, detection of leaks in integral fuselage tanks is more difficult. The point where the leakage is first noted rarely coincides with its internal origin. Tanks around air intakes usually have to be drip tight, as leaking fuel could severely damage an engine or cause an intake fire. Fuel leaking into 'hot' bays (e.g. auxiliary power unit (APU) or gun compartment) is a high-risk situation and requires redundant barriers or fire walls.

External fuel tanks

External tanks carry a high drag penalty with around 50% of the fuel they contain needed to overcome their drag. Moreover, aircraft with drop tanks will be subject to either dynamic pressure or g limits due to structural loading, and not all are designed for supersonic flight. Though it is common to jettison external tanks before merging into combat, this is expensive and may cause logistics difficulties during a prolonged conflict. The transfer of fuel from the external to the internal tanks is nowadays automatically sequenced, typically from the externals (wing then ventral), then wing tanks and finally the fuselage tanks. Emptying the wing externals first allows their jettison to reduce drag and the aircraft's rolling inertia.

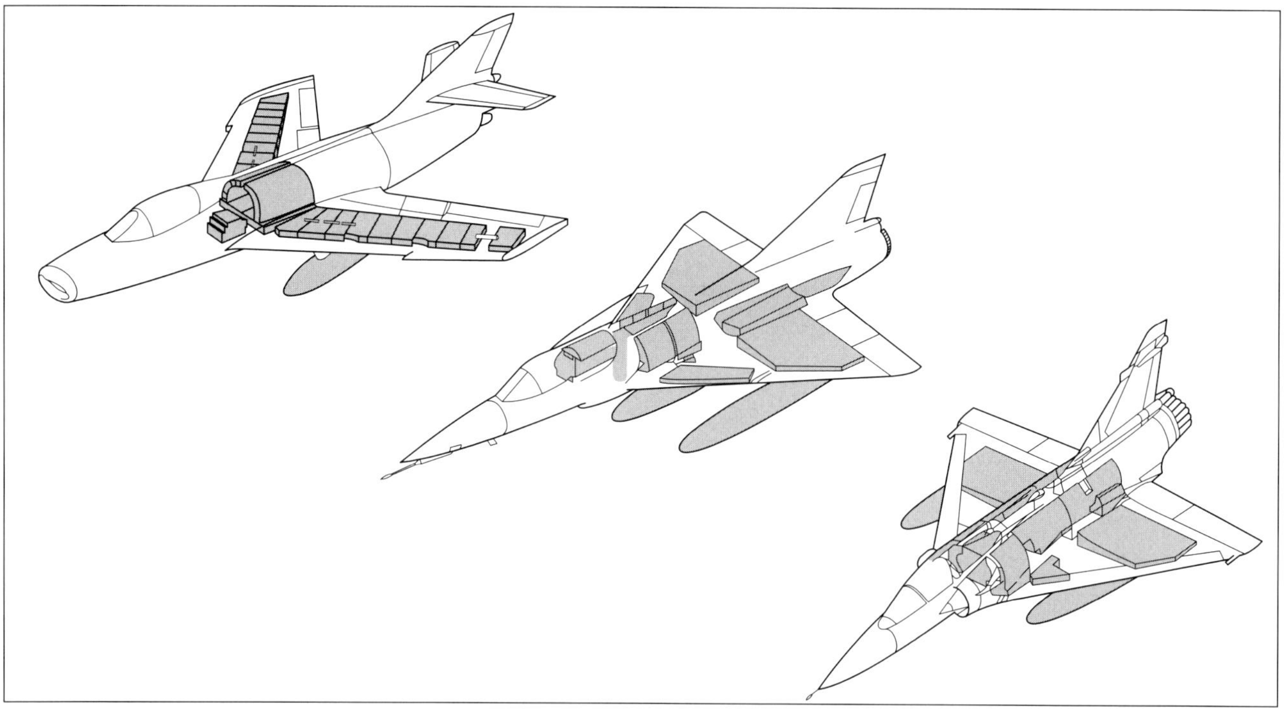

Evolution of typical fuel tank layouts.
TOP: Dassault Super Mystère B2 (1956) with flexible bag tanks (simplified). MIDDLE: Dassault Mirage IIIE (1961) with integral wing, flexible fuselage and ventral tanks. BOTTOM: Dassault Mirage 2000 (1978) with totally integral tankage. *(Mike Badrocke/AIR International).*

Fuel venting from the tip tanks (a common feature of early jet fighters) of a Grumman F9F Panther. Fuel constitutes a large fraction (1/3) of take-off weight and if an aircraft suffers an emergency or malfunction shortly after take-off it may be necessary to jettison a large amount of fuel to reduce weight rapidly to a level acceptable for landing, since most aircraft are not designed to land with a full fuel load. *(via author).*

Fighters are notoriously short of fuel volume, the British Aerospace Lightning especially so. Some F.6s (including this one from 11 Sqdn) were fitted with overwing ferry tanks on their strengthened, kinked and cambered leading edge wings. *(BAe via author).*

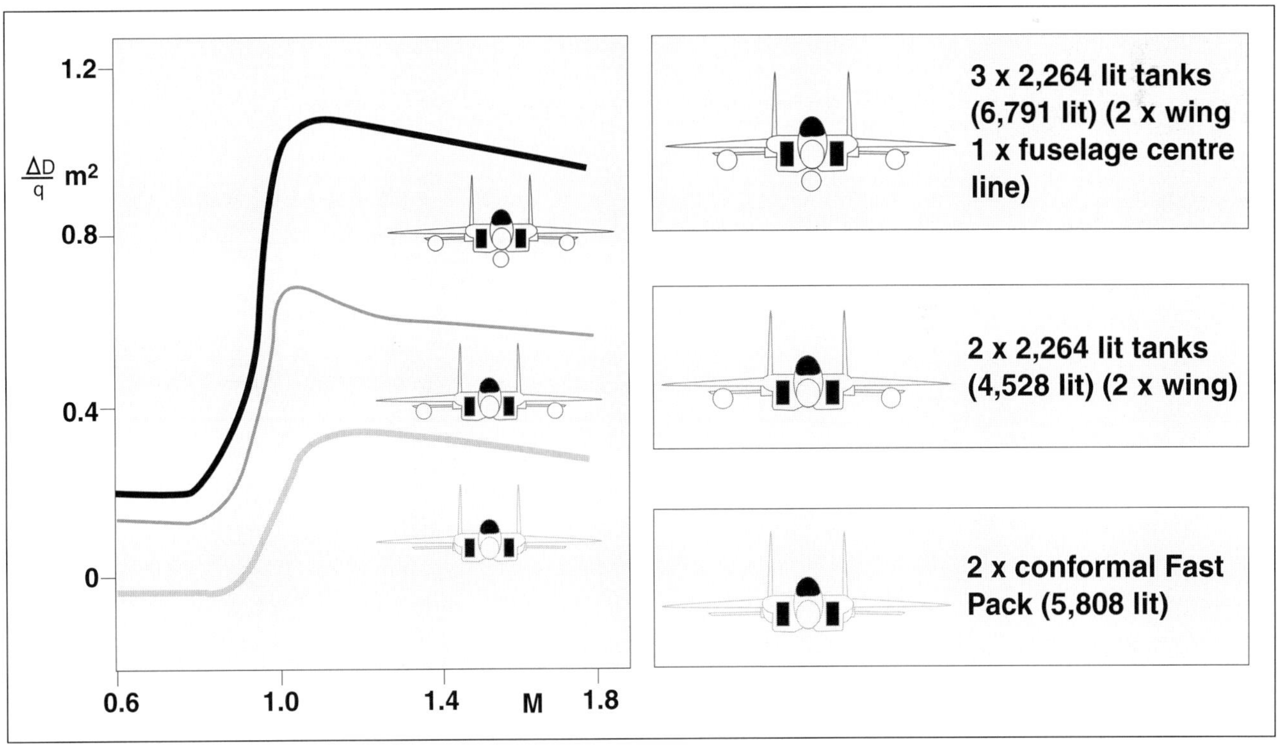

Traditional external fuel tanks add enormously to the drag of a fighter. The so-called Fast Packs (now called CFTs) demonstrated on the F-15A reduced subsonic drag (ΔD/q is drag increment/dynamic pressure), doubled the air superiority mission radius and almost tripled the interception loiter time (at 500nm) compared to the clean aircraft. *(Pete West/AIR International).*

Use of the 'buddy' refuelling system between two early production de Havilland Sea Vixen FAW.1s. The tanker is carrying a Flight Refuelling Ltd Mk 20 pod on its starboard wing and a 150 Imp gal fuel tank on its port wing. *(Flight Refuelling via author).*

A KDC-10 (DC-10-30-CF conversion) of the RNAF uses the standard USAF flying boom method to refuel one of four F-16s. The telerobotic system is equipped with TV cameras to provide 3-D displays to the dedicated refuelling operator. *(McDonnell Douglas via author).*

Air-to-air refuelling

The first demonstration of inflight refuelling was in 1934. However, the advent of jet aircraft rendered obsolete the 'looped hose' method that needed an operator in the tanker to handle the hose and an operator in the receiver to connect it. It was not until 1949 that Flight Refuelling Ltd in the UK developed the 'probe and drogue' method, viable for jet aircraft. It has since been very widely used by fighters because their limited fuel volume and high fuel consumption restrict range and endurance. The USAF holds out against the 'probe and drogue' method and instead uses the 'flying boom' system, which requires a dedicated operator in the tanker aircraft to 'fly' the boom into position. The main advantage of this is the much higher fuel transfer rate (about twice as fast but limited to one receiver at a time).

Fire/explosion suppression

Fire in the air, the historic killer of aircraft, has always been a hazard and great efforts have been made to combat it. Before World War Two engines were protected by extinguisher bottles, but the thermal detection switches used could result in the extinguisher contents being blown through the engine if the propeller was not first stopped. So the fire detector was used to flash a warning light, whereupon the pilot went through a propeller feathering (if available) and extinguishing drill.

The high-octane gasoline used in piston engines is explosively volatile and fighters went to war in 1939 with most tanks protected by self-sealing covers such as, for the RAF, the Henderson safety tank. Although the petrol-soluble interlayer of these tanks expanded to fill the bullet holes, a cannon shell tore them open. An addition to self-sealing is tank inerting. Such a system was used on the Yakolev Yak-1 of World War Two, whose tanks, as they gradually emptied, received burned engine gases, purified through a filter, to reduce the danger of an explosion.

Though the kerosene fuel for gas turbines is not as volatile, the occurrence of a fire remains extremely serious. Fuel tanks are a potential explosion hazard when partially full, since the volume above the fuel fills with fuel vapour. If air is present in sufficient concentration the resulting mixture is combustible. Penetration of the tanks by shrapnel or incendiary rounds can result in an explosion. To reduce the risk of this, tanks can be filled with a reticulating foam of low density (19 to 32kg/m^3 (1.2 to 2.0lb/ft^3)) which acts as a heat sink to reduce the temperature at a spark point. It also breaks up compression waves that precede the flame front in an explosion. The alternative to foam is an active fuel tank inerting system which fills the void in the tank with an inert gas, such as nitrogen.

The most likely place for a fire to occur is the engine bay. All the necessary ingredients for starting and maintaining a fire are readily available: plenty of fuel, air and hot ignition sources. Thus, fire detecting and extinguishing systems have to be provided and current systems include fire-wire detectors in the isolated hot zone of the engine. These activate circuits that close the air and fuel supplies to this zone before discharging the extinguisher bottles.

Table 14 Comparison of fuel systems of the F/A-18 and the F-4

	F/A-18	*F-4*
Fuel system		
Fuel tanks isolated from engines	✔	✖
Inlet duct/fuel tank damage control	✔	✖
Self-sealing fuel tanks 'get home' fuel	✔	✖
Self-sealing engine fuel feed lines	✔	✖
Fuel transfer lines inside tanks	✔	✖
Split system for crossover capability	✔	✖
Survivable external fuel tank	✔	✖
External void filler foam	✔	✖
Internal foam	✔	✖
Propulsion		
Firewalls between engine, AMAD and APU	✔	N/A
Fire detection and suppression system	✔	✔
Engine blade containment	✔	✖

Many lessons pertaining to survivability (the response of an aircraft given a hit) were learned in Vietnam. Redundant, separated, and damage-tolerant fuel and hydraulic systems are now the norm. Considerable survivability can be designed into an aircraft at little cost in weight or performance. Add-on fixes generally entail too high a weight penalty.

Undercarriage

Before World War One

Although skids were used on the Wright Flyer of 1903, a forerunner of the undercarriage as we know it today first appeared on Santos Dumant's 'Aeroplane-14 bis', which flew in 1906. Most aircraft were fitted with bicycle-type wheels carried in cycle forks, or had simple tubular or wooden skids. Many had castoring wheels to ease cross-wind landings, though when more robust gears were found necessary, non-castoring wheel layouts, suitably braced, took the place of the cycle-type arrangement. To prevent nosing over, a common problem, several aircraft of the period had forward projecting skids or wheels.

This Avro 504 (fitted with a Vickers 0.303in machine-gun) shows the use of laced-on wheel fairings and a forward projecting skid to prevent nosing over, a common problem for taildraggers. *(via author).*

The Sopwith Pup's undercarriage would not stand the slightest amount of drift on landing and it was not uncommon for the aircraft to finish up on its back or nose. *(Steve Jarrett).*

World War One

The fragility of wing structures required that the undercarriage was attached directly to the fuselage near the engine mounts, meaning a narrow track with poor stability but light in weight. The tailskid type with a pair of vee-braced mainwheels of bicycle type with wire spokes, for lightness, fitted with narrow large diameter section solid tyres, was the commonest arrangement. The poor lateral strength of the vee-type of undercarriage was considered a crashworthy feature that protected the airframe. The first advance was aerodynamic: the addition of laced-on fabric covers to reduce drag, which from 1915 was in use for more than a decade. Vertical descent velocities of up to 4.5m/s were not unusual and required reaction factors similar (4 to 8g) to the load factors of airframe design. The use of bungee cord wrapped around the axle, allowing a few inches' vertical motion to the wheel with respect to the bracing, absorbed initial shocks but lacked damping and led to 'hoppity-hop' landings.

Between the wars

It was realised that low shock-absorber efficiency due to the lack of damping caused high airframe reactions and undesirable recoil. Increased shock absorption became necessary to accommodate constantly increasing aircraft weights and sink speeds. Although shock-absorber stroke is not a function of aircraft weight, it was vital to increase the stroke in order to lower the landing load factors (vertical load÷aircraft weight), and thereby minimise the structure weight influenced by the landing loads. By the late 1920s, improvements in the quality of light alloy castings made possible the 'solid' wheel with a fat low pressure 0.2MPa (30psi) tyre. These provided some of the shock absorption, but size limitations and relatively low (47%) efficiency prevented a major contribution from this source.

Improved shock-absorbing support struts were devised, which ranged from rubber blocks and coil springs to leaf springs and oleo-pneumatic (oil-air) struts. The first application of the latter for aircraft by Bréguet in France stemmed from the use of gun recoil gear in several countries, a development that had begun many years before. In the UK, oleos were fitted to the BE2, FE2b and Armstrong Whitworth Siskin. With efficiencies as high as 90% in terms of minimising the stroke for a given sink speed and load factor, it remains an excellent device for absorbing the vertical kinetic energy. Unlike a coil spring, it does not store the energy then release it, causing the aircraft to bounce down the runway. Instead the oil returns to its normal static condition at a controlled rate so that rebound does not occur. Some pilots, however, complained that oleos were not good for 'getting off' because they did not 'bump' at all and full flying speed was needed before getting airborne.

Wheel brakes

Wheel brakes had been considered since the earliest times, but their principal disadvantages were the cost of fitting and maintenance, extra complication and weight. Brakes were useful for parking, but were not much use for stopping, since if they were powerful enough to be effective, they usually tipped the aircraft on its nose. However, they had occasionally been used before 1930, by which time increased landing speeds meant that they became mandatory and their adoption became widespread. So by the early 1930s, simple cable-operated drum brakes were coming into use, in conjunction with tailwheels instead of skids, to improve steering on the ground and eliminate the need for technicians at the wing tips. This meant, though, that the conventional vee-type undercarriage could not react brake torques, and the required additional members imposed significant weight penalties. This led to the internally sprung wheel that allowed a rigid structure of low weight to react the landing loads and brake drag, but it carried with it the penalty of large-diameter, low-pressure tyres to permit adequate travel. Nevertheless, the installation on the Gloster Gladiator, for example, showed the relative aerodynamic cleanliness of the concept.

Retraction

Increasing aircraft speeds made undercarriage drag of greater importance, but biplanes did not lend themselves to retractable undercarriages, and though airframe design changed from braced wood and fabric biplanes to aluminium alloy monoplanes, undercarriages lagged behind. Britain was slow to accept the change and persisted for several years with 'trousered' mainwheel legs. Many designers felt that the added complication and weight was not justified. Eventually they did become retractable, but the difficulties of achieving this by mechanical means were considerable and the general adoption of retraction had to await development in hydraulics. However, by the beginning of World War Two, with operational speeds of 300kts (550km/h), the hydraulically or electrically retractable undercarriage with power brakes was widespread. The single telescopic oleo leg was almost universally adopted for fighters, as exemplified by the Spitfire installation in its relatively thin wing, albeit with a narrow track.

Tailwheels versus nosewheels

The weight of early aircraft had to be kept to a minimum. Since the undercarriage is a dead weight, the lightest, simplest and cheapest type was the tailwheel, with the aircraft sitting down at an angle of attack slightly less than its stalling angle. The fixed-nosewheel layout never stood a chance of becoming commonplace on propeller-driven fighters since a nosewheel had to be larger than a tailwheel and would have produced far more drag and reduced propeller thrust. In most circumstances, furthermore, a retractable nosewheel would have been impractical because the engine fully occupied the nose.

The inter-war period saw the introduction of the first true nosewheel layout aircraft, the Bell P-39 Airacobra of 1938. The advantages of this type of undercarriage were the improved view over the nose, making it far less demanding, it could be taxied faster and straighter without needing to swing the nose from side to side to see ahead, and the protection of the propeller. Nosewheels only started to become more common in the 1940s (with differential braking for steering control) when the future of fighters was seen to lie with the turbojet.

Low pressure (35lb/in^2) tyres were typically used by World War Two fighters because they were restricted to operations from grass airfields. At a time when the art of aircraft hydraulics was in its infancy, Sydney Camm's emphatic dictum was that the undercarriage of the Hawker Hurricane (Mk 2A shown), when down, was to be a sound structure not dependent in any way on hydraulic pressure for its integrity. *(BAe via author).*

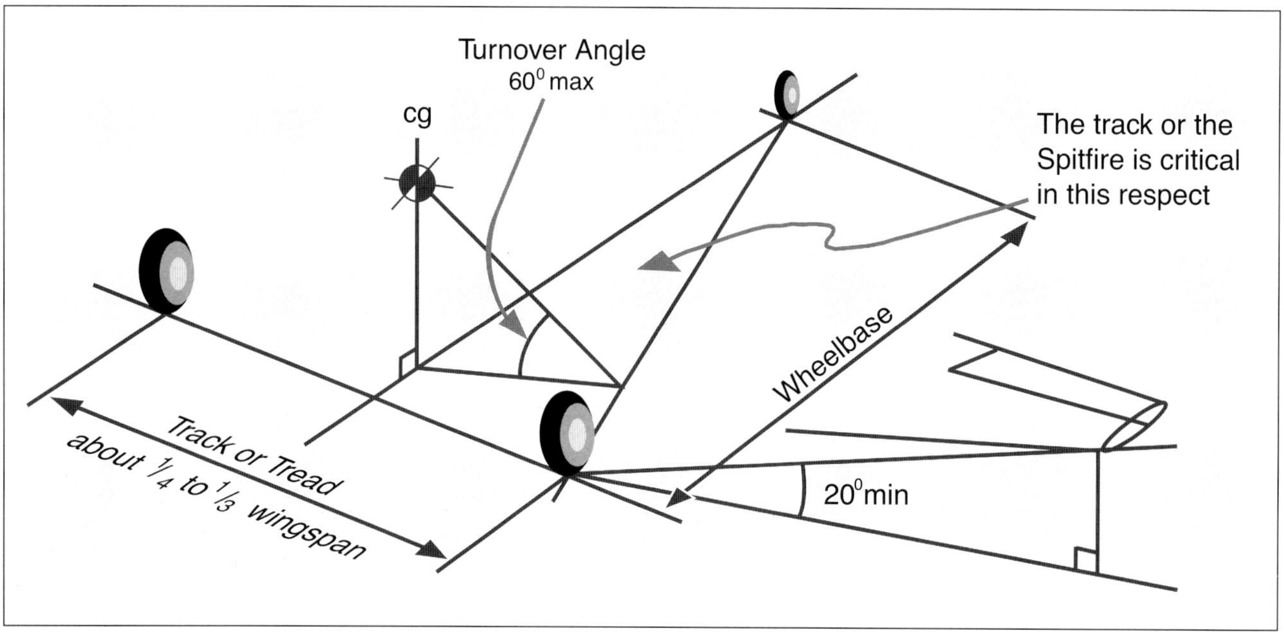

Lateral positioning of the undercarriage affects the turnover angle and ground clearance (with combined pitch and roll on landing) with flaps (they may need their deflection angle reduced or span reduced), underwing stores (diameter may be limited or very short pylons used) and wing tips. The turnover angle should not be more than 60° for a land-based aircraft (50° for carrier-based aircraft). The Spitfire had 60°, F-4E: 39°, F-104G: 36° and strenuous efforts are made to not exceed 55°. *(AIR International).*

Table 15 Tailwheel versus nosewheel

Parameter	*Tailwheel*	*Nosewheel*
View over nose	Poor	Good
Nose over tendency	High	Low
Control on ground	Poor	Good
Acceleration on take-off	Slow	High
Undercarriage weight	Light	Heavy
Deceleration on landing	High	Low

The disadvantages of taildraggers lay in ground handling. Because the CG of the aircraft is behind the mainwheels, there is a powerful tendency for the tail to swing outwards in a turn, leading to a ground loop. The nosewheel type's lower AoA made for faster acceleration on take-off and it restricted the damage from blown debris. Because nosewheel units have to take higher loads than tailwheels they are heavier, and lower mainwheel loads means reduced braking.

Positioning

The undercarriage represents a significant fraction of total aircraft weight (typically 3% of TOW for current land-based fighters). A complex retraction mechanism is heavy and a considerable maintenance cost item. Much skill is needed in making them light and simple, yet strong enough to absorb the repeated shocks of moderately heavy landings.

A simple rule for the modern tricycle undercarriage is to locate the mainwheels so that the aftmost CG is forward of a line drawn at 15° forward of the vertical drawn through the wheel centre. This avoids tail tipping by preventing the CG from moving over and aft of the main gear. The tail clearance angle for take-off rotation is usually about 15°.

The main gear needs to be placed to give efficient transmission of loads into the airframe. Retracting the undercarriage into a hole on the undersurface of the wing, between the spars, is undesirable, as it weakens the structure and displaces fuel volume. The alternative is the use of a stub spar aft of the rear spar with the wheel retracting between the two, though this makes for a narrow track.

For fuselage-mounted gears, used on high- or mid-wing aircraft, it was usual to have a main fuselage frame attached to the wing rear spar. This is an ideal structure for mounting the undercarriage, although a wing's mean aerodynamic chord moves aft with sweepback, and the CG, having moved aft as well, is often too close to the rear spar frame to suit mounting of the undercarriage at that point. In that event, a secondary frame must be added aft of the rear spar frame, with the undercarriage loads transmitted forward to the rear spar, through the fuselage skin. Modern fighters, however, with thin multicell wings having little trailing-edge sweep, have at least one of three hefty fuselage frames to pick up the undercarriage loads.

Nosewheel loads in the static condition vary from the extremes of 6% to 20%, though a preferable range would be 8% with CG aft to 15% with CG forward. Too small a nosewheel load leads to insufficient grip for steering, while too large a load means that nose-up rotation on take-off is difficult. It is preferable to have the nose gear retracting forward. In the event of a complete failure to extend it, the pilot can release the locks and allow the emergency blowdown system, gravity (with extra pulled g) and air drag to push the gear into the down and locked position.

Carrier operations

The undercarriages of carrier-borne aircraft have to be designed for twice the sink speed of land-based types (7.3m/s (24ft/s) rather than 3.6m/s (12ft/s)). Thus, the kinetic energy to be dissipated (which increases with

One of the earliest tricycle undercarriage fighters, the highly innovative Saab J 21 first flew in mid-1943. The aircraft had heavy control forces and the 1,475hp Daimler-Benz DB 605B pusher engine installation was prone to overheating. As a fighter it was obsolescent, but was used for ground attack. Only the second production fighter to be equipped with an ejection seat, this is a J 21A in the markings of Flottilj 8. *(Saab via author).*

Location of the main undercarriage has to be in a position that appears to be the most conducive to efficient transmission of loads into the airframe. To retract the undercarriage into a hole on the undersurface of the wing between the spars is undesirable. It compromises structural integrity where bending loads are reacted in the wing skins, and it uses up valuable fuel tank volume. Indeed, with thin wings it is impossible to stow a tyre inside. The BAe Hawk overcomes this dilemma by mounting its mainwheels on the front wing spar ahead of the wing box though blisters are required to accommodate them. *(BAe via author).*

Of mid-wing layout, the Yak-25 had its engines mounted at 40% span and so required its undercarriage to be of bicycle configuration with wing tip outriggers. This is a Yak-25K of the Soviet Air Force of the early 1960s. *(Alexei Mikheyev via author).*

The original McDonnell Douglas F-18 had its undercarriage so far aft that despite a big tailplane, the tail arm x download was insufficient to adequately lift the nose on take-off with the CG at its most forward position. The flight control system's software needed modifying to toe-in the rudders to give extra nose-up moment for rotation.

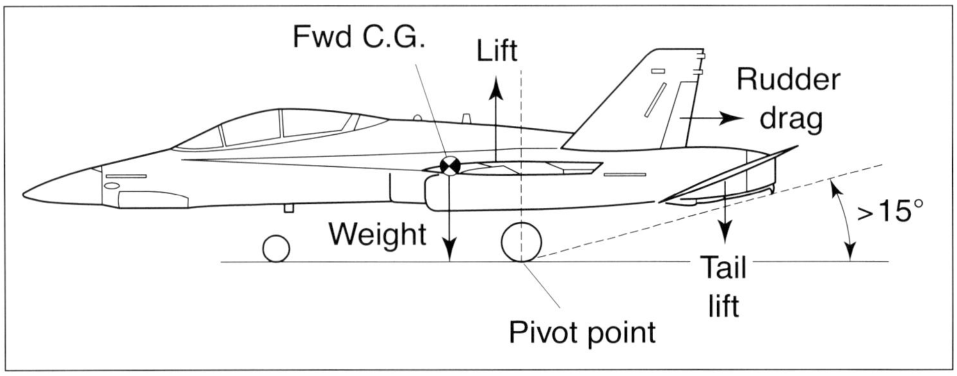

Nosewheel loads in the static condition vary from the extremes of 6% to 20% of take-off weight, though a preferable range would be 8% with CG aft to 15% with CG forward. Too small a nosewheel load leads to insufficient grip for steering, while too large a nosewheel load means that nose-up rotation on take-off is difficult and may be outside the tailplane's power.
(N. W. lift: nosewheel lift).

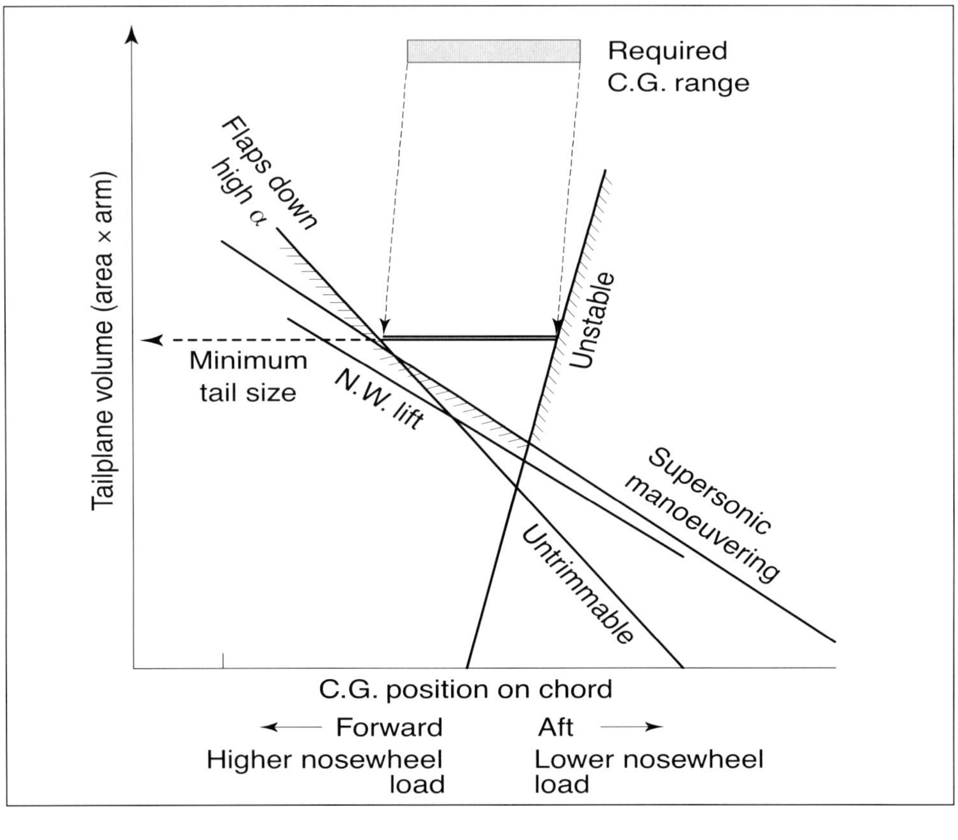

Many carrier-borne aircraft, with vertical kinetic energy levels four times those of land-based ones at equal weight, use levered suspension (on nosewheel) and triangulated support structures for their mainwheels. These are relatively heavy gear types, but they have one big advantage: they provide large strokes (of the wheel) while shock strut length is minimised, thereby providing a long stroke gear which is, at the same time, compact.

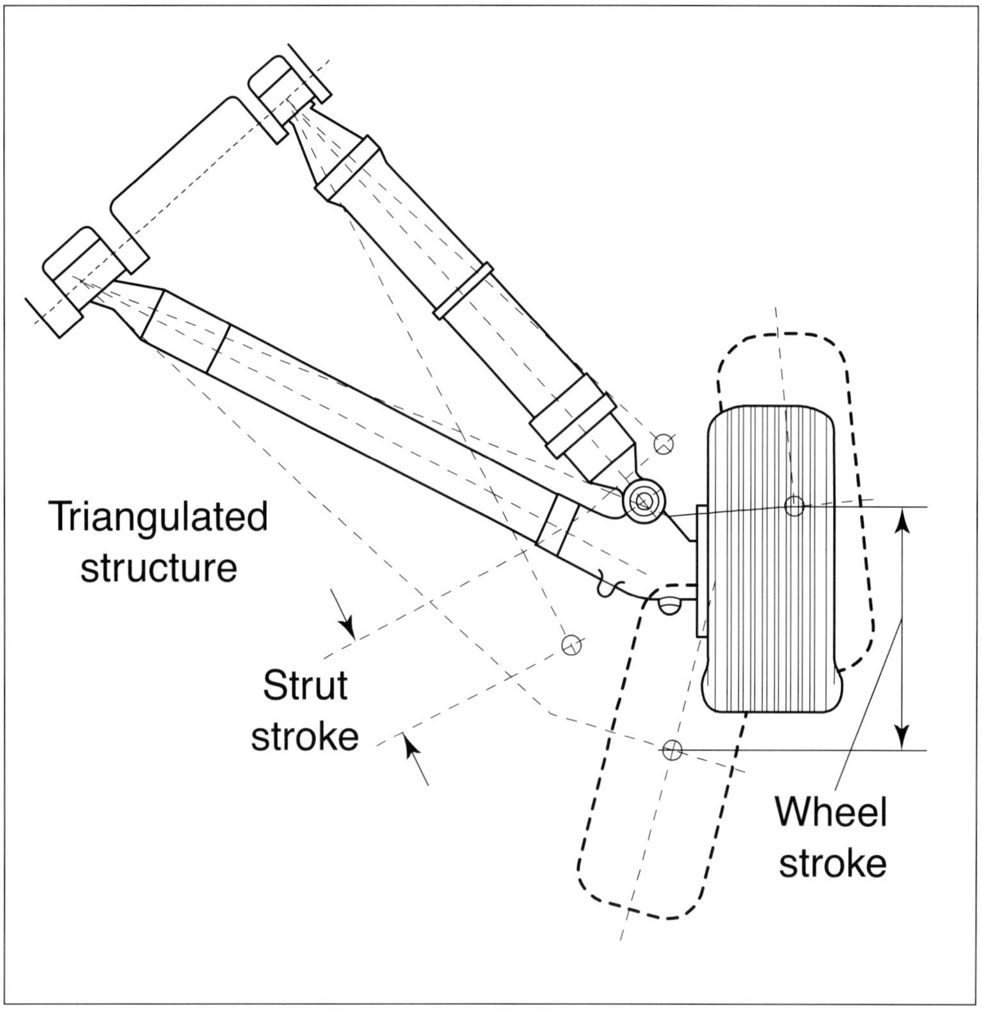

speed^2), is quadrupled for a given aircraft weight. Certain parts of the airframe structure then have to be designed by the 'controlled crash' loads experienced during carrier landings. Superimposed on this design case is the drag load caused by rotational inertia of the wheel about the axle during the time of very violent spin-up on touchdown. Additional loads come from a ship's roll and heave. To minimise these effects, many navy aircraft use levered suspension and triangulated support structures. These were major developments in shock absorption during World War Two that brought one big advantage: compactness. The bending load was removed from the shock absorber and at the same time a large wheel displacement was available with small piston movement, and hence short shock strut length. These are relatively heavy gear types, and though the narrow track of fuselage-mounted undercarriage renders them less stable for carrier operations, many naval fighters have them.

Skid control

Until the end of World War Two the drum brake was the standard, whereas today the disc brake is universal. In stopping an aircraft its kinetic energy is transformed into heat. To begin with, all brake heat sinks were made from steel, though beryllium was introduced and used on the Grumman F-14 Tomcat originally. In more recent years carbon has been used (e.g. McDonnell Douglas F-15 Eagle), and though it is expensive, its high specific heat at low weight, high thermal conductivity and low thermal expansion are essential. In an aborted take-off, the brakes can glow red hot (over 1,000°C (1,830°F)), sufficient to weld the stacks on steel brakes.

A scrubbing tyre in an undetected skid can burn through its many plies and blow out in seconds. In an even shorter time the tyre can flat spot and be doomed to be removed and probably scrapped. These problems were alleviated by anti-skid braking systems that started to appear in the late 1940s as cyclic on-off systems, and the Dunlop Maxaret system of the 1950s made the problem of stopping much easier. The current systems of skid control are now digitised with fibre optics, and provide selective deceleration, which the pilot pre-selects prior to touchdown (e.g. Lockheed Martin F-16).

Tyres

After World War Two, tyre pressures, which had been restricted to about 0.24MPa (35lb/in^2) by the need to operate from grass, increased rapidly as aircraft weight and take-off and landing speeds rose, so that 2.4MPa (350lb/in^2) and 260kts (480km/h) speed ratings have been produced. In selecting tyre sizes, an allowance is made for at least 25% growth in aircraft gross weight within the same basic size. This avoids the necessity of changing wheel and tyre sizes during service life. Tyres are normally designed to be 30 to 35% deflected under static loads, being fully squashed at about three times static load (i.e. 3g).

Carrier-based aircraft require very robust undercarriages with a wide track for no-flare landing stability on a deck that may be pitching and rolling.
ABOVE: Low-wing Dassault Etendard
ABOVE RIGHT: McDonnell Douglas F-4 Phantom with wing-mounted mainwheels.
RIGHT: High-wing McDonnell Douglas F-18 with fuselage-mounted mainwheels with levered suspension to widen the track. *(Author, IAT '97).*

An extendable nose leg is one way to get extra AoA for a carrier take-off, this shows another. Supermarine Scimitar F.1 of 803 Sqdn being launched from HMS *Hermes* in May 1960. Of note is the levered-suspension nosewheel and wing-mounted mainwheels, inboard of inner wing pylons carrying 150gal fuel tanks. *(Crown Copyright via author).*

The original high-wing MiG-23 'Flogger' had unusual undercarriage retraction kinematics that required forward then upwards and inward rotation into the fuselage. The MiG-27 derivative has fatter tyres designed to aid soft-field operations and required bulging of the aft fuselage. This is an Indian Air Force MiG-27M with nose housing a laser ranger and marked target seeker, while under the rear fuselage is the folded ventral strake. *(Hindustan Aeronautics via author).*

The Sukhoi Su-27 undercarriage is a simple, sturdy and rugged system, intended to allow operation from semi-prepared airfields. The nose leg has a single wheel mounted on a trailing arm and fitted with a mudguard to avoid projection of stones, debris, mud etc. into the air inlets, and against the lower surface of the aircraft. The nosewheel is equipped with a brake, in accordance with Russian practice. The long leg is braced by a telescopic bar which, when retracted, remains outside the fuselage within a fairing. This avoids complicating the lower fuselage structure with a long opening just to house a slim rod. The mainwheel legs are attached to the wing-carry through structure immediately outboard of the engine nacelles. Retraction is by forward rotation around a slant axis, so that each wheel is positioned flat, part inside the engine nacelle above the air ducting and part into the wing root. Tailboom scraping on take-off is apparently a common occurrence. *(Author, IAT '97).*

Because of its unique propulsion system the Harrier is forced to use a bicycle main landing gear with wing-mounted outriggers. These are GR.7s with FRS.2s in the background aboard a carrier. *(Crown copyright via author).*

Table 16 Tyre pressure and life

Aircraft	*Pressure MPa (lb/in²)*	*Landings*
F-100	1.5 (220)	29
F-105	1.5 (220)	68
Lightning	1.4 (200)	20
F-4E	1.8 (265)	53
A-7D	1.9 (280)	66
F-15E	1.5 (215)	50

Kinematics

There are almost as many kinematic concepts used in undercarriage retraction geometry as there are aircraft. Each type presents its own set of geometric and operational conditions, so that the undercarriage has to be custom-built in order to provide the best solution. Fighters, being volume-limited, exercise the ingenuity of designers, and in order to retract into the smallest space possible complex motions are often necessary. In many cases the wheels must be rotated about two axes to fit inside the available space, and as with the linkages there are many ways to do this, including shortening (e.g. F-14). If the nosewheel must be rotated during retraction, a radius link can be used (e.g. Saab JAS 39 Gripen, and F-16). Typical retraction/extension times are in the range of four to eight seconds, and because of the high acceleration of today's fighters, speed for retraction limits had to be raised from 250kts (460km/h) for the McDonnell Douglas F-4 Phantom to 300kts (550km/h) for the F-16 and F-15.

Hydraulic Systems

Between the wars

To make best use of the low-wing monoplane, coming into general use in the mid-1930s, the retractable undercarriage was introduced along with wing flaps. Both needed some form of actuation. Attempts were made to use mechanical hand cranks, but the gearing necessary to make the loads acceptable resulted in very slow operation. Both pneumatic and vacuum power systems were tried. These were light and clean, but were troubled by problems of leakage. The compressibility of air meant that the position of systems needing partial movements, such as control surfaces, could not easily be controlled with any precision. They also lacked instantaneous response, and the rate of actuation is highly load dependent so that even when the flow of air had stopped in the system, the actuator would move in response to load variations.

Hand-operated hydraulic systems were originally used because there was a reluctance to mount engine-driven power supplies. Early Supermarine Spitfires, for example, had a hydraulic handpump for the wheels and a vacuum system for the flaps. However, by the outbreak of World War Two, most RAF aircraft had comprehensive engine-driven 6.9MPa (1,000lb/in^2) hydraulic systems with components supplied by specialist manufacturers. They suffered shortcomings, but by the end of hostilities, hydraulics had been established as a moderately light and relatively trouble-free method.

The basic hydraulic system has one or more engine-driven pumps. Although these and the pipework needed to supply pressurised hydraulic fluid around an aircraft might seem to be heavy, the system could be lighter than an electrical or mechanical system. This was because the high pressures used produced small units of high power. Being almost incompressible, hydraulic fluid enabled very precise control of actuator position, and made it ideal for use in powering flight controls.

Since hydraulics may not be used all the time, accumulators, which store fluid under pressure for use during peak demands or failure cases, are incorporated in the system. Fighters have at least two independent hydraulic systems, arranged in such a way that if any system fails the others are able to maintain all of the systems with little or no degradation in function. It is mandatory for wheel brakes to have a standby accumulator. Cockpit canopies are operated hydraulically and emergency opening can be done via accumulators.

Jet age

A new feature of fighter aircraft of the 1950s was the use of much higher pressure 21MPa (3,000lb/in^2), made necessary because of the higher response rates required to match the exacting control demands at high speed. This means that a hydraulic jack capable of exerting a force of 10kN (2,300lb) need only be about 24mm (1in) in diameter. Modern fighters have undercarriage and braking (with nosewheel steering) and arrester hook, leading- and trailing-edge flaps, flying controls (ailerons, rudders, tailplanes, airbrakes, spoilers), air intake and exhaust nozzle, canopy and flight-refuelling probe, all operated hydraulically. With such a number of applications, the chief objection to hydraulics, even with modern 3,000 to 4,000lb/in^2 systems, is their bulk and weight.

Hydraulic fluid

Hydraulic systems use a mineral- or synthetic-based oil to transmit pressure. Being an oil the hydraulic fluid presents a fire risk, as any small leakage at the very high pressures used will cause a fine mist to be produced. Since the Vietnam War much research on the vulnerability of hydraulic systems to battle damage has been directed to finding a fluid with reduced flammability to improve aircraft survivability following a hit.

Whereas the take-off performance of the modern fighter with a high thrust/weight ratio can require just 300m, landing is another matter. Even with wheel and aerodynamic braking (from dorsal airbrake and foreplane deflection) the Eurofighter Typhoon uses a 4m (13ft) diameter brake 'chute. *(BAe via author).*

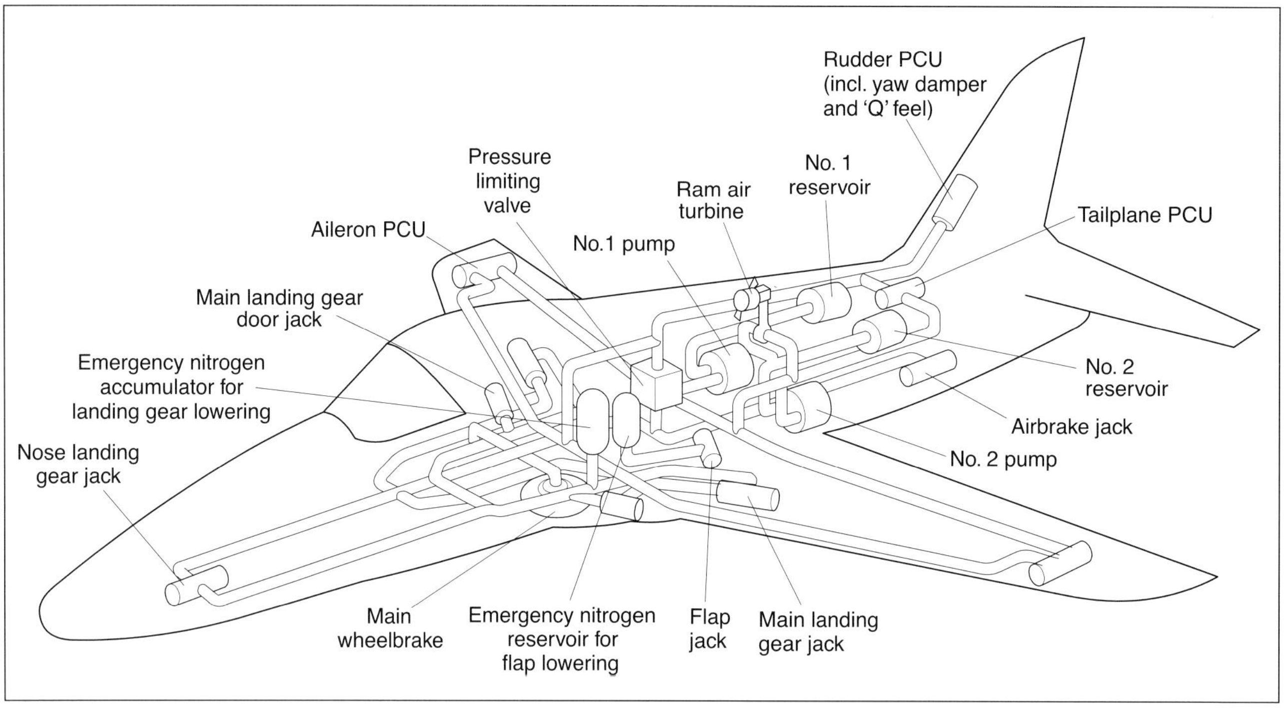

Simplified hydraulic system for British Aerospace Hawk 200. Power is supplied by two independent systems, each with its own engine-driven pump. Both systems power the flying controls by means of tandem actuators at ailerons and tailplane. In the event of engine or hydraulic pump failure a ram air turbine (RAT) automatically extends from the upper rear fuselage into the airstream to drive a pump to provide power to the flying controls down to landing speed. A pressurised nitrogen system is available to operate flaps and undercarriage in an emergency.

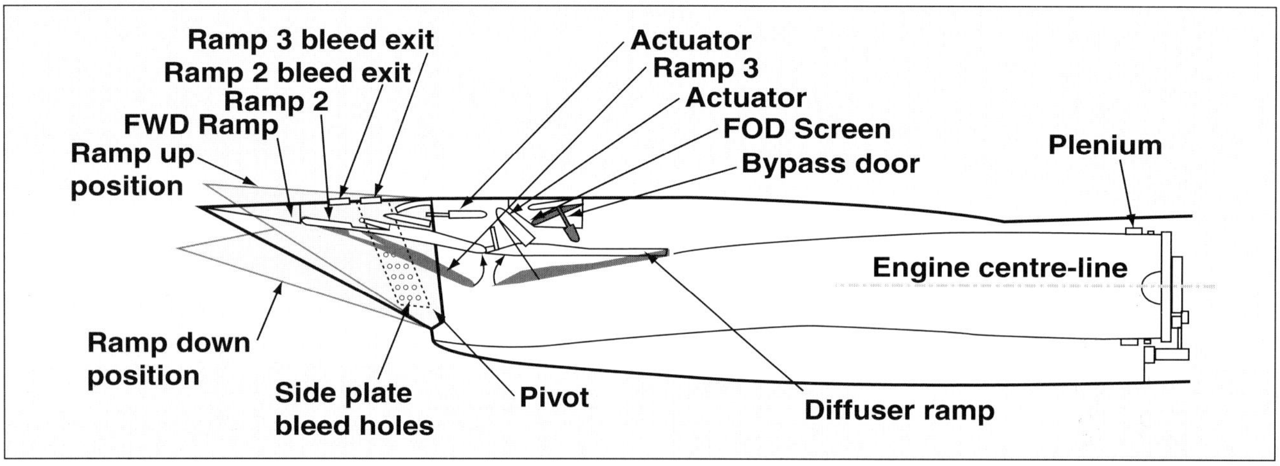

The automatic hydraulic control systems of the F-15 drive three-ramp variable capture area air intakes. The variable forward ramp changes angle with AoA and Mach number and nods up through 4° and down through 11°. Each intake is separately controlled via a system that senses AoA, stagnation temperature, pitot and static pressures and then positions the ramps and bypass doors. *(Pete West/AIR International).*

Electrical Systems

World War One

There was initially a certain distrust, mainly based on ignorance, of electrical equipment. Although electricity was accepted as necessary for some equipment, many individuals were inclined to the view that if systems could not be operated by air, cables, numerous mechanical linkages or fuel, then they were unnecessary. The earliest aircraft voice radios were massive sets, little changed from those for ground use. They needed a supply of electricity, provided by lead/acid batteries, which could be recharged by a strut-mounted windmill-driven generator. A single 12V lead-acid battery met the demands of aircraft of the period.

World War Two

The ongoing development of radio communication made further demands for electricity and by the outbreak of World War Two all fighters had electrical systems, simple but reasonably powerful. It was impossible to store, via batteries, the amount of electricity needed, so its inflight generation was required. Largely because of the needs of airliners during the early 1930s, lighter and more reliable generators were developed so that power became available for engine starter motors, eliminating the need for propeller swinging. During the war, operations became steadily more complicated; electrics had to supply ever more power for lighting, heating, radios, instruments, flap and, in some cases, undercarriage operation, arming and releasing weapons, and operating radars. By the end of World War

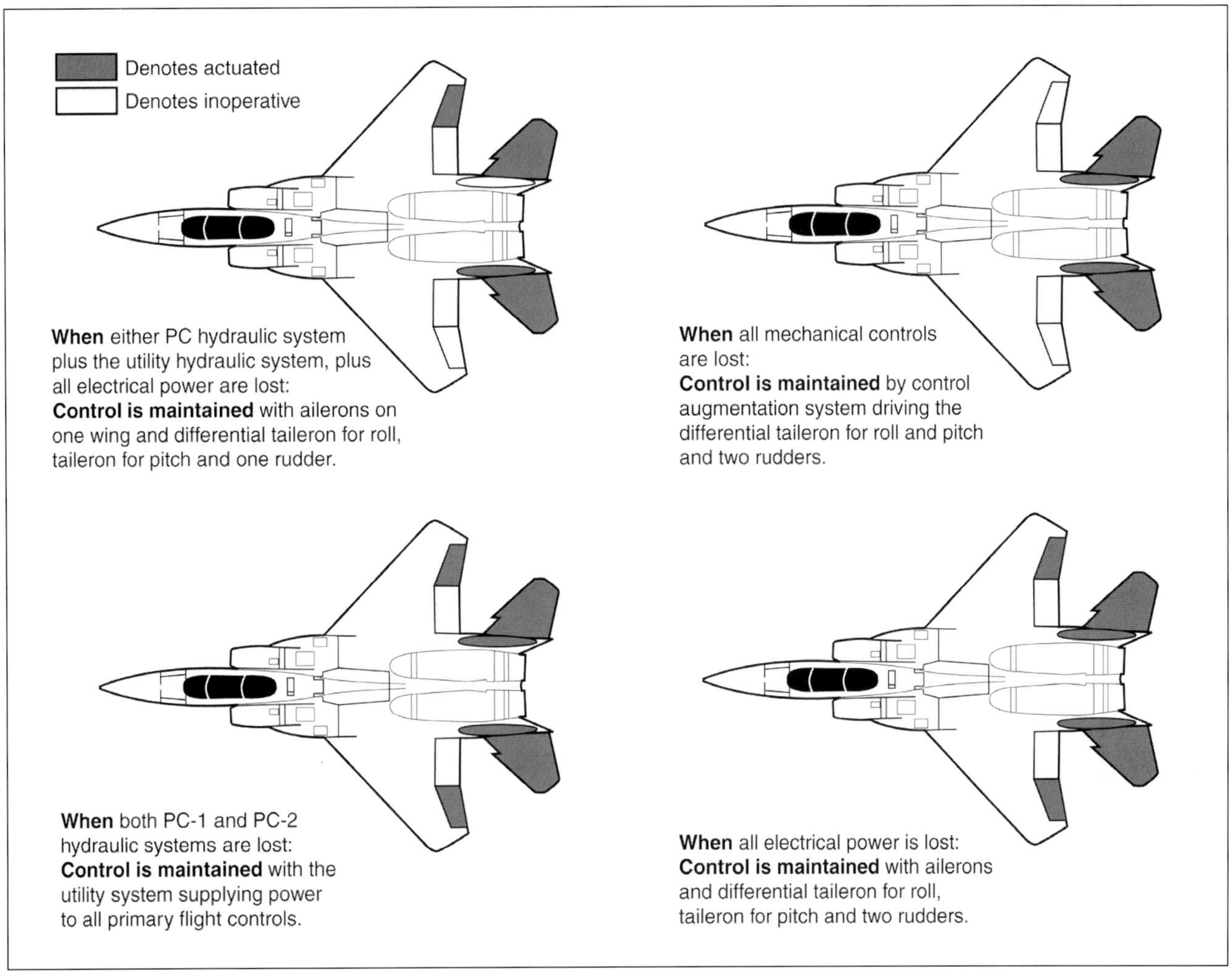

The primary flight controls on the F-15 consist of conventional ailerons, twin rudders, and differential tailplanes. Seconday flight controls are simple trailing edge flaps and airbrake. Apart from the electrically powered flaps, all flight controls are hydraulically powered. A highly survivable system was needed that could provide safe return to base and landing with major control, hydraulic, electrical, or stability augmentation system damage or failure. Combined with two engines and a highly redundant secondary power system, the F-15's flight control system provides high survivability in the event of a failure, as shown (N.B. PC: Power control). *(Pete West/AIR International).*

Space aboard aircraft carriers is at a premium and wing folding (via hydraulics) is required. This photo of a McDonnell F3H Demon on USS *Forrestal* (CVA59) in 1957 shows the large noseleg extension needed to get extra AoA on take-off. *(US Navy via author).*

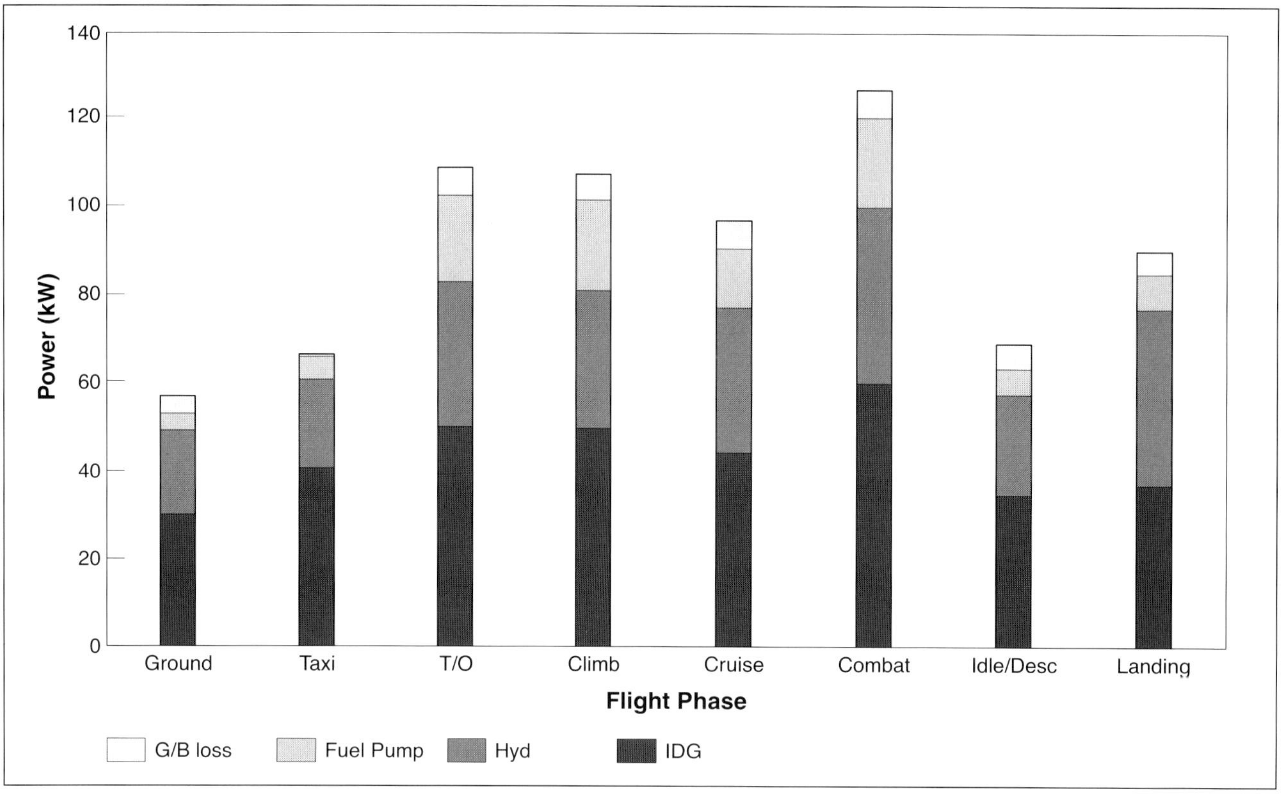

Apart from producing propulsive thrust a fighter's engine needs to provide power to onboard systems. This figure shows typically how engine power off-takes vary with flight phase (G/B: Gearbox; IDG: Integrated Drive Unit). *(AIR International).*

In many ways the Fw 190 was an 'electric aircraft'. Nearly all sub-systems were electrically operated through a 24 volt system. This approach had been selected early in the development process to enhance combat survivability in the event of battle damage, reduce weight and improve reliability. The wide-track undercarriage was actuated by small, powerful high-torque electric motors with a reduction gearbox of 3180:1 ratio of ingenious design and superlative workmanship. The motors proved highly reliable in operational service and eliminated many of the problems encountered with hydraulic actuation, a feature that would later be invaluable during the extreme cold of the Russian winter. *(Brian Birlison via author).*

Simplified electrical system for British Aerospace Hawk 200. Primary power is supplied by a 25kVA 200/115V 400Hz three-phase generator driven from the auxiliary gearbox mounted on the Rolls-Royce Adour engine. Two batteries provide sufficient capacity for engine starting and for 20 minutes of essential services should the primary supply fail. An auxiliary power unit (APU), driven from the gas turbine starter, provides electrical power at 12kVA and can be used on the ground to supply cooling air or for avionic equipment set up when no external supply is available. GSE is the alternative.

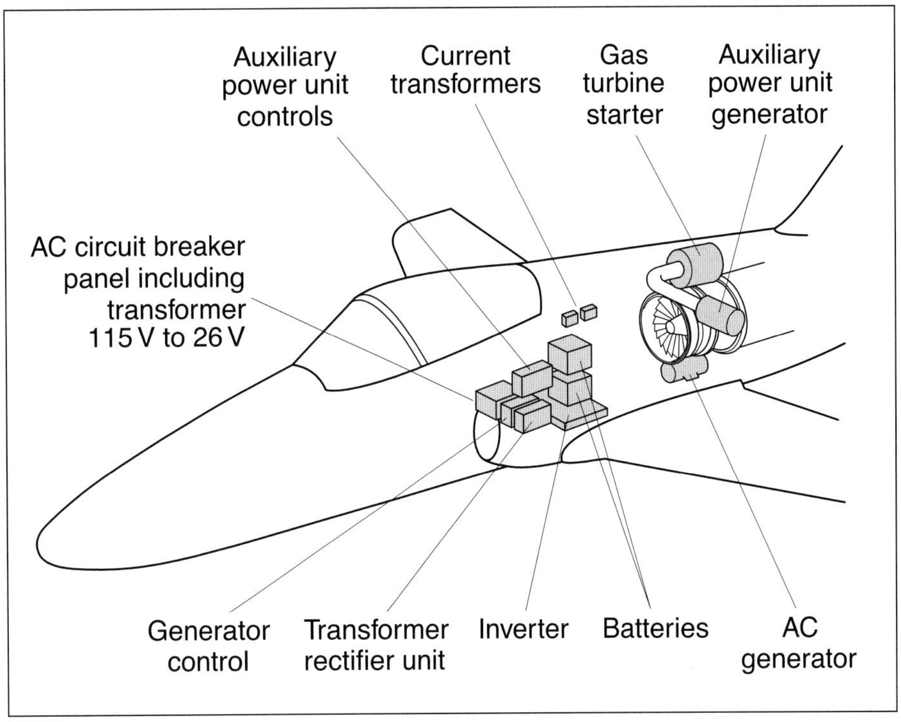

Two, 28V DC generators were straining to meet these demands.

Jet age

It was the great leap forward in the development of radar during World War Two that led to the substantial increase in the demand for electrical power. This growth has never slackened as the inclusion of more electronic equipment has been needed. Together with cockpit systems and anti-icing, this resulted in very large electrical generating loads which can reach high peaks, so that undercarriage and flaps were seldom electrically powered. There was a trend towards AC current, which is also required by instruments and radar equipment. The main advantage of AC power is that it operates at a higher voltage: 115V rather than 28V for the DC system. This is not an advantage in itself; indeed, higher voltage requires higher standards of insulation. It is in the transmission of power where the

McDonnell Douglas F/A-18E/F during its first sea trials from the USS *John C. Stennis* in January 1997. Unlike earlier F-18s the Super Hornet has dispensed with a dedicated airbrake and makes use of its FBW system (with no mechanical back-up) to work other control surfaces to provide the airbrake function. To allow the mechanical back-up controls to be removed, redundancy in electrical generation was increased substantially. Whereas the C/D has three independent power sources, the E/F has nine. In comparison with the C/D (described as an excellent carrier aircraft) the E/F is reportedly much less nervous on the approach. The C/D's leading and trailing-edge flaps are continuously in motion, during the approach, whereas the E/F's larger, higher rate surfaces do not appear to be so active, making the aircraft more solid, 'rather like running on rails'. *(McDonnell Douglas via author).*

advantage of higher voltage lies. For a given amount of power, a higher voltage relates to lower current. The lower the current, the lower the losses such as voltage drop (proportional to current) and power dissipation (proportional to current^2).

During the 1980s, the US Navy championed deriving 270V DC from the 200V/60Hz power readily available on its carriers. It allows output power of several kilowatts to be reached without excessively high currents being required. The use of 270V is an extension of the rationale for moving from 28V DC to 115V AC. However, there are a number of disadvantages associated with 270V DC: components are by no means commonplace, a large number of aircraft services will still require 28V DC or 115V AC supplies, and the use of higher voltages places greater reliance on insulation to avoid voltage breakdown. Nevertheless, a greater use of electricity, possibly to supplant conventional secondary and hydraulic systems or at least augment them to a substantial degree, is sought. The next generation of US combat aircraft (e.g. F-22, JSF) will have this imposed as a tri-service requirement.

'More-electric' aircraft

Hydraulic power is nowadays used almost exclusively as the 'muscle' or actuator function of many large loads. There has been the philosophy of 'better the devil you know' with hydraulics, as they have the benefit of millions of hours in all types of aircraft. Aircraft designers therefore have opted to continue using them, rather than breaking new ground. Until recently this seemed like reasonable caution; moreover, there were no realistic alternatives. However, recent progress in power electronics and motors (with high power/weight ratios over 10kW/kg) has changed this perspective. Furthermore, the increasing need for electrical power has revived the possibility that it could be a prime means of control and actuation within aircraft.

Many feel that the enabling technologies are now in place to permit tomorrow's aircraft to take advantage of all-electric systems to replace the costly, labour-intensive, prone to leakage and contamination hydraulic systems and eliminate the inefficient practice of using engine bleed air to power them. It is claimed that, to a great extent, the 'more-electric' aircraft would not only improve aircraft safety, but also reduce the design, development and test costs, in addition to direct labour and life-cycle cost.

At the heart of the 'more-electric' aircraft lie 'more-electric' engines, the source of power to all systems in flight. The problem is that many of the accessories driven by the engine require power only intermittently. For example, the undercarriage is only required twice during a flight – an inefficient use of power systems. In addition, the engine's electrical, hydraulic and pneumatic drive systems each have to be sized to their maximum demand. If a single type of power system was used for all the secondary systems, the overall maximum load would be significantly reduced, assuming that each individual system did not require maximum power simultaneously.

Secondary Power Systems

Aircraft require a source of power other than the primary need for propulsion, via a combination of hydraulic/pneumatic/electrical generators. It was normal for these to be driven by accessories attached to the engine. By the 1950s, the typical layout was that shown for the Hawker Hunter. The demand for weight/volume reduction (but still maintaining system integrity) led to early attempts at shared power-generation systems mounted on common gearboxes driven by both engines (e.g. de Havilland Sea Vixen).

It was not long before autonomous starting of combat aircraft was required, to make them independent of ground support equipment (GSE). From this came the air turbine cartridge starter, followed by gas turbine starters (GTS) or

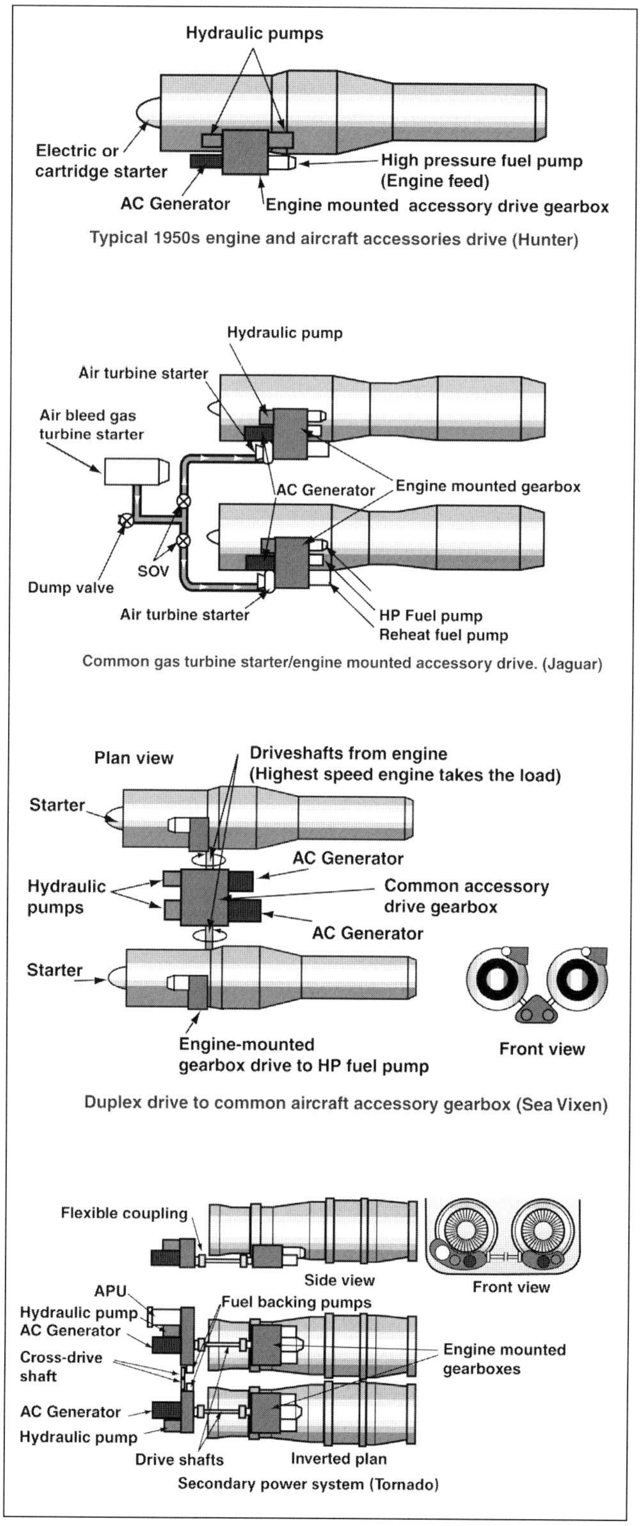

Comparison of secondary power systems. (*Pete West/AIR International*).

jet fuel starters (JFS). These were introduced during the 1960s (e.g. F-4, SEPECAT Jaguar). The GTS formed the basis of the continuously running auxiliary power unit (APU) by the addition of an air cooling system and the use of long-life materials. Many fighters carry an APU that powers the ancillary services such as electric, hydraulic, and often air conditioning on the ground and also provides emergency power in flight.

Finally, a desire for improved maintainability led to the 1970s requirements for checkout of systems without needing to run the main engine and replacement of engines without the need to breakdown the secondary systems. This led to the idea of integrating the secondary systems as a separate module, rather than mounting the accessories directly on the engine. Certain advantages accrue: reduced engine frontal area, simplified engine installation, ready access to the engine, accessories located away from engine hot spots, flexibility to drive the engine for starting and/or accessories from a separate source (e.g. Panavia Tornado). This was the basis of an integrated system of meeting aircraft power requirements that could be driven by either the engine, the APU or a ground power source.

For operational reasons it is highly desirable to have uninterrupted power supply during the transition from ground standby to engine-running mode. The aircraft is then independent of GSE and test sets on the flightline, so

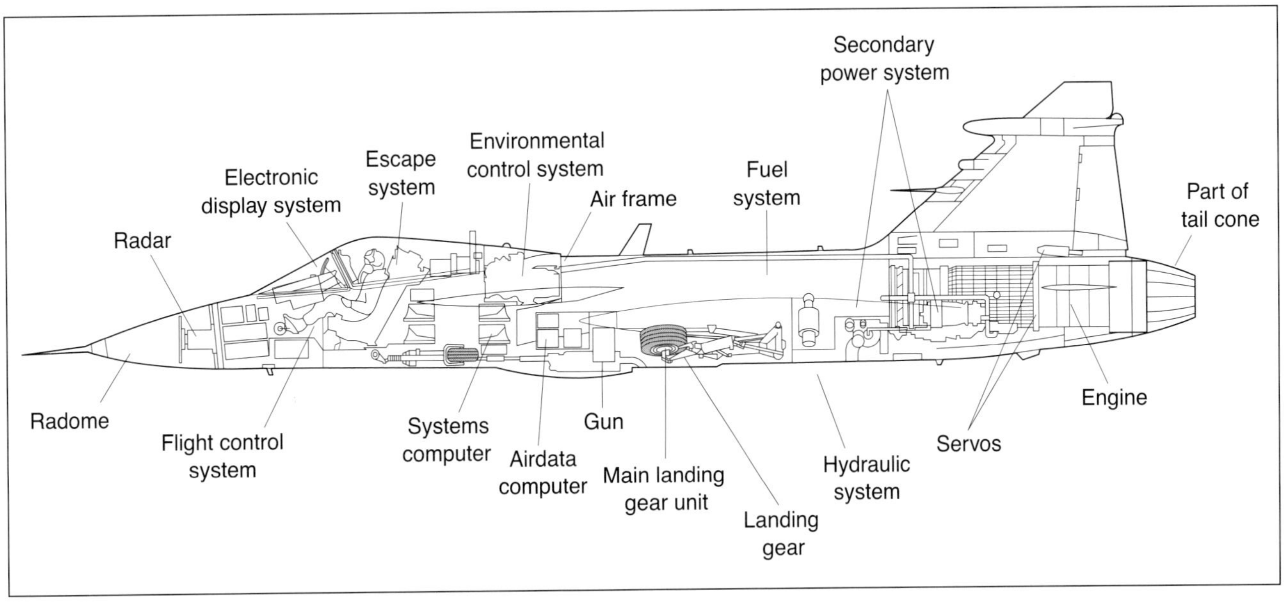

Current fighters are highly compact as shown by this side view of the Saab JAS 39 Gripen. Fuel tanks bracket the CG with the main wheels tucked under the wing box. The nosewheel now twists to lie flat under the avionics equipment bay aft of the cockpit and forward of the gun bay. In the 1950s the low density of fighters coped with the odd 20mm cannon shell which today would likely strike a control line or vital system. *(Saab).*

The need for an emergency power system aboard the single engine F-16 was fulfilled by a Garrett one-shot hydrazine-driven emergency power turbine located on the starboard fuselage side just aft of the cockpit canopy. This is a Lockheed Martin F-16C of 113th Fighter Wing, DC ANG based at Andrews AFB, MD. *(113FW/PA via author).*

only rearming and refuelling facilities are needed. The main benefits are: improved aircraft operations and maintenance, increased efficiency in power generation, flexibility in launching, operational readiness at short notice, remote base operation, reduced hazard of accident using GSE, fewer personnel required for operations and lower life-cycle cost of operation, less on-ground main engine operation, thus increasing life and reducing fuel consumption, and the reduced maintenance cost of an APU compared with GSE.

Emergency Power Systems

A highly manoeuvrable fighter, especially one with a single engine, requires emergency power in the event of an engine failure. When making extreme manoeuvres there will be some distorted airflow within the intake that can lead to engine flameout. This will cause a loss of engine-generated utility power. Emergency power can be provided by a ram air turbine (RAT) or an electro-hydraulic pump (EPU) to supplement the APU.

The RAT is extended when an emergency occurs. It furnishes the crew with sufficient power to fly the aircraft while attempting to restore the primary generators or carry out a diversion to the nearest airfield. It needs no source of power other than forward speed of the aircraft to drive its blades. Small hydraulic and electric generators are mounted immediately behind the turbine on the same shaft. However, space must be found to stow the device and a small accumulator is necessary to deploy it (once deployed the RAT remains extended). Because speed governing and blade feathering are employed, the mechanism is complex; nevertheless, RATs have proved their worth on many occasions.

An EPU consists of a turbine made to rotate by the release of energy from a sealed tank of a high energy/density monofuel such as hydrazine. A sensor will detect that the aircraft is in flight and the engine is below a predetermined speed. The quickly-starting turbine drives a gearbox that enables at least one hydraulic pump and one generator to be energised. This provides power for a limited time, sufficient to recover the aircraft and restart the engine (see Table 17).

Conclusion

Despite the vast amount of equipment under the skin of today's fighters, the aim is to achieve the lightest and simplest layout, reconciled, to the greatest extent, with the often conflicting requirements of reliability, maintainability and survivability.

Table 17 Requirements for APU and EPU in a typical fighter

Regime	APO *Ground/low altitude & speed*	EPU *High altitude/speed*
Output power	105kW continuous	55kW/200sec
Number of cycles	1,000 cycles of one hour	50 cycles of 200sec
Propellant	Kerosene/air	Hydrazine H70
Starting power	Electric motor/battery	Autonomous run-up
Starting time	15sec	3sec

Chapter 9

Armament and Tactics

One of the myths of World War One aerial warfare was that the early flyers did not want to commit mayhem on each other, but were instead amiably content to share the sky. The classic story has it that this pastoral state of affairs went on until one fellow, obviously no gentleman, attempted to drop a brick on an enemy. The tactics of air-to-air combat and the weapons employed have changed vastly during the last eighty years. History has shown that combat outcome is at least as dependent on tactics as equipment. Though aircraft speeds, weapon lethality and size of the battlespace have increased enormously, some fundamental principles endure.

For sustained turning (assuming that thrust = drag)

The radius of turn $R = \dfrac{2W}{S\rho C_L g\sqrt{n^2-1}} = \dfrac{V^2}{g\sqrt{n^2-1}}$

The turn rate $\omega = \dfrac{g\sqrt{n^2-1}}{\sqrt{\dfrac{2W}{S\rho C_L}}} = \dfrac{g\sqrt{n^2-1}}{V}$

To minimise turn radius and maximise turn rate (degrees/second), the wing loading (W/S), speed (V) and altitude $\left(\dfrac{1}{\rho}\right)$ need to be low, whereas lift coefficient (C_L) and load factor (n) need to be high.

World War One

Pilots were patently inexperienced but eager to learn and come to grips with the enemy. Reconnaissance (scout) pilots found they had to fight for their information and there began a crude form of duelling in the air. Initially there was no lack of desire to inflict harm, merely a limited means of doing so.

Armament

The British War Office was opposed to the procurement of foreign-made guns, so lacking political influence, the Royal Flying Corps had to use indigenous single-shot weapons. Among the assortment of weapons used in the first months of the war were 0.45in (11.4mm) carbines, 0.303in (7.7mm) Lee Enfield rifles and pistols. The Lewis machine-gun, used extensively, was fed from a rotating drum that could be changed in the air and held forty-seven rounds of 0.303in ammunition; later the double drum held ninety-six rounds. Because they could not fire through the propeller, the Lewis was mounted on the top wing. The gunner had to stand to aim and fire, or to clear one of the far too frequent stoppages – not a simple matter in a gale of 50kts, and when lives depended on clearing stoppages within seconds.

For use against airships, Hales grenades (ordinary 'pineapple' fragmentation types with a steel rod for penetration) were fired from rifles. But they were short-range, inaccurate and not guaranteed to explode at the right time. Later, incendiary, explosive and even chain-shot cartridges were used. It was hoped that the incendiaries would be effective against Zeppelins, though it was guessed, wrongly, that the airships' gas bags were surrounded by an envelope of inert gas. A significant improvement in aircraft armament came in mid-1916 when explosive bullets for the Lewis were used with good effect.

A height advantage was essential if a fighter was to use its armament against an airship. However, a BE2c, for example, required an hour to climb to 12,000ft (3,660m), where it was virtually stalled and barely controllable, often enabling the airship to escape. To compound the problem at altitude, airships could climb much faster than an aircraft. Nevertheless, during 1916, BE2cs shot down five Zeppelins during night operations.

Though the fighter of tractor layout was much cleaner aerodynamically, it was the pusher type that was first to get guns into the air and to use them. In 1914 the Vickers FB 5 Gunbus was introduced and, although slow, they were effective against slower aircraft. In single-seaters, almost universally tractor types, there was no safe structural spot to mount a machine-gun where it was accessible, and no way to fire a gun through the propeller.

Forward-firing guns

By installing steel deflector plates on the propeller blades of tractor aircraft, the French had a machine-gun that could be mounted directly in front of the cockpit. Bullets passed between the two blades of the propeller, which spun at slightly more than 1,000rpm when it was out of the way, and into the propeller when it wasn't. It worked fairly well, but it was obviously hazardous to shoot at your own propeller, regardless of the deflector plates. That said, shooting was greatly simplified because it enabled pilots to approach their quarry from any direction and aim by flying their aircraft at it. These *avions de chasse* were soon taking toll of their German opponents, and the air war swung in favour of the French. Because of the ability to fire through the propeller disc, the tractor layout with its higher efficiency quickly made the pusher fighter obsolete.

In April 1915 a Morane-Saulnier Type L fitted with a Hotchkiss machine-gun to fire through its armoured propeller fell into German hands. Copies of the propeller deflector plates were made but the firing tests were almost fatal for onlookers. Anthony Fokker was asked to study the problem, whereupon arose the myth that the Dutchman originated the interrupter system that revolutionised air

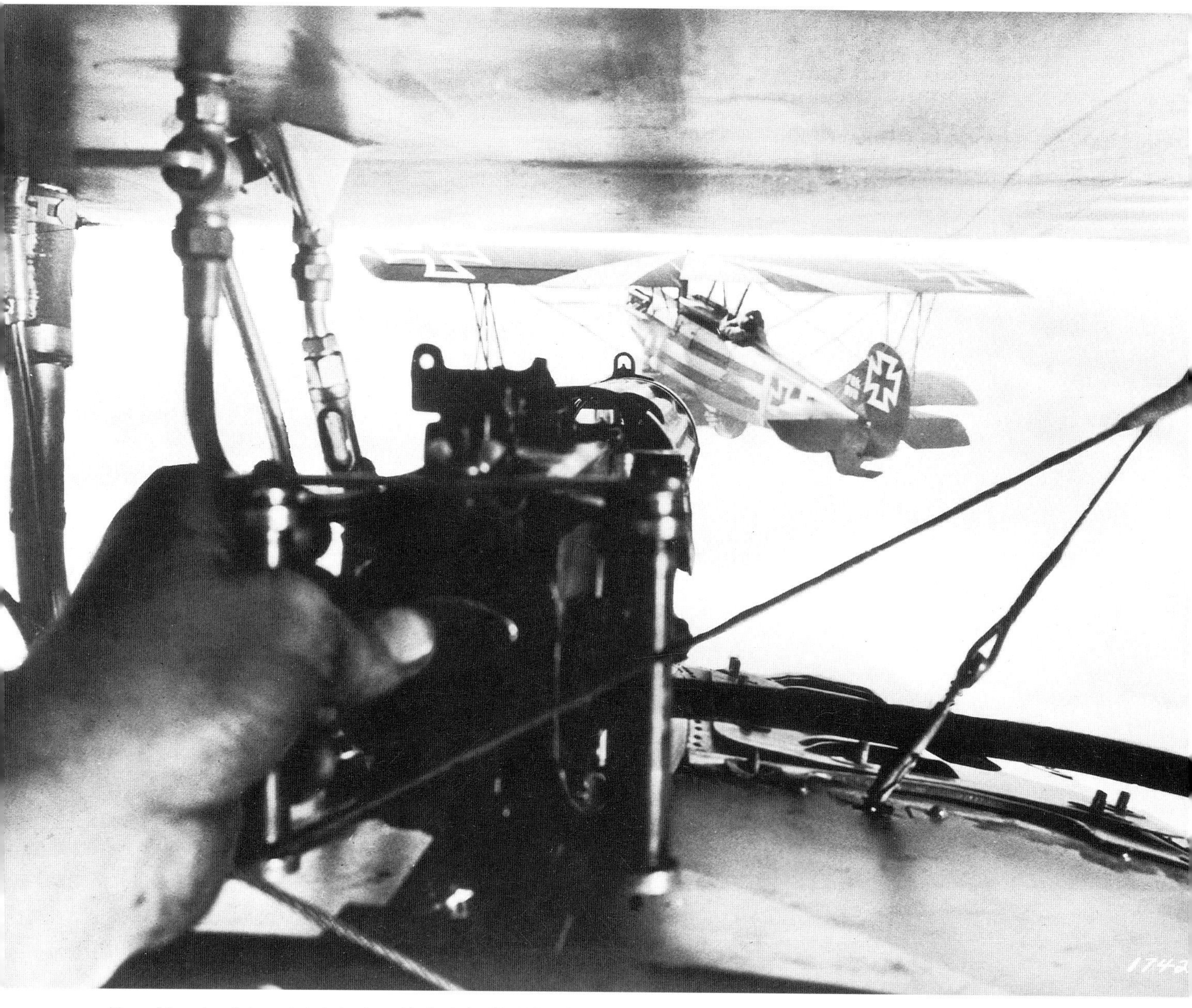

Many of the rules of air combat, derived empirically during World War One, have stood the test of time and are just as relevant today for all but beyond visual range engagements. One example, from Oswald Boelcke (40 victories); is "fire only at close range and only when your opponent is properly in your sights". For Boelcke, the ideal machine-gun range was 18m, where it was hard to miss. *(Philip Jarrett).*

warfare. However, the idea of a mechanism to allow a gun to fire through a rotating propeller was not new. Indeed, Saulnier, the French aircraft designer, made firing trials with such a system in 1914. Moreover, a German patent had been granted in 1913 but had been rejected on the grounds that aeroplanes would never become fighting vehicles. Nevertheless, the 'Fokker' interrupter gear synchronised firing one bullet with the passage of each propeller blade.

The new system on the Fokker Eindecker (monoplane) E1 began to play havoc with Allied aircraft over the Western Front from August 1915 onwards. Not until April 1916 did the Allies obtain an enemy aircraft equipped with synchronising gear intact. Equality was reached by mid-summer, followed by alternating superiority as new designs and modifications were introduced by both sides.

(SEE TABLE 18 ON NEXT PAGE)

Weapon aiming

The weapon-aiming systems of the time were typically ring-and-bead gunsights set to compensate for trajectories, provided the attacking aircraft was within range. The guns most suited to the synchronising mechanisms of both Allied and German air arms were the various versions of the Maxim gun. The Vickers Maxim became the standard British pilot's fixed gun, remaining in service until 1935.

Tactics

Survival in air fighting is to see the other aircraft first, and this is a question of experience and training, of knowing where to look and what to look for. Experienced pilots always see more than newcomers, because the latter are more concerned with flying than fighting. Halfway through the war an RFC pilot was required to have twenty-eight hours' solo flying before joining a squadron in France.

Table 18 Major advances in World War One fighters

Date	*Aircraft*	*Armament*	*Speed*	*Ceiling*	*Remarks*
Oct 1914	Morane-Saulnier Type L	Deflector plates on prop, Hotchkiss or Lewis gun	62kts @ 6,500ft	11,000ft	First tractor with forward-firing gun
June 1915	Fokker E1 (monoplane)	Interrupter gear fitted to Maxim MG08/15 firing through prop	70kts	11,000ft	First with interrupter gear – 'Fokker Scourge'
Apr 1916	Nieuport 16	Lewis over top wing or Lewis via interrupter through prop	90kts	13,000ft	First Allied aircraft to exceed 100mph
May 1916	Sopwith 1½ Strutter	Vickers via interrupter gear through prop	88kts @ 6,500ft	13,000ft	First British fighter with synchronised gun
Dec 1916	Albatros DIII	Twin MG08/15 through prop	91kts	18,000ft	Streamlined but suffered lower wing failures
Apr 1917	SE5a	Vickers through prop, Lewis over top wing	108kts @ 10,000ft	20,000ft	One of the finest scouts, when flown like a single-seater
May 1917	Sopwith F.1 Camel	Twin side-by-side Vickers through prop	100kts @ 6,500ft	22,000ft	Bizarre handling; 5,490 delivered
Sept 1917	Fokker DR I (Triplane)	Twin synchronised MG08/15	89kts @ 10,000ft	22,000ft	Highly manoeuvrable; von Richthofen mount
Apr 1918	Fokker DVII	Twin synchronised MG08/15	100kts @ 3,000ft	20,000ft	Stable, highly manoeuvrable and deadly
July 1918	Sopwith Snipe	Twin 0.303s	104kts @ 10,000ft	20,000ft	4,500 ordered, 1,100 delivered

Over the period of the war, endurance of fighters doubled from around 1.5hr to 3hr.

Wherever possible the novices were given more training with their squadrons, but sometimes, during periods of heavy fighting, they were thrust immediately into the arena, having scant knowledge of any combat techniques and possessing little air experience. It was not surprising that the life expectancy of newcomers sometimes averaged only a few hours in the air. A golden rule was 'Always above, seldom on the same level, never underneath'. Leaders always tried to get in at the top of a fight because it gave them time to decide how to attack and a good speed when they dived. The first dive was all important. The half-roll was useful, for if they missed on the first dive they could zoom higher, half-roll and return to the attack. Elaborate manoeuvres, such as loops, were to be avoided, and to make steep turns and keep turning without losing height until there was a chance of a shot was often successful.

The introduction of faster aircraft meant that the prevailing westerly wind, although still to be reckoned with, was no longer such a menace to Allied pilots. It was important for pilots to have a sound knowledge of the engine, so that they could carry out simple repairs if forced down. Moreover, they had to know best speeds at all altitudes, fighting ceiling, radius of action, limiting speed in a dive, turning qualities, and how to go into a 'recoverable spin' (spiral dive) to retire from a fight.

Big formations

The early air fighters flew alone, but soon realised that a lone pilot could not guard his own tail, so they began to hunt in pairs. The trend moved towards ever bigger formations. Pairs grew into sections of four or five aircraft, sections into squadrons. A good squadron formation was thought to be three flights each flying its usual vic pattern, the leader in the lowest flight. The other flights were echeloned 600 to 900ft (180 to 275m) away on either side, the highest flight being the reserve that only came down when a fight was going wrong.

By 1917 there was increased emphasis on team fighting with German Wings of forty to fifty aircraft. These big formations were difficult to handle, and when in a good attacking position it was difficult to remain together in any subsequent fight. They looked frightening but could be easily seen, so were used only in good weather during the late evening, when there was less danger from Allied patrols tucked away in the sun.

Strength had to be met with strength and the RAF (formed on 1 April 1918) often assembled fighting formations of two or three squadrons to counter the enemy. Henceforth large formations became fairly common, though the great majority of RAF missions were flown by five or six aircraft, either encountering too many enemy aircraft or none at all. There was no satisfactory means of ground-to-air communication for the greater part of World War One and interception remained a matter of chance sighting.

Between the Wars

Following the Armistice it became difficult for the RAF to obtain defence funding. There was little incentive for

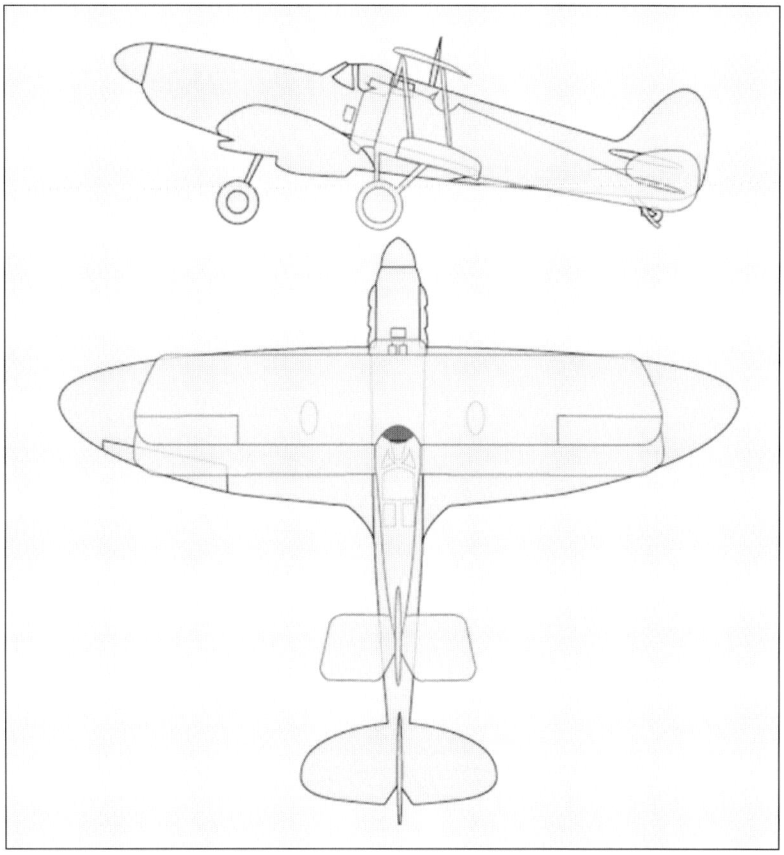

The Hawker Demon was the first two-seat fighter to be introduced by the RAF following World War One and was a version of the Hart day bomber. Powered by a supercharged Rolls-Royce Kestrel engine of 485hp, it was fitted with twin forward-firing 0.303in Vickers machine-guns and a canted rear cockpit coaming, to improve the field of view of a ring-mounted Lewis gun. The Demon was finally withdrawn from front-line RAF service during 1939. *(BAe via author).*

Size and configuration comparison between the Sopwith Camel and Supermarine Spitfire Mk IA

	Spitfire/Camel
Wing area	1.0
Take-off weight	4.0
Military load	4.0
Power available	7.7
Total drag	0.5
Maximum speed	3.0
Airfield speeds	1.5

What made it possible for the Spitfire to carry so much added weight, when the wing areas were roughly the same?
The answer is in four parts: i) Development of flaps giving much higher maximum lift (extra 65%); ii) Engines of much higher power-to-weight ratio which for the Merlin was almost twice the 1917 Clerget's value, largely due to high octane fuels which enabled superchargers to be used to their utmost; iii) Constant speed propellers to enable that power to be fully available over a wide speed range; iv) Larger and better airfields permitting higher take-off and landing speeds (increased by 50%).
(Pete West/AIR International)

technical progress and many designs were no more than maintained throughout the 1920s. The UK's air fleet was reduced from 185 squadrons in 1918 to 28 by 1921. Because of the recognised value of deterrence – that is, threatening destruction of an enemy's industries and civilian morale – British investment in bombers took precedence, with roughly three bomber types being introduced for every new fighter.

Experience showed that aircraft hit by numerous rifle-calibre bullets could survive without major damage and numerous arms experts advocated the use of 0.5in (12.7mm) guns. An Air Staff committee investigated the most effective calibre of future guns and protracted trials were carried out. One problem was that no really suitable British heavy weapon was available. By 1929 it was decided that although the range and hitting power of a heavier gun was superior to the 0.303in weapon, the latter's higher rate of fire and lighter weight were more suited to fighters.

The threat

There were conflicting attitudes within the British Air Ministry in the mid-1930s as war with Germany loomed. The much-publicised dictum that 'the bomber will always get through' had persuaded a number of staff officers that it was better to continue concentrating on the offensive weapon. However, the 'bolt from the blue' concept, that an enemy would strike without warning and that a strong fighter force was therefore essential, gained a firmer footing.

The first problem was to catch the bomber in the new 250kts (460km/h) monoplane fighters and, because of shortened gun-firing opportunities, kill it in a matter of seconds. There would indeed be little time to catch the

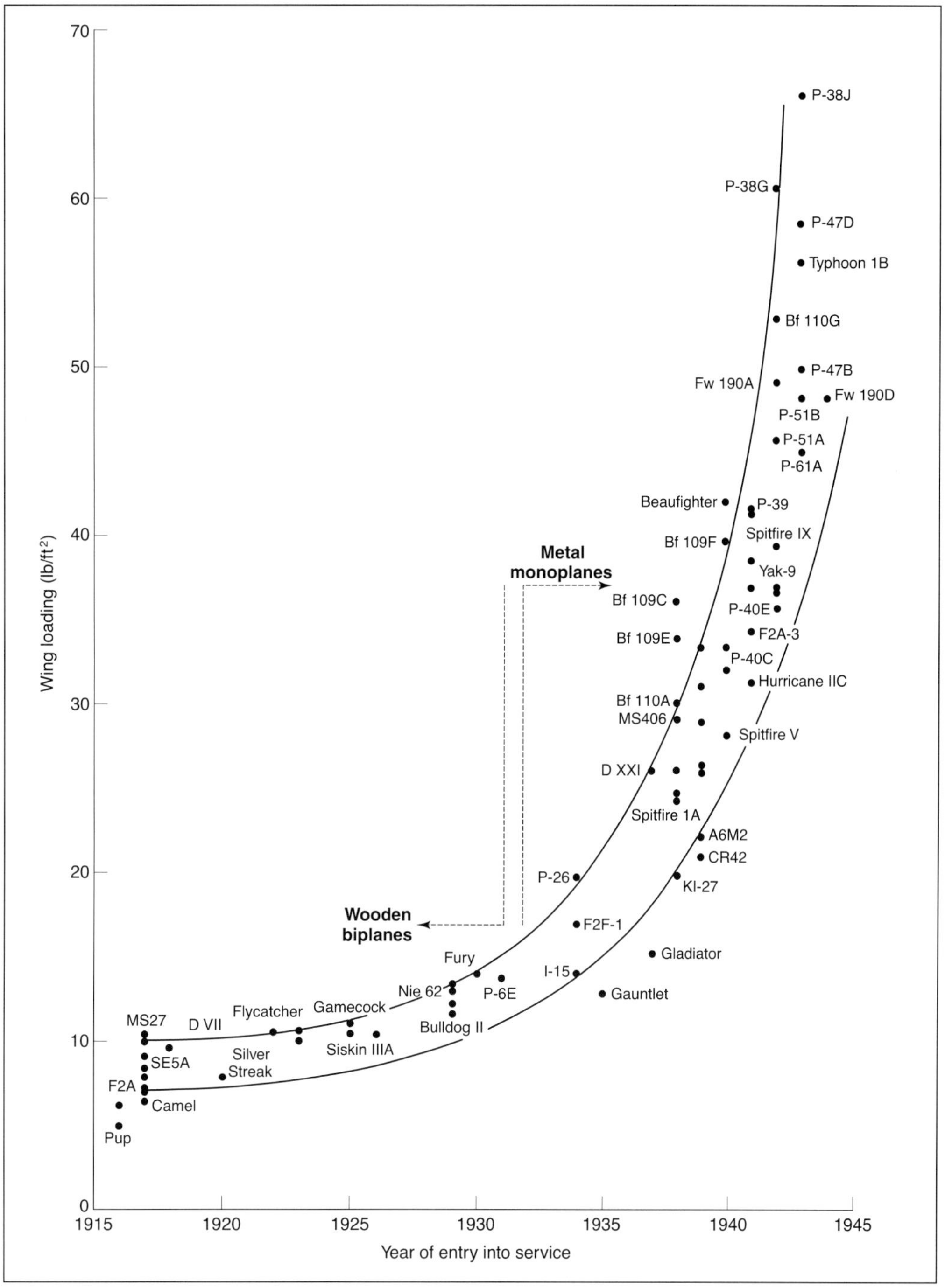

This figure shows the very gradual increase in wing loading (W/S) of biplanes until the mid-1930s. The much higher speeds attainable by the new monoplane fighters combined with their inevitably smaller wings and higher W/S, led to lower turn rates and larger turn radii. *(AIR International).*

bomber at all. Moreover, it was felt that if a target could be held in the gunsight for two seconds, at that sort of *immense* speed, as it seemed at the time, the average pilot would be doing everything he could. Having got that right, what sort of lethal density could be delivered in two seconds? Overcoming the growing structural and defensive strength of contemporary bombers, the most important target, was a difficult problem.

Choose your weapon

It was decided in 1933 to find a replacement for the 0.303in Vickers. Though it had proved to be an excellent gun in many ways, it was perennially prone to stoppages and had reached the end of its development potential. After comparative trials with the Vickers, Hotchkiss, Darne, Madsen and two Colt-Browning types, the MG40 and MG40/2, the latter (0.303in Browning) was chosen for development. By going for a 'new' gun (its origins lay in 1890) altogether, and getting the rate of fire accelerated to its maximum, *and* by not firing any guns through the propeller (which slowed the rate of fire), the whole concept became clear. If the guns could go in the new monoplanes' wings, this would minimise the frontal area of the aircraft, and eliminate the archaic business of loading guns in the air and clearing stoppages, to which pilots had long become accustomed. It could make a simple and workable, if unorthodox, armament system.

However, it was essential for higher lethal density, and not just to maintain hitting power, to increase the size of the gun battery. The prevailing wisdom, at the time when the Hurricane and Spitfire were being designed, was that six 0.303s were adequate. This was abandoned for the

Hurricane, Hawkers being able to fit eight Brownings. The Spitfire was intended from the start to use eight machine-guns, located as quartets in each wing outside the propeller disc. The arrangement was not initially popular with armament officers. Not content with putting eight guns in the wings with a new method of mounting, the monoplane designers had chosen to have new guns as well, about which the RAF knew nothing.

The Browning was the first in RAF use to have the facility of adjusting the barrel in relation to the breech-block. Some armourers adjusted the barrel too far forward leaving some of the case protruding so that the end of the round was blown off, causing a 'separated case' stoppage. During the Battle of Britain, if a fighter returned from a sortie with such a stoppage the armourer responsible was disciplined. Excessive fouling of the muzzle attachment arose, with the guns seizing after firing 200 rounds, but recoil operation with safety features assured nearly trouble-free operation. The most usual stoppage was caused by rogue ammunition or badly made-up belts.

Cannon

Although the British Air Staff had opted for the 0.303in Browning, it was known that other air forces were introducing 20mm (0.79in) cannon to counter the armour protection being built into the latest aircraft. Both RAF fighters delivered a high rate of fire, but it lacked the punch to penetrate enemy armour. A British team had visited the Hispano-Suiza factory in Paris in 1936 to see a demonstration of its newly designed and secret 20mm Type 404 cannon. In its very early stages of development it needed a very rigid mounting, and the engine was to be used as the baseplate for the gun. But, since it fired through the propeller, its rate of fire was limited and demanded more aiming skill to achieve decisive damage. Nevertheless, after tests at the Aircraft and Armament Experimental Establishment at Martlesham Heath the Hispano was adopted for use in future RAF fighters. However, with the Munich crisis of 1938, all new production fighters were to have it. A factory to make it was set up in the UK just before France was overrun, and the gun proved to be one of the most successful of the air war.

World War Two

There was even more resistance to fitting the Hispano cannon than the Browning machine-gun, despite its much greater destructive potency (ten times in hitting power alone, not counting the explosive effect). Enormous disruption to production was caused in June to July 1940 by the conversion. The Spitfire's wing had to be redesigned to accommodate one cannon with two Brownings outboard, on each side. The first guns were available for Spitfire IBs and achieved spectacular results, but despite this, the gun's debut was marred by faults with ammunition feed and it was temporarily withdrawn from service. The feed problems were mainly due to the bulky sixty-round drums that could only be housed with the gun mounted on its side. Gradually the stoppage rate was reduced to one in every 1,500 rounds.

Ammunition

It had been thought that explosive 20mm cannon shells would destroy an aircraft with very few hits, but this was not borne out by events. The larger available amount of solid ammunition did as much damage as the fewer explosive rounds that tended to explode directly on contact without penetrating. It was realised that the ideal round

The Hawker Hurricane's canopy was made up of a number of separate panels, whereas the Spitfire made use of the latest technology moulded Perspex. The Hurricane went through the war with only minor changes to its much fenestrated canopy, even so visibility over the sloping nose was better than that of the Spitfire. The Hurricane was robustly built and could take a lot of punishment. The two banks of guns were mounted closely together just outside the propeller disc. As a gun platform it was as steady as a rock, though while the Hurricane I hardly shuddered when firing its eight Browning 0.303in machine-guns, when on the Hurricane IIC the four Hispano 20mm cannon fired it was a different story. Marking the end of production of 12,975, the final Hurricane (a Mk IIC) was named 'The Last of the Many'. *(BAe via author).*

On the Spitfire Mk IB only 60 rounds per gun (rpg) of the Hispano cannon's 20mm ammunition could be carried, as drum-type magazines were the only ones available. These necessitated wing blisters, which remained to the end of Spitfire production in spite of design efforts on Supermarine's part. Later, belt feed became available allowing ammunition capacity to be increased to 160 rpg. Eventually, it became evident that 20mm ammunition was the only really effective type and the provision of four cannon was made standard on Spitfire Mks 21, 22, and 24 with 140rpg. To ensure the guns would operate at temperatures down to -40°C, the gun bay was heated by warm air from the engine radiator duct. This is a Spitfire Mk 24 with enlarged tail surfaces and powered by a Rolls-Royce Griffon of 2,375hp driving a five-blade propeller. *(via author).*

should first penetrate the airframe and then ignite fuel or oil. In 1942 the semi-armour piercing/incendiary (SAP/I) shell with a tungsten nose was introduced. On impact the head penetrated the armour followed by a flash of flame that ignited anything within 0.3m (1ft). From mid-1942 two types of round were used, belts being made up equally of high explosive/incendiary (HE/I) and SAP/I.

Gun harmonisation

Fighter pilot training assumed that the air threat to Britain would be hordes of German bombers flying in close formation, and RAF fighters would never meet, let alone engage, enemy fighters. Since the Bf 109 could not reach Britain from Germany, stern attacks on unescorted bombers were on the agenda. Apparently, what was not foreseen was the possibility of more adjacent airfields becoming available to the *Luftwaffe*, or that its fighters might be able to increase their range by fitting drop tanks.

The standard of practice shooting in Fighter Command before the outbreak of war was not high. Too little attention had been devoted to gunnery instruction and the 'kills' in any squadron always fell to the same three or four pilots. The remainder had to be content with a 'probable' or a 'damaged', because they 'hose-piped' their guns from skidding aircraft and fired beyond range.

Thus the 'Dowding Spread' method of gun harmonisation was to be used against bombers. It was calculated to give a poor marksman a wide enough bullet pattern to compensate for aiming error with sufficient lethal density to ensure a kill. Moreover, setting the guns to converge on a spot 1,200ft (365m) in front of the aircraft put the fighter outside effective enemy defensive fire.

Many of Dowding's pilots, however, were not seduced. Even if, at 1,200ft, the bullet velocity was still high enough to prevent them tumbling, still maintain accuracy and penetrate armour, the spread produced by aiming, shooting and random errors combined would be unlikely to kill a bomber. This was even more so against a small target like a fighter, which RAF pilots were unconvinced they would never meet. Moreover, it handicapped good marksmen, who often closed to assured killing ranges only to find that the heavy concentration of firepower was not particularly lethal.

Fighter Command dismissed its pilots' theory, but during shooting practice in the spring of 1939 Hurricane pilots of No.1 Squadron secretly harmonised their guns on a spot at 750ft (230m) range. The excellent results obtained proved the point and the method was adopted for the RAF's fast fighter squadrons.

Weapon aiming

The first 'head-up displays' were installed in fighters dating from before World War Two. This was a basic reflector gunsight in which a lamp-illuminated reticule was projected into the pilot's line of sight using an unsophisticated optical system. Though it had been invented before the Wright brothers left the ground, it came into use only in the late 1930s. At the base of the Mk II sight were knurled rings by which the pilot set the required range and target wingspan. When the gap between the movable bars, projected onto the sight, spanned the wing tips of the enemy aircraft, the pilot knew he was at maximum range at which to open fire. Though the reflector sight was a great improvement over the ring-and-bead sight, the most difficult problem facing pilots was that of allowing the correct angle of deflection or 'lead' needed to hit the target when it was moving across the line of fire. The average pilot could hit an enemy aircraft when he overhauled it from dead astern, but a deflection shot usually missed.

The gyro-gunsight

Within squadrons there was much debate on deflection shooting since the top speeds of the Spitfire I and Bf 109E,

With its twin Allison V-1710 engines mounted on the wings, the nose of the Lockheed P-38 Lightning was free to fit four 12.7mm machine-guns and a 20mm cannon plus ammunition. It was fast, being able to reach 360kts at 25,000ft and had a maximum range of 1,960nm. This is a P-38J-10 shown in the markings of the 55th Fighter Sqdn, 20th Fighter Group, 8th Air Force in the UK, 1944. *(Lockheed Martin via author).*

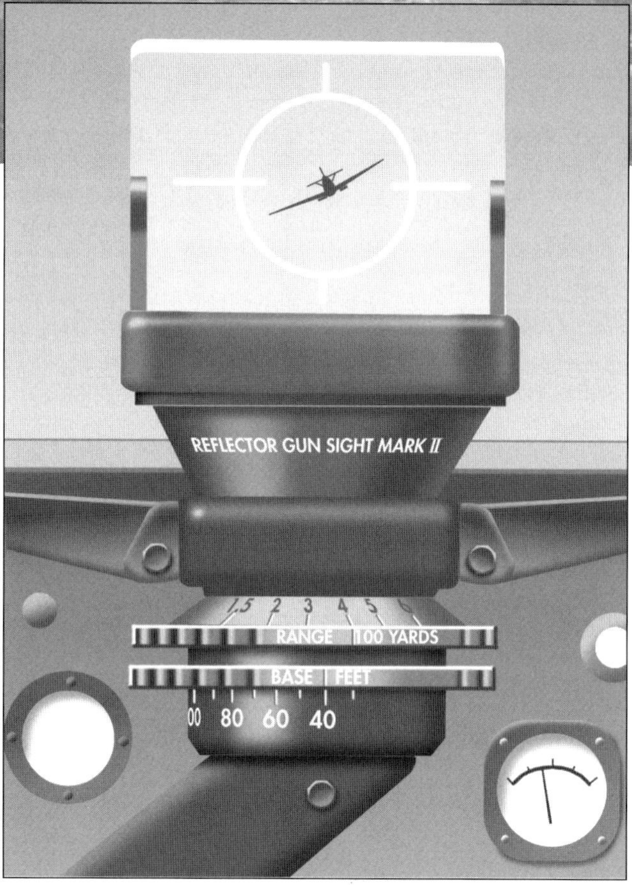

At the base of the Mk II reflector gunsight were knurled rings by which the pilot set the required range and target wingspan. When the gap between the movable bars, projected onto the sight, spanned the wing tips of the enemy aircraft, the pilot knew he was at the maximum range at which to open fire. *(Pete West/AIR International).*

at about 300kts (560km/h), gave little time to adjust one's aim. The first successful attempt to provide a lead-computing gunsight was the gyro-gunsight, devised at RAE Farnborough in 1939. These were sights that could compensate for range, deflection and gravity drop of the projectile by setting up a lead angle using inputs processed from a fighter's gyros. An illuminated circular reticule, which moved to the correct position allowing for deflection, was projected onto the reflector plate. The pilot set the type of enemy aircraft on a dial, enabling the sight to compensate for range, and operated a twist grip on the throttle lever to set the size of the circle of diamonds over the target's wingspan. Produced by Ferranti, they became a dominant feature of RAF cockpits during 1944, much larger than the reflector gunsight. Despite this vastly improved method of aiming guns, armourers still put a proportion of tracer ammunition into gun belts to provide a real-time cue of aiming errors.

Later on, some versions of the Hawker Tempest dispensed with the familiar sloping glass combining plate (so called because it combined the pilot's view ahead with the lens-generated reticule) on the gunsight. Instead, the images of the optical system were reflected directly off the windscreen. This improved the pilot's forward view, but as the airspeed increased so the airloads distorted the windscreen throwing the sight out of alignment. This was reduced when Grade A armoured windscreens became available.

Tactics

German fighter pilots of the Condor Legion, which fought in the Spanish Civil War, found that owing to their increased speeds, greater turn radius (five times that of World War One) and restricted view, it was essential to use the smallest fighting unit. Flying in pairs, for a lone pilot was more vulnerable than ever, a Bf 109 squadron stretched 1½ miles (2.4km) across the sky, at varying heights. The starboard group, deploying down-sun from the leader, could search into the sun and guard the rest against surprise attack. As the leader manoeuvred before attacking, he could keep his team fully informed by the use of the radio telephone (R/T). This was a vast improvement over the previous methods of wing rocking or firing coloured lights to signal intentions.

An evasive tactic that could be used by pilots of Bf 109s

was to bunt with considerable negative g directly into a dive, since its fuel-injected engine did not cut out immediately. This was unlike the carburettor-equipped Rolls-Royce Merlin engines of RAF fighters, though the limitation was initially overcome by rolling inverted before commencing a dive (a split-S). Speed quickly built up in the dive, and the inferior manoeuvrability of the Bf 109 at high speeds quickly became known. On several occasions Bf 109s were coaxed to self-destruction; the RAF pilot would do a half-roll and quick pull-out from the subsequent steep dive. In the heat of the moment the German pilot would follow, only to find that he had insufficient height for recovery, owing to heavy elevator loads, and would fly into the ground without a shot being fired.

The most important qualities in manoeuvrability were reckoned to be the ability to hold a high-speed dive with precision and ease and to break away quickly in the horizontal plane, and the minimum turning radius in the horizontal plane had to be as small as possible. Thus, a high acceleration into a roll (up to 20% better against the propeller than with it, especially under g) was required so that a turn could be started quickly. This gave the ability to wrong-foot an adversary by getting him established in a turn in one direction, then snap-rolling the other way faster than he could follow. Remarkable aileron control was one of the most outstanding features of the Focke Wulf Fw 190.

Once an aircraft has banked and is turning steadily in the horizontal plane, aileron effectiveness is no longer a decisive factor. The degree of manoeuvrability in this condition (turn rate) is inversely proportional to wing loading (W/S) and speed. Thus, for any given W/S, there is a definite minimum radius of turn, when the manoeuvre is flown at maximum lift. There is also the limitation of the maximum g that the pilot and the aircraft could withstand. High load factors could be designed into the aircraft, but a highly trained and physically fit pilot can only withstand an acceleration of around 6g without blacking out. G-suits appeared only later.

Aircraft comparisons

There was much argument over the merits and demerits of the Hurricane and Spitfire. The Spitfire was a graceful aircraft, and as fast as the later versions of the Bf 109E. However, it was more difficult to fly and costly to produce and repair than the Hurricane, nor was its narrow landing gear as robust. Nevertheless, it was repeatedly updated and re-engined, and was the only fighter to remain in continuous production throughout the war. On the other hand, the Hurricane was slower than the Spitfire and its best altitude was lower. That said, it was easier to fly, more manoeuvrable, immensely strong and could take a lot of punishment. Much of its battle damage was repairable on the station or at repair centres to which patched-up Hurricanes could be flown. Its serviceability during the Battle of Britain was remarkable: 63% against 37% for the Spitfire. The Spitfire proved to be 6% better at shooting down aircraft (based on the claims of the time), but since the Hurricane formed the backbone of Fighter Command during 1940, for every two aircraft downed by Spitfires, three fell to Hurricanes. Because of the better performance of the Spitfire at altitude, it was generally used against the *Luftwaffe*'s fighters, whereas the Hurricanes looked after the bombers. It was an excellent partnership.

At the beginning of World War Two the frontline British

Built in substantially greater numbers (5,216) than any other version of the Curtiss Warhawk, this P-40N sports the shark's mouth marking popular with many user-squadrons, has six 12.7mm (0.5in) machine-guns and carries a 43.5 Imp gal fuel tank. Powered by the unsupercharged Allison V-1710 it was severely limited in its altitude performance. *(via author).*

In terms of visibility the Bf 109 had almost complete absence of view to the rear but because of its canopy's large flat panels (much more angular than the British ones), the German fighter had a good view forward with little distortion and distressing highlights. This, however, had to be paid for by increased drag which, at the Bf 109E's top speed, was roughly equivalent to 3kts (6km/hr). *(Philip Jarrett).*

fighter, the Hurricane I, was inferior to the Bf 109E whereas the Spitfire I was matched in top speed and the altitude at which top speed was attained. Both British fighters were inferior, however, in speed above this critical altitude and in climb at all altitudes. The reason was simply that though the Daimler-Benz 601A engine of the Bf 109E developed about the same power for about the same weight as the Merlin it was installed in a considerably lighter aircraft. The maximum loaded weight of the Hurricane I in April 1940 was about 6,700lb; that of the Hurricane II in November 1940 was 7,000lb. The Spitfire IIA of 1940 weighed 6,300lb whereas the weight of the Bf 109E in 1939–40 was 5,500 to 5,800lb.

The bigger-winged and thus more lightly-loaded Hurricane I and Spitfire I with W/S of around 120kg/m^2 (25lb/ft^2) had the advantage over the Bf 109E with its W/S of 160kg/m^2 (32lb/ft^2). From mock dogfights with RAF fighters, the good points of captured Bf 109s were seen as good control at low speed and gentle stall even under g, high top speed, and excellent rate of climb. When put into a full throttle climb at low airspeed, the Bf 109 climbed at a very steep angle. RAF pilots had difficulty in keeping their sights on the enemy, even at a height when their climb rates were comparable.

At full throttle at 12,000ft (3,600m), British test pilots flying Bf 109Es established a minimum sustained turn radius of 900ft (275m) compared with the 700ft (210m) of early Spitfires and Hurricanes. Although the more heavily loaded Bf 109 was at a considerable disadvantage, it is important to bear in mind that the minimum radius was obtained by going as near to the stall as possible. In this respect the Bf 109 scored, due to its excellent control near the stall and innocuous behaviour at the stall. This was conferred by its automatic leading-edge slats and good spin-resistant rudder/tailplane design that gave its pilots the confidence to get the last ounce out of their aircraft's turning performance.

In terms of visibility the Bf 109 had almost complete absence of view to the rear, but because of its canopy's large flat panels (much more angular than the British ones), the German fighter had good view forward with little distortion and distressing highlights. This, however, had to be paid for by increased drag which, at the Bf 109E's top speed, was roughly equivalent to 3kts (6km/h). The Bf 109 was also too lightly built to take much punishment. That said, it was still a formidable mount, though its pilots were sometimes accused of 'standing off and picking off'. They were wise but no less bold men, for the characteristics of the Bf 109 were ideal for such tactics. To 'mix it', except under favourable circumstances, was to invite trouble.

In designing the Fw 190, Kurt Tank called a new tune. For example, the canopy of the Fw 190 was a great advance and rear visibility was exceptionally good. The armour plating behind the pilot was cut away so that the pilot could see around it and the shoulder harness was designed to allow the pilot to twist freely. Its high wing loading of 205kg/m^2 (42lb/ft^2) gave the Fw 190A a poor turning circle at low airspeeds. Manoeuvrability was sacrificed for speed and rate of climb. In combat against fighters of inferior speed and climb performance, but better manoeuvrability, the Fw 190 could gain height by its superior climb performance, then dive, open fire at short range and climb away. As a result of its short, stiff wing and superb control harmony, manoeuvrability of the Fw 190 at high airspeeds was excellent. *Luftwaffe* pilots exploited these qualities to the full and made slashing attacks, without indulging in low-speed, circling dogfights. With its four 20mm cannon (two firing through the propeller) and two machine-guns the Fw 190 was far superior to the Spitfire V, especially at low altitude.

Long-range escort

During 1943, daylight raids by Boeing B-17 Flying Fortresses of the USAAC's 8th Air Force on Germany suffered enormous losses. Republic P-47 Thunderbolts, later used as escorts, had limited range and often had to turn back before the bombers' targets were reached. From inspections of downed B-17s, *Luftwaffe* fighter pilots noted that its firepower was much greater at the back than at the front. To make successful attacks the German defenders were instructed to fly parallel and to one side of the bombers, until they were about three miles out. Sections of four fighters were to turn in for head-on attacks, aim at the cockpit, open fire at 2,000ft (600m) and fly over the bombers. Few pilots had the courage to do this, and most of them half-rolled ahead of the targets and dived away. This, however, reduced firing time and gave the rear gunners a simple deflection shot (a box of twenty-seven B-17s, each with nine gun positions, could defend their rear with a lot of crossfire), and delayed a repeat attack until height was regained.

The single cannon and two machine-guns of the Bf 109F did not have enough hitting power to bring down the big, well-armed bombers. However, Fw 190s of the so-called *Sturmgruppen* (assault troops) assigned to heavy bomber interception carried two fuselage-mounted 13mm MG131 machine-guns, inboard wing-mounted 20mm MG151/20 cannon and two MK108 30mm short-range cannon and were fitted with plenty of armour plating. These would make mass stern attacks very close to the bombers, while the Bf 109s held off any escorting fighters. The attrition suffered by B-17s during 1943 due to such tactics rose to more than 20%. Eventually, the USAAC bombers' targets were restricted to those within the range of the escort fighters.

Design of the North American Mustang to meet British needs started in early 1940, and as the Mustang I with its unsupercharged Allison V-1710 engine, it first entered RAF service in February 1942. However, because of its disappointing high-altitude performance the aircraft was used only for reconnaissance and close support. When the P-51, as it had become known in USAAC, was given, at the suggestion of Rolls-Royce, the Merlin 61 (later produced in the US as the Packard-built V-1650), the P-51B's performance at altitude was much improved (380kts at 35,000ft (708km/h at 10,670m)). Clever engineering provided a large amount of fuel, including two large (108 USgal) wing drop tanks made from laminated paper. By spring 1944, P-51Ds with the 1,450hp Packard V-1650-7, slimmed down aft fuselage and tear-drop canopy were accompanying bombers on the seven-hour, 1,100-mile (1,600km) round trip to Berlin. With the new escort fighters, 8th Air Force bomber attrition dropped to around 5% from enemy air action.

Table 19 Comparison of Spitfire and Mustang

Aircraft	*Range ext. fuel*	*Empty weight*	*Loaded weight*	*Empty W/S*	*Maximum W/S*	*Remarks*
Spitfire IX	1,577km	2,555kg	4,300kg	23.28lb/ft^2	39.25lb/ft^2	Fuel in fuselage, undercarriage in wings between spars
(1942)	(850nm)	(5,634lb)	(9,500lb)			
P-51D (1944)	2,655km (1,430nm)	3,463kg (7,635lb)	5,488kg (12,100lb)	32.8lb/ft^2	52lb/ft^2	Fuel in wings and aux. fuselage tank, undercarriage ahead of front spar

Despite the close similarity in size of the P-51D and Spitfire IX, the slightly faster American fighter had higher empty and maximum wing loadings (40% and 32% respectively) than the British design. The Mustang's large fuel capacity gave it an almost 70% greater range.

The Yak-9U of early-1944 normally carried a 20mm cannon that fired through the hub of the propeller (which restricted visibility while manoeuvring on the ground) and two 12.7mm machine-guns in the wings. Its empty weight was 350lb heavier than that of a Spitfire IX and, powered by a Klimov M-107A V-12 of 1,650hp up to 6,000ft, its maximum speed was 320kts at sea level and 360kts at 18,000ft. Some of the Yak-9P variants saw service in the Korean War. *(Alexei Mikheyev via author).*

By the time battle was joined by bomber-escorting Republic P-47s and North American P-51s their pilots had already been in the air several hours. Thus the levels of noise, vibration and comfort in the cockpit while never unimportant assumed particular significance in ensuring that the pilot arrived comparatively fresh and ready for the fight. A tired and cramped pilot whose head is buzzing from a noisy engine is not likely to fight as effectively as one who has spent the previous three hours or so in a comfortable seat in a quiet cockpit. These considerations must be put in the context of the primitive radios and indifferent sound attenuation qualities of 1940s helmets. Developed specifically for operations in the Pacific Theatre, where the principal requirement was range, the Republic P-47N (shown) was by far the heaviest of all Thunderbolts. It saw extensive service in the Pacific, frequently escorting B-29s since it had the ability to fly the return trip of Saipan to Tokyo. Its armament comprised eight 0.5in (12.7mm) machine-guns. *(via author).*

The Jet Age

Korean experience

The swept-wing MiG-15, which made its combat debut in the Korean War in November 1950, had been designed to intercept the Boeing B-29 Superfortress, a job it performed well. It also quickly showed that it outclassed the straight-winged American fighters (e.g. North American F-51 Mustang, Lockheed F-80 Shooting Star, Republic F-84 Thunderjet and Grumman F9F Panther) in action at the time. To intercept the United Nations' (UN) fighter-bombers, the Chinese MiGs (often flown by Russian pilots) needed to fly south and operate at medium to low altitude. The task of the North American F-86A Sabre, which arrived shortly after the Soviet fighter, was to prevent the MiGs from doing this. Thus, the F-86s regularly flew north to the Yalu River border with China and patrolled. Any determined attempt to intercept the fighter-bombers would have put the MiGs at a severe tactical disadvantage below the Sabres. Because this was against the doctrine prevailing at the time, the MiGs were forced to use their superior altitude performance to gain a height advantage over the Sabres, thereby placing them well above the bombers and giving little opportunity for their interception.

Whereas World War Two had seen the USAAC fighting in sixteen-strong squadrons, the F-86 did not have the endurance to assemble such formations before flying north. They had to fly at high engine rpm to keep the speed up and fuel consumption and aircraft vulnerability down. A squadron leader could not have held together sixteen Sabres at high speeds since the F-86 did not have a sufficient thrust margin to allow one to catch up once it fell behind in combat. Survival could not be found in tight turns and it was vital for wingmen to stay close. Sections flying in 'finger-four' groups therefore staggered their take-off at five-minute intervals, and once over the Yalu they could stay on patrol for only twenty minutes.

Having reached the Yalu River, the F-86s flew between 27,000 and 33,000ft (8,200 to 10,000m) in order to stay just below the usual contrail height and be able to see those made by the higher MiGs. A tactic used by the MiGs was to send in two decoy aircraft below the F-86s while the others waited up-sun for an opportunity to bounce. Most of the American pilots had fought in Europe and were not fooled; instead they waited for the MiGs to come down.

Once a fight was imminent, the Sabres jettisoned their drop tanks, increased their speed and then took the initiative by turning into the MiGs. They fought in fours and pairs until their low fuel state obliged them to turn south. During high-speed combat of eight to ten minutes the high stick forces and wide turning circles usually made aiming with the F-86's gyro-gunsight so difficult that the average pilot managed to get only one burst of fire. Sometimes, Sabre pilots exceeded Mach 1 in a steep dive, and recovery was a delicate affair. Typically, the Sabre's controls would freeze and the stick would jump back to the full aft position, causing wrinkles in the aircraft's skin.

Aircraft comparison

The North American F-86E, with water-methanol injection on its General Electric J47 turbojet, started to replace the A-model Sabre in the autumn of 1951. It had a speed advantage of 16kts (30km/h) at low altitude, was roughly equal at 30,000ft, whereas at 40,000ft (12,000m) the MiG-15 was 18kts (33km/h) faster and turned quicker. This was partly due to the relative ineffectiveness of water injection at altitude. Another factor was the lower wing loading of the MiG-15 at 220kg/m^2 (45lb/ft^2) compared to the F-86E's at 300kg/m^2 (60lb/ft^2), both aircraft having similar internal fuel capacity. Although the F-86 was bigger (30% larger wing but a fuselage only 3% longer) than the MiG-15bis, there was a disproportionate 33% difference in empty weight. This was mostly due to the heavier axial-flow J47 engine compared with the MiG's VK-1 centrifugal-flow type (a copy of the Rolls-Royce Nene). The result was a significant advantage in combat thrust/weight (T/W) ratio (0.48 for the MiG-15 and 0.37 for the F-86), which provided the Soviet fighter with a 40% better initial climb rate.

The US engine's lower specific fuel consumption, allied to greater fuel capacity (albeit with drop tanks), gave the F-86E its 560km (300nm) combat radius, needed to operate over 'MiG Alley'. The MiGs were often able to operate within sight of their home bases just north of the Yalu River. Whereas the Russian design philosophy produced small, short-legged, lightweight, highly manoeuvrable but austere fighters, US aircraft were much more lavishly appointed. Indeed, F-86 pilots often complained that the US should stop building training aircraft and start making combat ones. They believed that a lot of the safety features for peacetime flying became combat hazards, and around 1,100kg (2,400lb) of equipment could be stripped from the aircraft.

Though the F-86A achieved good results, its pilots were convinced that they needed a powerful cannon, as the MiG-15 could take a lot of fire from the Sabre's six 0.5in (12.7mm) calibre machine-guns. The new F-86E, however, carried the same armament though deflection shooting was made easier with its radar-computing gunsight, and its increased thrust made it superior to the latest MiG-15. The overall kill to loss ratio during Korea was about 7:1 in favour of the UN forces.

Armament

Mauser, the well-known German arms manufacturer, had produced the MG151/15 cannon in 1936 and the MG151/20 in 1940. These were electrically operated, high-velocity weapons of exceptional quality. The 20mm version was the most important *Luftwaffe* gun throughout World War Two, but by 1942 it was realised that a larger calibre was required. Thus arose in 1944 the revolutionary 30mm

The success of the North American F-86A over the MiG-15 rested chiefly on the ability and aggressiveness of its pilots. The inadequate armament (six 12.7mm [0.5in] machine-guns with 267rpg to give 15 seconds firing) of the American fighter explained the high number of MiGs 'damaged and probably destroyed' against those 'definitely destroyed'. This armament deficiency persisted in the F-86E that followed the F-86A into production and combat, but was eventually replaced by four 20mm cannon. The F-86A's weight was also criticised (16,000lb against the MiG's 12,000lb). Some of this derived from such 'luxuries' as emergency fuel pumps, self-sealing fuel tanks that did not hold up against the MiG-15's large calibre and slow-firing cannon (two 23mm each of 80rpg and one 37mm cannon with 40 rounds) and an unreliable, radar-ranging gunsight that was hard to maintain and disliked. A USAF ace in Korea was quoted as saying 'I'd rather stick a lump of chewing gum on the windscreen'. These are F-86Hs that operated from Nellis AFB in the 1950s. *(via author).*

Table 20 Aircraft cannon

Model	*Length (m)*	*Weight (kg)*	*Cal/Wt (mm/kg)*	*Velocity (m/s)*	*Rate of Fire (rpm)*	*Firepower* (MW)*	*Aircraft*
Giat DEFA 553	1.66		30/.22	765–820	1,300	1.4–1.6	Fiat G91
Giat DEFA 554	2.01		30/.22	765–820	1,200–1,800	1.3–2.2	AMX, Rafale
Mauser Bk 27	2.31	100	27/.26	1,025	1,000–1,700	2.3–3.9	Tornado, Gripen, EF2000
Oerlikon KCA	2.69	136	30/.36	1,030	1,350	4.3	Viggen
ADEN Mk 4	1.59	87	30/.22	740	1,200–1,400	1.2–1.4	Hawk 200
ADEN 25	2.28	92	25/.18	1,050	1,650–1,850	2.7–3.1	Harrier GR.7
General Electric M61A1	1.875	120	20/.1	1,040	6,000	5.4	Most US fighters
GSh-23L	1.537	72	23/.175	740	2,800–3,400	2.2–2.7	Su-22, MiG-21
GSh-23–6			23/.175	740	8,000	6.4	MiG-31
GSh-301			30/.4	890	2,000–4,000	5–10	MiG-29, Su-27

Advantages of the intermittent single-barrel cannon are two-fold: higher-calibre ammunition offers a bigger explosive effect coupled with faster instantaneous (0.05sec) rate of fire, whereas the General Electric Gatling gun requires more time (0.4sec) to overcome inertia, spin up to speed and attain the full rate of fire. However, Gatling guns (M61A1 and GSh-23-6) have far fewer moving parts with much higher reliability (mean rounds between failure for the M61A1 gun is over 100,000) and less barrel wear.
*Kinetic energy only, explosive effect not included.

MG213c with a five-chambered breech. Though it was never used operationally, early tests at 1,200 rounds per minute (rpm) proved it had outstanding potential. This was realised by the Allies, and following their capture of the Mauser development complex France, Britain and the US developed it as the DEFA, ADEN and M-39 respectively.

The absence of a propeller on jet fighters meant that guns could be mounted in the nose of the aircraft or near the centreline to achieve optimum weight and balance. That said, gun firing from this location has been a cause of engine stalls and flameouts since the introduction of the turbojet. Ingestion of gun gases by the air intake enriches the fuel/air mixture and causes pressure and temperature transients to occur so rapidly that engines cannot respond fast enough to maintain stable operation. Most of the early installations used fuel derichment and/or exhaust deflection systems to prevent gun gas entering the intake. Later engines were even more sensitive to gun gas effects since the methods used to increase their T/W led to stall margins decreasing. The trend to larger-calibre (heavier), higher-muzzle-velocity ammunition aggravated the problem. This was because the newer 30mm ammunition produces about three and a half times as much muzzle gas as the 1960s General Electric M61 Vulcan 20mm gun. To avoid gun gas ingestion modern fighters have their gun port behind the inlet and to one side of the aircraft with automatic rudder deflection to counter the offset recoil.

Mach 2 considerations

In the quest for high performance, most fighters designed in the 1950s and early 1960s were given low-drag cockpit canopies. Good cockpit visibility, one of the prerequisites of a fighter, was lost sight of in spite of the excellent example set by the F-86 and MiG-15. The Korean War, in representing the high point of the day-fighter, was the one in which superior altitude performance was the ultimate tactical advantage. It also reinforced the view that human vision was no longer enough. At high altitude, the air is very clear and the sun illuminates the cockpit harshly, which in turn reflects strongly on the inside of the canopy. In addition there are few external cues on which to focus, which makes visual detection of targets with closing speeds beyond 1,000kts (1,850km/h) even more difficult.

Lack of excess thrust (i.e. thrust-drag) meant that the first generation of Mach 2 aircraft could not be afforded good rearward visibility via a high-drag bubble canopy. Moreover, it was not thought to be strictly necessary for the second generation. Radar warning receivers would show up rear-hemisphere threats, and anyway, was not dogfighting a thing of the past? Experience gained in the Vietnam and Middle East conflicts shattered many allusions about air combat tactics, showing forcibly that dogfighting had not passed into history. Good cockpit visibility is still important.

During the Vietnam War some pilots flying McDonnell F-4s stated a preference for removing the weapon system operator's (WSO) station and substituting extra fuel. The contrary view was that the WSO's eyes were invaluable both as an extra set and also to operate the radar system. A pilot who put his head down in the cockpit to operate the radar during combat deep into enemy airspace would be lucky to survive. Thus, pilots came to accept rear blind spots and manoeuvred their aircraft to compensate. Around 40% of all visual sightings were first made from the back seat. The new fighters, which appeared in the 1970s, embodied bitter lessons of actual combat, including teardrop canopies which restored pilot visibility to a standard not seen since the 1950s. The belief among pilots that 'If you can't see out of it, it isn't a fighter' was taken seriously by fighter design teams of the late 1960s.

North American F-86D Sabre firing 2.75in folding-fin Mighty Mouse spin-stabilised rockets (each with the explosive power of a 75mm shell) from its retractable tray of 24 such rockets. Larger and longer ranged guns were impractical, on the other hand sufficient range and hitting power could be had from quite small rocket projectiles. Starting from the German R4M the US developed the Mighty Mouse, an FFAR of 2.75in diameter, which was theoretically effective out to 9,000ft, although it was far from accurate as can be seen by the initial dispersion shown. To overcome this Mighty Mice were fired in a salvo, blanketing a large area, giving a good chance of one hit which could be devastating. Initial trials indicated that the optimum range was 4,500ft. *(via author).*

The introduction of air-to-air missiles

Even before the Korean War, zenith of the gun-armed fighter, it was realised that though dogfights would continue to arise in modern aerial warfare, increases in aircraft speed meant that the gun alone was becoming inadequate. A new air-to-air weapon was required. Many aircraft entering service in the 1950s were equipped with folding-fin air rockets (FFAR) developed from the German R4M, such as the 70mm (2.75in) diameter Mighty Mouse with a range of up to 9,000ft (2,700m), as an interim measure. Because of their inherent dispersion and inaccuracy, the use of these 'shotgun effect' weapons was typically restricted to beam attacks on bombers. Such attacks during training were always scary for fighter and bomber crews alike, as the risk of collision was high because timing was crucial after getting to within 800m (2,500ft) of the target. What was needed was a longer-range, but still fire-and-forget boost-coast missile to supplement the machine-gun and cannon.

The year 1950 marked the beginning of a decade of rampant technological progress. It was thought that the upcoming breed of 'wonder' missiles would rule the skies and make piloted fighters redundant. The prospect was that all future killing would be done by pushing buttons to launch long-range weapons at an adversary who was beyond visual distance. All capability would be built into the weaponry and systems, leaving the aircraft as a mere carrier and the pilot as a systems manager. Such was the perceived potential of the new technology that Convair's F-102A and F-106A and McDonnell's F-4 were designed for missile armament only.

These were the very chilly days of the Cold War when large enemy formations of high-flying bombers were projected as the main threat to the USA. An area weapon like the Douglas MB-1 Ding Dong (AIR-2A Genie) unguided rocket was seen as 'more bang for the buck'. Entering service in 1957, it had a range of around 7 miles (11km) and a nuclear warhead of 1.5 kilotons (1,500 tons of TNT) yield. This fearsome weapon could be fired in a snap-up attack at bombers flying at a higher altitude. Mission planners, after explaining the weapon's capability, needed to be upbeat to pilots in describing evasion tactics after firing it. For the defence of North America against Soviet bombers the Northrop F-89J Scorpion was the first to carry the AIR-2A. Genie-equipped F-106As flew singly with the ground controller able to watch for other threats. However, if the threat of fighter opposition was expected, two-ship formations were used with the wingman watching the six o'clock position.

Missile guidance

BEAM-RIDING With beam-riding missiles, the fighter used a simple ranging radar that sent out a very narrow pencil beam which the missile followed. Launch range was typically about 2nm (3.7km) and the pilot had to hold aim to within 0.25° during the full duration of the missile's flight, about eight seconds. Against a non-manoeuvring target this was moderately successful, but even quite gentle evasive action could easily render the opponent untrackable by the average squadron pilot. British experiments with the beam-riding Fairey Fireflash were abortive.

Originally developed by the US Navy's Bureau of Aeronautics and Sperry, the Sparrow I (AIM-7A) was the main air-to-air missile (AAM) for the US Navy from the mid-1950s. This medium-range beam-rider was operationally deployed with McDonnell F3H-2Ms Demons during the Lebanon and Quemoy crises of 1958. However, the limitations of beam-riding were such that a new form of guidance needed developing for the big 3.6m (12ft) long US missile.

SEMI-ACTIVE RADAR HOMING A simpler alternative to beam-riding, semi-active radar homing (SARH) involves the fighter's radar illuminating the target throughout the engagement with the missile homing on the reflected radar energy. The Hughes-developed GAR-1 Falcon for the F-102A was the first SARH missile to become operational. After launch it accelerated rapidly to Mach 3, whereupon it would coast out to an effective range of 4nm (7.4km). However, Falcon was overshadowed by the much longer-range Sparrow III (AIM-7) family, the later vastly improved variants of which all used a bigger rocket motor, and reached out to between 22 and 54nm (40 and 100km). The extended-range AIM-7E introduced a new dimension in air combat in 1963. First, it theoretically allowed attacks to be made from beyond visual range (BVR), even in clear weather. Secondly it extended the range at which shots could be realistically made at a fast-moving retreating target.

INFRA-RED HOMING This is a far simpler and more reliable system than radar. It relies on the missile seeker homing on the heat at infra-red (IR) wavelengths of 2 to 3 microns, emitted by the target whose hot parts will be around 1,000°C (1,830°F). They were designed so that the seeker acquires the target prior to launch while held in the boresight (straight-ahead position). The seeker is then uncaged so that it can track the target to hold its locked-on condition as it is launched (lock-on before launch). This influenced the type and location of its carriage, since the seeker had not to be shadowed by the wing or fuselage of the launch aircraft.

Without doubt, the Ford-Philco AIM-9 Sidewinder has been the most successful IR missile and over 160,000 of all types have been produced. It was first fired in 1953, entering service three years later as the AIM-9B with a maximum speed of Mach 2.5 and range 1.75nm (3.2km). As it was not able to 'see' through cloud, its use was restricted to clear air conditions. It also originally had the disadvantage that it was distracted by alternative heat sources including those on the ground, earning it the dubious name of 'Groundwinder', so was of little use below 5,000ft (1,500m). Its seeker sensitivity, speed and range were inferior to both the IR Falcon variants and de Havilland Firestreak and Red Top. However, it was very much simpler and cheaper, having originally only nine moving parts and seven thermionic valves, and did not require special handling. Early models of all types of IR missile could be used only in the rear quadrant, thus restricting attacks to the traditional curve of pursuit.

Vietnam experience

Despite the widespread view in the late 1950s that the day of the manned fighter aircraft was ending and dogfights belonged to a past era, the US Navy wanted a fighter for fleet defence. Accordingly it ordered the McDonnell F4H Phantom in 1955. Flown by the Navy, Air Force and Marine Corps it became the most versatile combat aircraft in south-east Asia. However, like the Spitfire of World War Two, the F-4 (as it was re-designated) was vulnerable to ground fire. Cockpit visibility was restricted and there was a rear blind spot. Moreover, its General Electric J79 engines produced heavy black smoke that helped enemy pilots and ground gunners to see, identify and track it.

The ageing MiG-15 'Fagot' was of little importance, but

The Hughes GAR-1 Falcon was standard armament for late models of the F-89, F-102A and F-106A. The Falcon was eventually designated the AIM-4. The AIM-4C, on the left, had an infra-red seeker, while the AIM-4A was radar guided. In Vietnam the IR Falcon proved a minor disaster since it had a long preparation time and once activated but not launched within 30 seconds, could not be switched off for re-use. Cooled seeker heads for the infra-red versions came in the 1980s. The aircraft is F-102A (53-1813) and when launching its AIM-4Cs care to avoid them chasing the radar-guided versions was necessary. *(Hughes via author)*.

The North American FJ-4 Fury of 1954 was a complete structural redesign of the FJ-3 to produce a wholly new aircraft with 50% increased internal fuel capacity. Four wing stations were provided capable of carrying AIM-9B Sidewinders (shown) in addition to the integral four 20mm cannon. *(via author).*

Aware of the shortcomings of the MiG-15, the Mikoyan design bureau set out to rectify them with the MiG-17, which first flew in December 1949. It demonstrated the Soviet philosophy of taking a good simple design and improving it. Very similar in appearance to its predecessor, the main difference lay in its wing which was of broader chord and swept at 47° inboard and 43° outboard. The wing loading remained the same, while the T/W (with afterburning) at 0.63 was 25% higher. Although merely transonic it posed problems to Mach 2 aircraft in the skies over Vietnam, 20 years later. Typical armament was one 37mm N-37D (starboard) and paired 23mm NR-23 (port) cannon, though this MiG-17F also has a pod under the port wing containing eight 55mm unguided rockets. *(A. Maxeeb via author).*

the MiG-17F 'Fresco', with a maximum take-off weight (MTOW) of only 6,200kg (13,700lb) and wing loading (W/S) of only 270kg/m^2 (55lb/ft^2), was a lively fighter with excellent manoeuvrability. It was designed as a dogfighter, and with its low wing loading could turn well inside the F-4 (W/S roughly 490kg/m^2 (100lb/ft^2)). Armed with one 37mm and two 23mm cannon the MiG-17 left little smoke and was hard to see. They flew in 'finger-four' formations with spacing between a pair of aircraft being 2,000ft (600m), with a mile (1.6km) between pairs. After several encounters US aircrews noticed that a favourite MiG-17 tactic was to try to lure F-4s into low-level dogfights when they were returning south and low on fuel. The F-4 kill ratio against MiG-17s was about 4:1.

The MiG-21 'Fishbed', introduced into the Vietnam War in 1967, was the best enemy fighter with a standard armament of one 23mm cannon and four AA-2 Atoll IR missiles. It was superior to the F-4 in acceleration and manoeuvrability at all heights at low speeds and at high subsonic speeds above 25,000ft (7,600m) (see Table 21).

To use a beam-riding missile, the launching fighter used a simple ranging-radar that sent out a very narrow pencil beam which the missile followed. To avoid interference between the ionised gas from a rocket motor and the reception of the launch aircraft's radar beam the AAM was boosted by upper and lower cordite boost motors that quickly burned out and were jettisoned. The remaining unpowered dart, with a guidance receiver antenna in the blunt tail, was then free to pursue the target. British experiments with the beam-riding Fairey Fireflash (shown here being carried in 1960 by one of the 12 Supermarine Swift F.7s used at RAF Valley for weapon firing trials) were abortive. *(via author).*

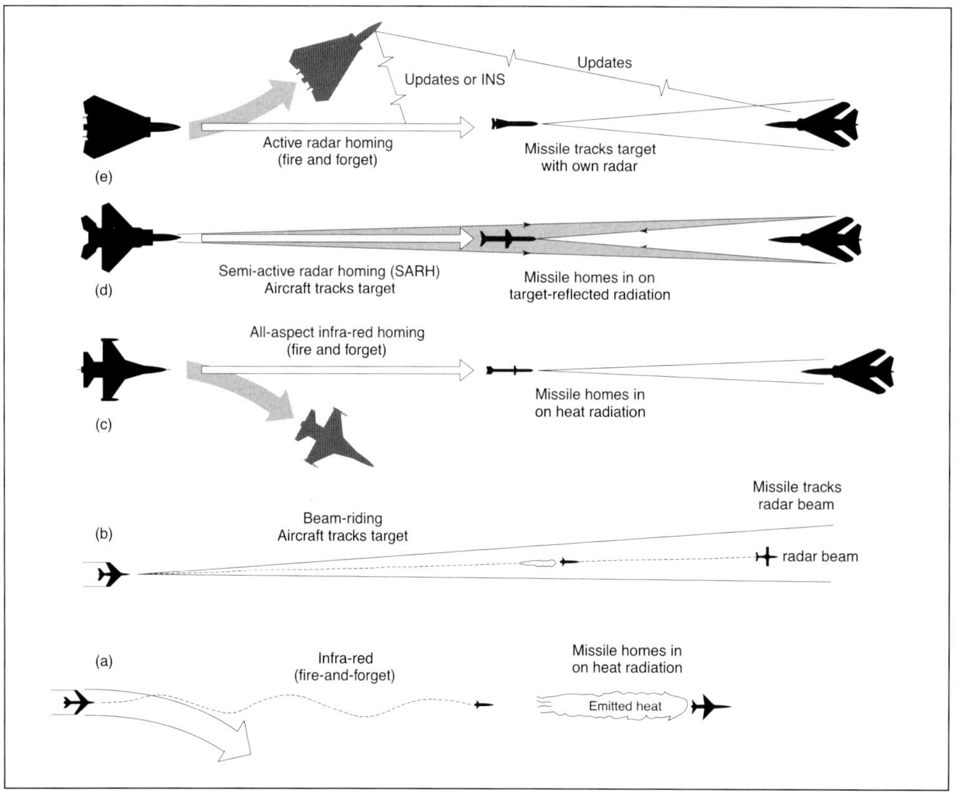

There are five basic homing methods for AAMs. Because of their low sensitivity, early IR seekers could respond only to the hot rear section of the target's engines. As a result, they could only be fired from directly behind the target, in the traditional curve of pursuit attack. With the improved all-aspect IR seekers of the 1970s and beyond, head-on shots could be taken. Beam-riding and semi-active radar homing methods require the target to be continuously painted by the launch aircraft until missile intercept. The IR method was the only 'fire-and-forget' technique until the advent of active-radar AAMs which can be wholly autonomous. *(Pete West/AIR International).*

Thus F-4 pilots were unwise to engage it in sustained dogfights. Operating in pairs, MiG-21 pilots were vectored by their ground controllers to a position behind a strike force from which they made brief high-speed (Mach 1.3) hit-and-run attacks. The F-4 kill ratio against MiG-21s was around 2:1.

The Vietnam experience during Operation 'Rolling Thunder' (1964–68) with an overall US kill/loss ratio of 2:5:1 was a salutary one for US fighter forces. Though it never endangered US air superiority in south-east Asia, when transposed to the European scenario it was immensely disquieting. One highly favourable and enduring reaction was the setting up in 1968 of the US Navy's 'Top Gun' Fighter Weapons School. By 1972 the results of dissimilar air combat training were beginning to show and the US Navy's kill/loss ratio had risen, while the Air Force's remained at 2:1.

Energy-manoeuvrability

The energy-manoeuvrability concept for measuring combat performance across a spectrum of speed, altitude and load factor was developed in the 1960s for indicating an aircraft's capability and for comparing aircraft. The parameter used at a given speed and altitude is specific excess power (SEP). It is used to assess acceleration or rate of climb at a given load factor (g). In units of metres/sec (or ft/sec), it is derived from (excess thrust × speed)/weight. For air-to-air combat, in which high manoeuvrability is essential, there are five main objectives, as indicated in Table 21.

An adjacent figure shows a typical plot of SEP contours on the axes of altitude versus Mach number for the F-4 and MiG-21, giving an idea of where you wanted to drive the fight depending on what you were flying. Such graphs apply only to the given aircraft configuration, weight, load factor and thrust setting. The effect of pulling g is to collapse the contours inwardly.

Self-protection

Before 1966 US fighters did not carry radar warning receivers (RWRs) and surface-to-air missiles (SAMs) took their toll if the pilots were unable to see the missiles coming, as when flying above overcast. US pilots realised that it took more than speed and manoeuvrability to evade the SAMs, so chaff and jamming pods were carried. By 1967, after initial teething troubles, the electronic countermeasures (ECM) pods worked well, and to lose use of a pod meant that a pilot had to abort. Radar-homing and warning equipment told a pilot that a SAM radar was activated and missiles about to be launched. Once a missile was launched a cockpit warning told the pilot to take evasive action and/or jam the incoming missiles. By 1972 more than 150 SAMs were fired for each aircraft brought down; previously it had taken fifty-five. Most fighter-

Originally code-named 'Blue Jay' the de Havilland Firestreak was an IR-homing AAM that reached IOC in 1958 after starting development in 1951. It first entered service with the Sea Venom of 893 Squadron RN. The Firestreak and later Red Top IR-homing missiles were the only AAMs to enter service in the UK in the 1960s. This is a Firestreak being test fired by a de Havilland Venom NF.2 in the late 1950s. *(via author).*

Table 21 Cumulative evolution of performance requirements and its effect on design requirements

Objective	*Performance requirement*	*Design requirement*	*Era*
To gain an energy advantage over opponent at the beginning of the engagement	High acceleration or rate of climb (high supersonic SEP)	High afterburning T/W and low supersonic drag	1950s
To convert energy advantage into an angular position advantage	Minimum deceleration during a turn	High L/D at high g, high usable lift coefficient	1960s
To achieve an advantageous angular position in sustained combat	High sustained turning capability (high subsonic SEP)	High dry T/W, high subsonic L/D, high g tolerance of pilot	1970s
To force an overshoot if opponent has an energy or heading advantage	High deceleration and maximum instantaneous turn capability	High maximum lift coefficient and low W/S	1980s
To point and shoot	High agility	Helmet-mounted sight, thrust vector control, high-agility weapons	1990s

The MiG-21J 'Fishbed' was the most commonly used Soviet fighter in the latter stages of the Vietnam War. Though highly manoeuvrable, its dogfighting capability was restricted. Its radar had a range of only 12 miles but since it was used only under GCI conditions this was of no consequence. However, that its gyro gunsight toppled at around 3g was critical as was its limited usable fuel capacity of only 365gal. The common MiG tactic, therefore, was to come in at supersonic speeds, often Mach 1.6, behind the US forces and launch their Atoll IR missiles, then dive for the deck and run. This is a MiG-21bis of the Indian Air Force with four wing points for carriage of R-13 or Matra Magic IR missiles. *(Hindustan Aeronautics Ltd via author).*

Differential specific excess power for F-4 versus MiG-21. At Mach 0.9 at 15,000ft, the instantaneous turn rates of the F-4 and MiG-21 were almost identical at 13°/sec, but at Mach 0.5 the MiG-21 could manage 11°/sec, whereas the F-4 was capable of only 8°/sec. In sustained turns, which this plot can be used to evaluate, the F-4 had a slight edge at high subsonic speeds and low altitude. Such graphs apply only to the given aircraft configuration, weight, load factor and thrust setting. The effect of pulling g is to collapse the contours inwardly. *(AIR International).*

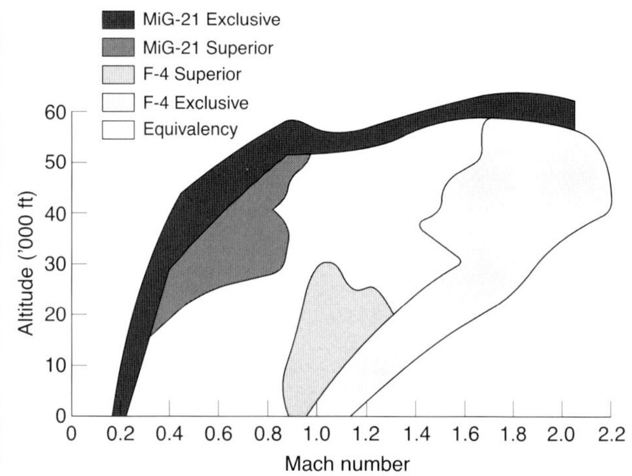

bomber missions were flown at 12,000 to 15,000ft (3,600 to 4,500m) where SAM effectiveness was reduced because the missile (SA-2 Guideline) was still accelerating, not reaching its maximum speed of Mach 2.4 until 25,000ft (7,600m). At the lower altitudes the aircraft were more vulnerable to anti-aircraft artillery (AAA), particularly 57mm guns.

Misleading missiles

Experience with AIM-9E Sidewinders and AIM-7E Sparrows in the Vietnam War showed that they failed to perform in dogfights, since they had been designed for use against less manoeuvrable bombers. They achieved kill rates of only around 15% and 8% respectively, so were often launched in pairs. This was wasteful but may appear more likely to achieve a kill. The main limitations were due to the AIM-9 originally being optimised for medium- and high-altitude operation so it could not be used successfully against targets close to the ground, nor for head-on attacks, and it had a short range. In addition, both missiles were restricted to 2.5g launch manoeuvres. Ten years later via a series of upgrades and development, though, the all-aspect AIM-9L, with reliability vastly improved by semiconductors compared with the original thermionic valves, a cooled-seeker head, improved fusing and a higher-thrust, longer-burning motor, was used with great success during the Falklands campaign of 1982 when out of twenty-seven fired by Harrier FRS.1s, twenty-four made hits with nineteen kills.

The gunfighter returns

One of the major lessons to come out of the Vietnam War was that dogfighting there was surprisingly similar to experiences in World War Two and Korea. Despite their limitations during the Vietnam era, AAMs gave US fighters greater capability compared with the MiG-17, which

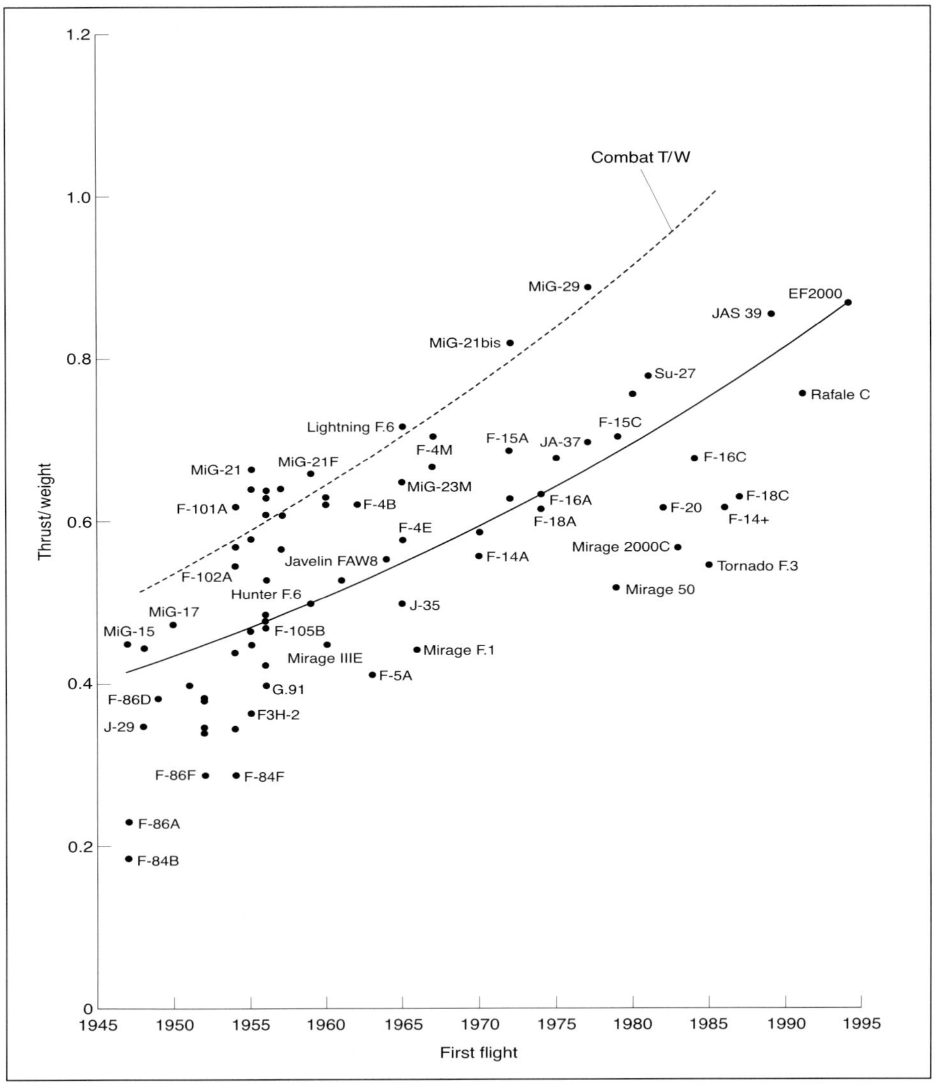

Aircraft thrust/weight ratio versus time. Values shown are based on MTOW, the upper dotted line indicates combat T/W. Noticeable are the high values for the limited-endurance (low fuel fraction) Lightning, MiG-21 variants and MiG-29. *(AIR International).*

Air-to-air missiles tend to be relatively low density stores and may be subjected to significant disturbances (roll and pitch rates) as they pass through the flow field around the aircraft. Consequently, fuselage-mounted missiles are ejected as swiftly as possible through the flow field. Many stores have shown a distinct reluctance to leave the environs of their home (e.g. aftmost AIM-120 beneath an F-15). *(Pete West/AIR International).*

Missile armament of the Lockheed F-104A (this one of ARDC) was two wing tip mounted AIM-9B Sidewinders, early versions of which proved to have serious shortcomings, not the least of which was a low launch-g limit of a little over two. Because of their unreliability they were often fired as pairs as shown. The Starfighter was the first aircraft to be equipped with the General Electric M61 Vulcan 20mm Gatling-type cannon which has, through a series of upgrades, grown in reliability and a version of it is to be used on the Lockheed Martin F-22. *(via author).*

The Westinghouse AWG-10 fire control system of the F-4J included the first multi-mode pulse-Doppler AI radar (Westinghouse APG-59) to enter service on an aircraft; the first transistorised one and the first with a digital computer. Furthermore, all weapon data was displayed on the HUD. The F-4J was a USN aircraft with a 32in radar antenna, that for the USAF was the F-4E which had the APQ-120 radar, similar in many ways to the AWG-10 but with smaller antenna, adopted partly to allow the installation of a 20mm M1 61A Vulcan cannon in the nose. When fitted with an under-fuselage pod-mounted Gatling gun the F-4 Phantom regained the essential armament of a fighter. The later, nose-mounted M61 gun of the F-4E reduced the inherent flexibility of the pod and improved accuracy. Two rails for AIM-9s are carried on each wing pylon of this *Luftwaffe* F-4F. *(via author, IAT'97).*

carried only cannons and unguided rockets. Nevertheless, there were numerous reports of F-4 pilots, having used all their missiles, finding themselves in positions where an opponent could have been downed if they had had a gun. American pilots were quick to recognise that fighting a MiG, on its terms, in a gunless F-4 was less than desirable. Some F-4Cs were fitted with a centreline pylon-mounted SUU-16/A cannon pod, though the flexibility of the mounting caused dispersion. Eventually, the F-4E entered service in 1968 equipped with a chin-mounted General Electric M61 Vulcan that could fire at 6,000rpm. With only a few seconds to fire in any MiG engagement, the very high firing rate was more than adequate and the gun jammed very little. The F-4E achieved 50% of its kills using its gun.

It was this addition that enabled US forces to adequately match the agile, if admittedly low-endurance, MiG-21. The Russian Mach 2 aircraft had both manoeuvrability and excess thrust, despite its single engine. It was small, and carrying missiles and guns it was versatile, and whatever its shortcomings in other areas it was well suited to the task. In most cases the US forces had to concede the initiative and await the arrival of the MiGs and then fight on the MiGs' terms. Fighting under adverse circumstances, for example, the visual identification rules of engagement (RoE), the wonder is that the US fighter force performed as well as it did.

HOTAS

Inevitably, the workload involved in combat while monitoring threat warnings, countering radar and communication jamming, operating an advanced multimode radar and fire control system and maintaining situational awareness, became enormous. McDonnell Douglas evolved a new concept to deal with it for its F-15 Eagle that entered service in 1974. This is hands-on throttle and stick (HOTAS), in which all the radar and weapon function controls likely to be needed in combat are grouped under the pilot's hands. It has been adopted for all modern fighters but needs high manual dexterity to operate effectively.

Weapon aiming

In air-to-air gunnery the gyro-gunsight of World War Two presented a single aiming point and required the pilot to track his target to establish the lead angle before firing. A head-up-display (HUD), by contrast, can present the loci of aiming points for different target ranges, enabling the pilot to fire when the target and appropriate aiming point coincide, without tracking. This 'snapshoot' capability is valuable in air combat against a target that is manoeuvring evasively. Of great importance to today's fighter is its maximum attained manoeuvrability rather than its ability to sustain, during dogfighting, several turns. In essence, 'Spot your enemy, abruptly point your aircraft and fire'.

Newer than the fixed HUD is the helmet-mounted display (HMD), whose electro-optic system tracks head movement to give it look-and-shoot capability. One was first fielded on F-4s in the 1970s but could not be fully exploited due to limited missile capability. Pilots wearing a simple HMD were seen flying MiG-29s in 1985. During exercises against USAF Lockheed Martin F-16s in 1994, *Luftwaffe* MiG-29 'Fulcrum' pilots using the system were able to get the first shot most of the time in close-in encounters. Advanced systems are planned for the Dassault Rafale, Eurofighter Typhoon and Lockheed Martin F-22 Raptor.

AAM development

Active-radar guidance for guided weapons was, in theory, a promising alternative to SARH since it could provide a fire-and-forget weapon. This would free the launch aircraft from the limitation of having to keep its radar pointed at the target throughout the missile's flight time. The fighter could then turn away either to avoid close-in combat or to engage another target. Using such a technique, the missile would fly, using autopilot/inertial control guidance, to a point calculated by the launch aircraft where the missile could engage its terminal-guidance active-radar seeker for a good chance of lock-on.

An active-radar seeker needed its own pulse-Doppler (PD) transmitter, greatly increasing complexity, and remained impractical in the pre-microchip era. This was for two reasons: firstly, the difficulty in producing a unit small enough to fit inside an AAM, and secondly, radar detection range is proportional to the size of the antenna. To a rough approximation, the power of a radar can be higher the longer the wavelength (the lower the frequency), and conversely is lower at the high frequencies, at which an AAM's radar had to work. However, in the mid-1960s the US Navy began flight-testing the Hughes AIM-54A

The McDonnell Douglas F-15 first flew in 1972, 20 months after the Grumman F-14 and its design was influenced by the largely mythical performance of the MiG-25 but there were other considerations. In many predicted scenarios, the F-15 would be heavily outnumbered by agile lightweight fighters such as the latest MiG-21 variants. It therefore needed to combine the virtues of a high performance interceptor and daylight air superiority fighter while retaining full night and adverse-weather capability. To meet these demands, armament consisted of a single M61A1 20mm cannon (with 940 rounds) and eight AAMs, a mix of Sparrows and Sidewinders, combined with a one-man operable radar. Inevitably the workload involved in flying the aircraft while operating an advanced multimode radar, navigation and using communications was enormous, and McDonnell Douglas evolved a new concept to deal with it. This was HOTAS in which all the controls likely to be needed in combat were grouped under the pilot's hands. It needs high manual dexterity to operate effectively. Hughes, which developed the AN/APG-63 radar for the early McDonnell Douglas F-15A, evolved the design to form the AN/APG-70 that is installed in USAF F-15E and some of its later F-15C/Ds. Here, the two-seat F-15E is flight testing inert JDAM/Mk 84s in 1995 at Eglin AFB test range to provide structural and environmental data. *(Boeing via author).*

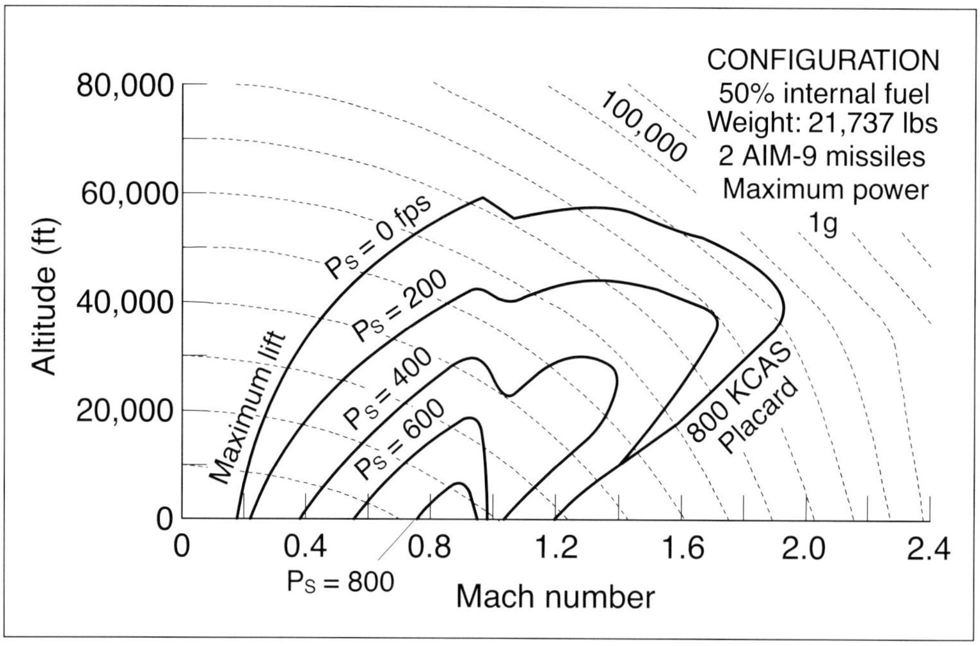

Contours of constant values of specific excess power for F-16A under the given conditions. The contours indicate 'sustainability' (on or inside the Ps=0 boundary) within which the aircraft can accelerate and/or climb. By unloading g (often to zero) all the contours blossom, whereas pulling g collapses the contours. *(AIR International).*

Phoenix for the Grumman F-14 Tomcat. This million dollar 450kg (1,000lb) very-long-range weapon uses SARH for the fly-out and its limited-range active radar only when close to the target.

By the early 1980s work started on a smaller, lighter and faster replacement for the AIM-7M Sparrow and became known as the Advanced Medium Range AAM (AMRAAM), later designated AIM-120. It is fired in the general direction of its target, which has been previously detected and its range and velocity measured by the launch aircraft, far beyond visual range. There are two basic modes: radar-supported targeting with datalink support (greatest probability of success), and autonomous with or without initial radar targeting. For most of its flight the missile flies on a strapdown inertial navigation system, which keeps it flying in the right direction, towards the predicted future position of the target. In some attack modes, especially if it is thought that the target is aware of an attack on it, the launch aircraft has to occasionally paint the target with brief sweeps of its radar. Its processor then checks that the echo still belongs to the original target. After each paint a coded digital update of future target position is sent from the launch aircraft, which results in a small alteration of the missile trajectory until the missile's radar goes active. However, the radar sweeps of the target needed for mid-course correction throws away the chief advantage of having an active missile, since the target is warned of an imminent attack. When it is within range, the missile's radar seeker goes active and counters attempted jamming using its home-on-jam mode. If launched without a radar target the missile will go active immediately and intercept the first detected target. From the mid-1980s active radar has come to be used on the Vympel R-77 Amraamski (AMRAAM equivalent) and R-37 Amos, AIM-120 (8 to 10GHz seeker) and Matra Mica (10 to 20GHz seeker).

Beyond-visual-range combat

Despite the predictions made over the last four decades, the era when the air superiority mission would consist of launching long-range missiles at an adversary beyond visual range (BVR), consistently failed to materialise. In every conflict from World War One to the end of the Gulf War of 1991, the overwhelming majority of air-to-air engagements occurred within visual range, albeit artificially aided, and have involved manoeuvring to a position for using short- to medium-range weapons. However, of the Iraqi aircraft downed during 'Desert Storm' just 28% fell to Sidewinders (nineteen fired) and the rest to Sparrows (seventy-one fired). This showed that medium-range missiles (e.g. AIM-7M) had finally achieved long-promised BVR capabilities, though the bulk of in-service AAMs still have very low kill rates.

To make BVR a safe option, positive identification is essential. This is one reason that prevented F-14s using their long-range Phoenix missiles during the Gulf War of 1991. That is not to say that BVR combat is unrealistic, since not only would this deny the main *raison d'être* behind the F-22 and Typhoon, it would disregard non-cooperative target recognition. The new AI radars have the ability, given the opportunity, to look down the intakes of a 'bogey' and to use PD matching to recognise an engine from its compressor's signature. Moreover, airborne warning and control system (AWACS) aircraft and joint tactical information distribution system (JTIDS) are other long-range sources to aid target identification. The use of safe corridors flown through at an agreed speed is another piece in the jigsaw towards identification. Despite these resources, available to only a handful of countries, there will be occasions when operating rules of engagement (RoE) do not allow engagement without a visual identification. The ability to detect, track and identify threats early enough either to avoid or engage them without their knowledge will be cardinal in obtaining the vital 'first look, first shoot, first kill' advantage claimed for the F-22.

US complacency: WVR overlooked

Because within-visual-range (WVR) combat was thought to be outdated, much emphasis was put on Cold War-type BVR shoot-outs at the expense of short-range weapons (reminiscent of the gunfighter versus missile fighter controversy of the 1950s). However, given restrictive RoE combined with reliance on an inadequate identification

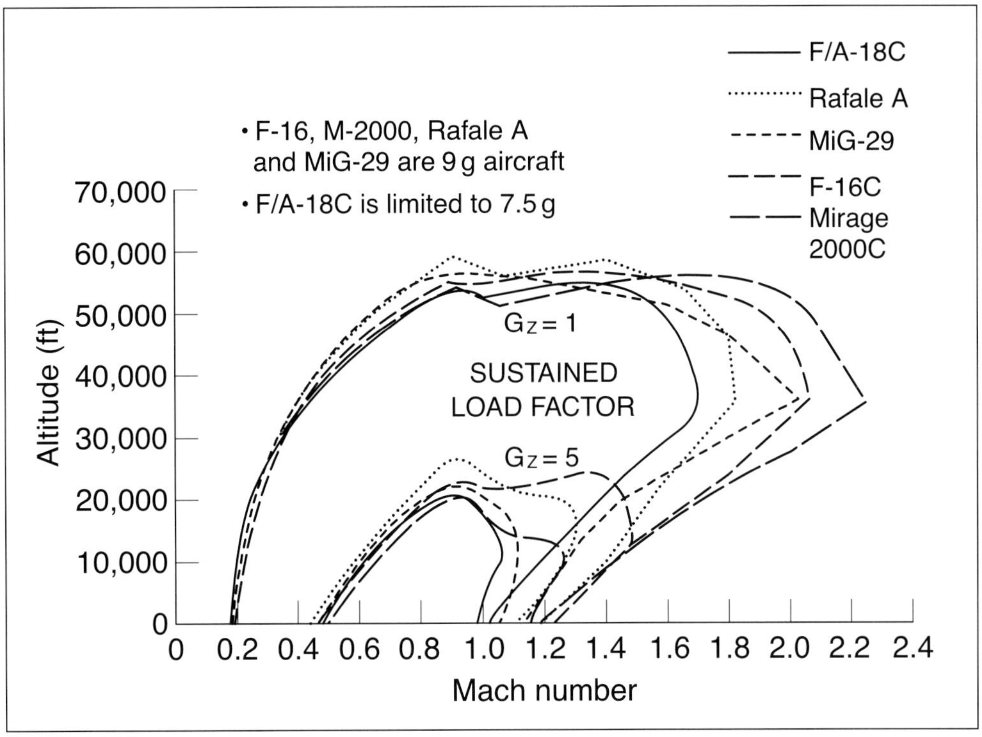

Comparative sustained g manoeuvre envelopes, illustrating the effect of pulling g on manoeuvrability. A missile suffers more severely since its g level is typically now more than four times that of the aircraft it is designed to attack. Moreover, once its motor has fired, its level of sustained g is constantly dropping. *(Lockheed Martin/AIR International).*

For the MiG-29 the IRST is an integral component allowing medium-range interceptions to be made in radar silence. On the other hand, if the target vanishes in cloud, thereby losing the IRST, the radar automatically takes over. Gun attacks are mainly with IRST coupled by a laser ranger, a combination said to give exceptional shooting accuracy, while the helmet-mounted display allows missiles to be fired at large off-boresight angles. Pilots wearing a simple HMD were seen flying MiG-29s in 1985, and using the system *Luftwaffe* pilots who inherited the East German aircraft were able to get the first shot most of the time in close-in encounters during exercises against USAF F-16s in 1994. The early MiG-29 was regarded as a superb point-defence dogfighter but severely handicapped by fuel capacity. This is a hunched-back MiG-29 (9-13) 'Fulcrum'-C. *(Alexei Mikheyev via author).*

friend or foe (IFF) system, long-range AAMs' low-kill probabilities against modern aircraft having multiple sensors that are often able to detect a missile attack and agile enough to evade, and defensive electronic warfare effects, WVR combat remains an inherent scenario for the crop of fighters in current development.

Moreover, using Vympel R-73 Archers obtained by the US Department of Defense, it was realised that the Russian missile had much higher manoeuvring performance than first thought. It totally outclassed the standard AIM-9M in acquisition, speed, resistance to countermeasures and turning capability. Indeed, in simulations the AIM-9M only won one in fifty engagements compared with the R-73. Short-range combat simulation of an F-15 with AIM-9M versus a MiG-29 with helmet-mounted sight and R-73 found that that the MiG could acquire targets in thirty times the volume of sky of the F-15: the MiG was overwhelming. However, Soviet-style tactics did not make best use of the MiG-29/HMD/R-73 combination since ground-controlled intercepts gave it little chance in close-in combat where the combination excels.

The realisation that R-73 was such a step forward gave added impetus to the development of smarter and more agile short-range AAMs. There was also resentment in the US regarding the time (seven to ten years) lost by waiting for Europe to live up to the Memorandum of Understanding it signed in 1980 (but which died in 1991), that allocated it short-range missile development while the US dealt with medium-range that led to AIM-120. The intention is therefore to field the Raytheon AIM-9X Evolved Sidewinder in 2003, eighteen years after the first appearance of R-73. Until that time the US will have an inferior WVR missile to that of Russia, Israel (Python 4) and Britain (ASRAAM).

Advances made with light surface-to-air IR missiles such as Stinger and Mistral enabled the use of more sophisticated scan patterns and dual band sensors (3 to 5 microns and 8 to 12 microns). Non-imaging seekers simply react to a point source of heat, but the more recent imaging-infra-red (IIR) sensors create a thermal image of the whole target, locking on to a portion of it by means of signal processing. Moreover, IIR sensors are more resistant to countermeasures since they are able to recognise flares by comparing differences in their frequencies, intensity and velocity with aircraft. They are more complex and expensive, and one of the critical issues is their affordability since the focal plane array seekers they use are pushing the purchase price closer to that of seekers in medium-range SARH missiles.

It is contended that the group of AAMs now entering or about to enter service will leapfrog R-73. When it was introduced it greatly expanded off-boresight capability to ±60° over the previous ±15°, offering far more firing opportunities than the tailshot of previous types.

This is the Dassault Electronique AD4A active radar (10-20GHz) seeker for the 160mm (6.3in) diameter Matra/BAe Mica BVR air-to-air missile. The ability to fly close to the target before engaging a terminal active radar homer is already present in the AIM-54 Phoenix, AIM-120 AMRAAM, Vympel R-73 and the first active Mica firing took place May 1997. Short-range firings of Mica will involve locking the seeker onto the target before launch.
(Dassault Electronique via author).

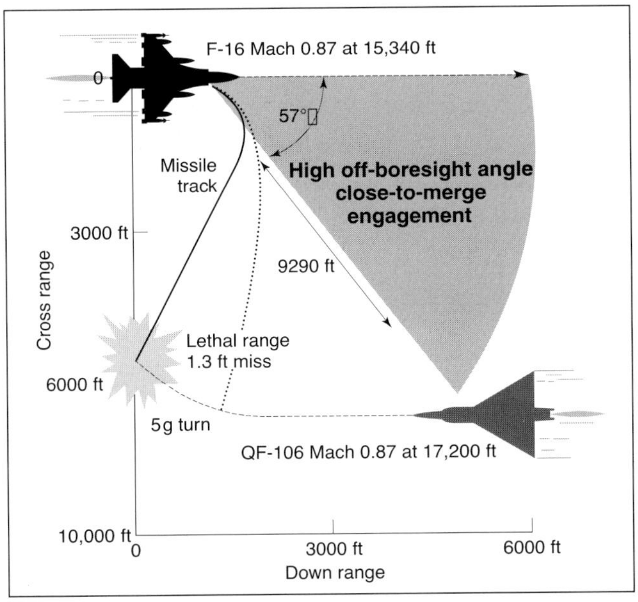

During a test in July 1994 an F-16 pilot, wearing an HMD to align the missile seeker with the target, fired a thrust-vectoring version of the Box Office missile at a QF-106 drone. Shortly after clearing the F-16 (within 100ft) the missile initiated a 30g turn towards the target, flying at up to 63° AoA. The missile passed through the drone's afterburner plume at a miss distance of 3.8m, well within the lethal radius of the warhead. The dotted line indicates the approximate trajectory ASRAAM would have flown. Whereas Box Office took 6.4 secs to kill the target, ASRAAM would have killed it 2–3 secs earlier. *(Pete West/AIR International).*

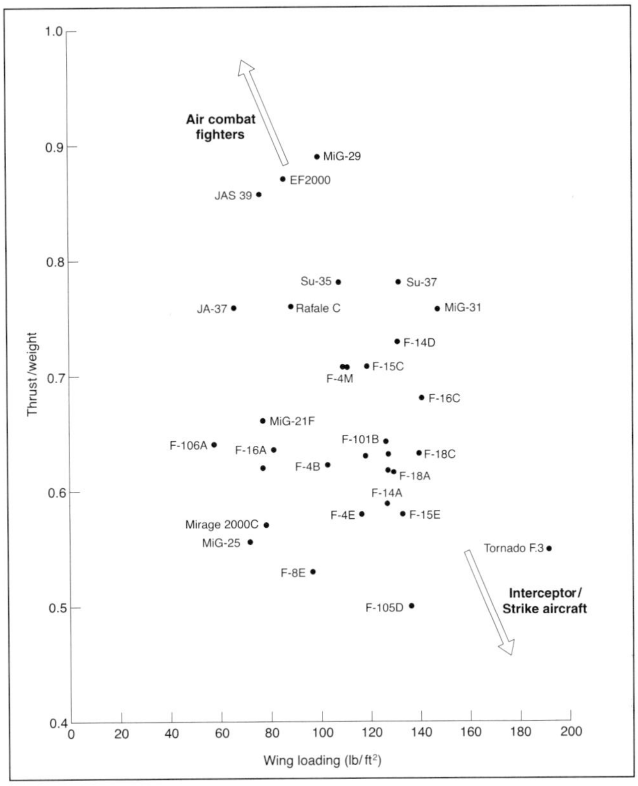

Thrust/weight ratio versus wing loading are the main parameters used in constraint analysis (known in the 1950s as limitations analysis) for conceptual design. High T/W and low W/S are typically sought for the highly manoeuvrable air combat fighter, whereas lower T/W and high W/S have traditionally been the domain of the interceptor type. *(AIR International).*

Raytheon's FMRAAM response to the RAF's requirement for an active-radar BVR missile (and one that could replace the AIM-54 Phoenix) is a development of its very successful 30nm (56km) range AIM-120 AMRAAM. As can be seen it employs a rocket for boost to about Mach 2 where the two ramjets can perform the 55nm (100km) sustained flyout to the target to a terminal Mach 3. The ramjets are Aérospatiale liquid fuel types with minimal RCS plume signature. *(Raytheon via author).*

Infra-red missiles are designed such that the seeker acquires the target prior to launch while held in the boresight (straight ahead position). The seeker then locks-on the target and is uncaged so that it can track the target to hold its locked-on condition as it is launched. This influences the type and location of its installation. The missile seeker must not be shadowed by the wing or fuselage of the launch aircraft. Rather than use up valuable underwing pylons to mount its self-defence AIM-9 Sidewinders, an unusual position had to adopted for the Jaguar. The photo shows a launch from the Sidewinder-dedicated overwing pylon. *(BAe via author).*

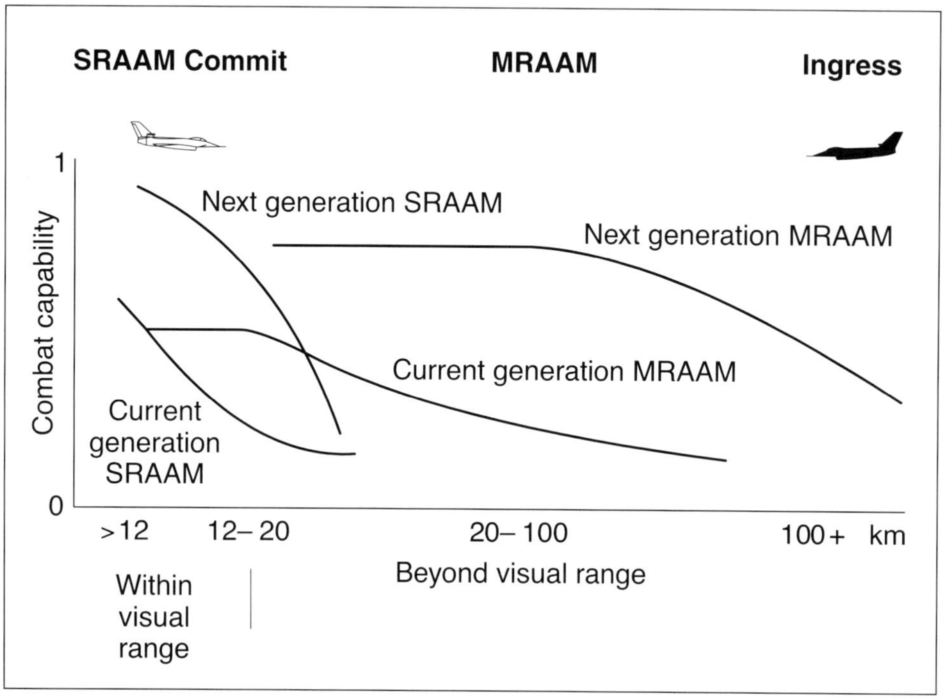

Effectiveness of missiles versus flight time. The current generation of MRAAMs have maximum combat capability out to around 15km. Due to emerging threats, shorter missile fly-out times and improvements in tailchase engagement capability of future BVR weapons are required. It is predicted that the use of ramjet cruise propulsion will take the next generation MRAAMs far into the BVR arena, engaging targets at ranges of about 80km. *(Pete West/AIR International).*

Furthermore, the provision of lock-on after launch capability with inertial guidance combined with low-drag aerodynamics and improved rocket motors allows head-on, pre-merge engagements at maximum range.

Billions of dollars have been spent on making modern fighters more manoeuvrable, but putting the manoeuvring performance in the missile and relaxing that of the aircraft could pay dividends. Manoeuvrability drives T/W and W/S and determines the required load factors. Some not very manoeuvrable fighters may be able to turn the tables by having the missile do most of the manoeuvring. Ultimately, however, if the weapon is not lethal in the endgame, all else is wasted.

Conclusion

History and simulations have shown that combat outcome is at least as dependent on tactics as on equipment. The purpose of tactics is to exploit one's own strength and the opponent's weaknesses. Recent computer modelling has shown 'cliff-edge' effects where large tactical variations have had little effect until a certain point is passed. This is typically when two or more optimums of a huge variety of combat parameters become matched. Whatever the future holds, we can be sure that the endless measure versus countermeasure battle will continue.

Matra BAe ASRAAM successfully destroys a QF-4 Phantom drone after being launched from an F-16 at Eglin AFB in 1998 on its fifth development firing. This photo shows its low signature Royal Ordnance motor designed to minimise the chance of detection at launch or in flight. The Aim-132 ASRAAM has a 10kg blast-fragmentation warhead and is effectively a hittile but with more fragments than early model AIM-9 warheads. The idea is to hit the target enough times that even double and triple-redundant systems are caused to fail. The missile's IIR sensor gives a crisp enough image of the aircraft to hit either cockpit, wing root or midfuselage so that much of the damage comes from the kinetic energy effect. The kinetic effect of its warhead is said to be equivalent to the impact of a one ton automobile doing 500mph. *(Matra BAe via author).*

Index